AN INTRODUCTION TO GAUGE THEORIES

AN INTRODUCTION TO GAUGE THEORIES

Nicola Cabibbo
University of Rome, La Sapienza, and
INFN Sezione di Roma, Rome

Luciano Maiani
University of Rome, La Sapienza, and
INFN Sezione di Roma, Rome

Omar Benhar
INFN Sezione di Roma, and
University of Rome, La Sapienza

Translated from the original Italian by Geoffrey Hall, Imperial College, London

CRC Press
Taylor & Francis Group
Boca Raton London New York

CRC Press is an imprint of the
Taylor & Francis Group, an **informa** business

This book was first published in Italian in 2016, by Editori Riuniti University Press, under the original title: *Introduzione alle Teorie di Gauge.*

CRC Press
Taylor & Francis Group
6000 Broken Sound Parkway NW, Suite 300
Boca Raton, FL 33487-2742

© 2018 by Taylor & Francis Group, LLC
CRC Press is an imprint of Taylor & Francis Group, an Informa business

No claim to original U.S. Government works

Printed on acid-free paper

International Standard Book Number-13: 978-1-4987-3451-6 (Hardback)

Library of Congress Cataloging-in-Publication Data

Names: Cabibbo, N., author. | Maiani, L. (Luciano), author. | Benhar, Omar, author.
Title: Introduction to gauge theories / Nicola Cabibbo, Luciano Maiani, Omar Benhar.
Description: Boca Raton, FL : CRC Press, Taylor & Francis Group, [2017] |
Includes bibliographical references and index.
Identifiers: LCCN 2016041647| ISBN 9781498734516 (hardback ; alk. paper) |
ISBN 9781498734516 (E-book).
Subjects: LCSH: Gauge fields (Physics) | Field theory (Physics).
Classification: LCC QC793.3.G38 C33 2017 | DDC 530.14/35--dc23.
LC record available at https://lccn.loc.gov/2016041647

Visit the Taylor & Francis Web site at
http://www.taylorandfrancis.com

and the CRC Press Web site at
http://www.crcpress.com

Contents

List of Figures

Preface

Introduction to Gauge Theories completes the series of three volumes based on lecture courses in relativistic quantum mechanics, electroweak interactions and gauge theories, taught by the authors to first-year master's degree students in physics and astronomy, and astrophysics (*Laurea Magistrale*) of the University of Rome, "La Sapienza", over a period of several decades.

The principal objective of this volume is to introduce the basic concepts of renormalisation in quantum field theory and the fundamentals of modern gauge theories. Even though it is connected to the previous volumes, the book can be read independently; it assumes only a general familiarity with special relativity, second quantisation and the phenomenology of electroweak interactions.

The basic tool is the Feynman path integral, introduced in the early chapters and systematically employed in what follows. The exposition follows a pedagogic course, which begins with the simple case of the quantum mechanical transition amplitude to lead to the generating functional of the Green's functions of field theory. The same pedagogic approach is used in the chapter in which equations of motion, symmetries and the Ward identity are discussed. The analysis of the path integral formalism is completed by a discussion of anticommuting variables and the quantisation of fermion and electromagnetic fields.

The perturbative expansion of the generating functional of the Green's functions, and the Lehman, Symanzyk and Zimmermann reduction formulae—which allow the latter to be related to the scattering matrix elements—are illustrated, first in the simple case of a scalar field theory with $\lambda\phi^4$ interactions and subsequently with quantum electrodynamics (QED).

Renormalisation in QED is treated in the central part of the book. The appearance of ultraviolet divergences, from which the necessity of introducing a regularisation procedure follows, is illustrated by taking as examples corrections of order α to the photon and electron propagators and to the interaction vertex. The fundamental Ward identity of QED is also proven to the same perturbative order.

The discussion of QED to second order is supplemented by the detailed illustration of several important applications, which include analysis of the infrared divergence, detailed calculations of the Lamb shift and the vacuum polarisation tensor, and Schwinger's calculation of the correction to the anomalous magnetic moment of the electron, extended in a subsequent chapter to the full Electroweak corrections in the Standard Theory. Introduction to the renormalisation group completes the part dedicated to QED, with discussion of the Gell–Mann and Low equation, and the Landau pole.

In the following part of the book, quantisation of non-Abelian gauge theories is analysed, and the evolution of the coupling constant with the energy and momentum scale. Asymptotic freedom in quantum chromodynamics, the fundamental theory of strong interactions, is demonstrated to second perturbative order through explicit calculation of the β function. The evolution of the coupling constant of the fundamental interactions to high and very high energies is discussed as an introduction to the hypothesis of grand unification.

The recent discovery of the Higgs boson has reopened discussion of the naturalness, or non-naturalness, of scalar fields in field theory, also in connection with the possible presence of new physics at energies higher than the critical energy of the Standard Theory. The final chapters are dedicated to the study of effects of scalar fields in the Standard Theory, analysis of limits on the mass of the Higgs boson, the calculation of quantum corrections to the effective potential of scalar fields and the problem that naturalness of elementary scalars poses for the structure of ultraviolet divergences of the potential.

A large part of this work is due to Nicola Cabibbo with whom we have been linked in a collaboration that lasted several decades. Nicola's contributions, several of which are on subjects not usually treated in introductory texts on gauge theories, are unmistakable for their originality and clarity.

In conclusion, we would like to express particular gratitude to Guido Altarelli, Riccardo Barbieri, John Iliopoulos, Gino Isidori, Giorgio Parisi, Antonio Polosa, Veronica Riquer, Massimo Testa, and to many other colleagues for numerous conversations and comments on the subject, and also express our affectionate thanks to Paola Cabibbo, for the patience with which she has followed the slow process of writing. Finally, we would like to acknowledge the fact that this book owes much to the comments we have received from our students, over the course of the years.

Luciano Maiani and Omar Benhar

Rome, August 2016

The Authors

Luciano Maiani is emeritus professor of theoretical physics at the University of Rome, "La Sapienza", and author of more than two hundred scientific publications on the theoretical physics of elementary particles. Together with S. Glashow and J. Iliopoulos, Maiani made the prediction of a new family of particles, those with "charm", which form an essential part of the unified theory of the weak and electromagnetic forces. He has been president of the Italian Institute for Nuclear Physics (INFN), director-general of CERN in Geneva and president of the Italian National Council for Research (CNR). He has promoted the development of the Virgo Observatory for gravitational wave detection, the neutrino beam from CERN to Gran Sasso and at CERN has directed the crucial phases of the construction of the Large Hadron Collider (LHC). He has taught and worked in numerous foreign institutes. He was head of the theoretical physics department at the University of Rome, "La Sapienza", from 1976 to 1984 and held the chair of theoretical physics from 1984 to 2011. He is a member of the Italian Lincean Academy and a fellow of the American Physical Society. For his scientific work, he has been awarded the J. J. Sakurai Prize, the Enrico Fermi Prize, the Dirac Medal, the High Energy and Particle Physics Prize of EPS and the Bruno Pontecorvo Prize.

Omar Benhar is an INFN research director and teaches gauge theories at the University of Rome, "La Sapienza". He has worked extensively in the United States as a visiting professor at the University of Illinois and Old Dominion University, and was an associate scientist at the Thomas Jefferson National Accelerator Facility. Since 2013, he has served as an adjunct professor at the Centre for Neutrino Physics of Virginia Polytechnic Institute and State University. He is the author of more than one hundred scientific papers on the theory of many-particle systems, the structure of compact stars and electroweak interactions of nuclei.

Nicola Cabibbo (1935–2010) was professor of Theoretical Physics and Elementary Particle Physics at the Rome Universities *La Sapienza* and *Tor Vergata*, and held research and teaching positions in prestigious institutions such as Harvard University, the Institute for Advanced Studies, Princeton, CERN, Geneva, University of California at Berkeley and Université Paris VI. In 1962, Cabibbo discovered the phenomenon of quark mixing, described by a new natural constant, the *Cabibbo angle*, measured with great accuracy in semileptonic weak decays of hadrons. According to a recent analysis, Cabibbo's paper on quark mixing was the most influential article published in the journals of the American Physical Society during 1893–2003. In the 1980s, Cabibbo provided important momentum to the applications of numerical techniques to theoretical physics, notably the gauge theories of strong interactions, promoting and leading the development of the family of APE (Array Processor Experiment) supercomputers. He served as a member of a number of learned societies: Accademia Nazionale dei Lincei and Accademia delle Scienze di Torino, in Italy, National Academy of Science and American Association for Art and Sciences, in the United States, and Accademia Pontificia delle Scienze, which he chaired from 1993. An internationally reputed science manager, Cabibbo was president of Istituto Nazionale di Fisica Nucleare (INFN) and of Ente Nazionale per le Nuove Tecnologie per Energia e Ambiente (ENEA). He was the recipient of the J.J. Sakurai Prize (APS), the Medaglia Matteucci (Accademia Nazionale dei XL), the Dirac Medal (ICTP Trieste), and the Benjamin Franklin Medal.

CHAPTER 1

INTRODUCTION

CONTENTS

1.1   QUANTUM ELECTRODYNAMICS

In this course we will largely focus on quantum electrodynamics, the field theory which describes the interaction of charged particles with an electric field. Quantum electrodynamics, or QED, is an incomplete theory, given that all the elementary particles are also subject to the weak interaction and, in the case of quarks, also to the strong interaction.

A more complete theory, which takes account of both electromagnetic and weak and strong interactions, is provided by the so-called Standard Theory (or Standard Model). The Standard Theory is also incomplete, given that it does not take account of gravitational interactions.[1]

Despite its limitations, QED is very interesting for several reasons:

- QED has a wide range of interesting physical applications, from the interactions of photons and electrons to the fine structure of matter at the atomic level.

- QED was the first field theory to have been studied in detail, mainly using the method of Feynman diagrams, to address fundamental issues such as the occurrence of infrared and ultraviolet divergences and renormalisation.

- QED is a gauge theory, and is therefore the prototype of the Standard Theory. A study of QED is a valuable introduction to the study of the Standard Theory.

---

[1] Although at the level of classical (non-quantum) physics, gravitational interactions are described very successfully by Einstein's general theory of relativity, and in many cases even Newton's theory can be satisfactorily adequate, there still does not yet exist a universally accepted quantised form of gravity. The most widely held opinion is that it will be necessary to progress from a field theory to a string theory, an argument beyond the scope of this course.

| **FERMIONS** | | | *matter constituents* | | |

Leptons spin = 1/2			**Quarks** spin = 1/2		
Flavor	Mass [Gev]	Electric charge	Flavor	Approx. Mass [GeV]	Electric charge
ν_e electron neutrino	<1×10⁻⁸	0	u up	0.003	2/3
e electron	<0.000511	−1	d down	0.006	−1/3
ν_μ muon neutrino	<0.0002	0	c charm	1.3	2/3
μ muon	0.106	−1	s strange	0.1	−1/3
ν_τ tau neutrino	<0.02	0	t top	175	2/3
τ tau	1.7771	−1	b bottom	4,3	−1/3

Figure 1.1 Fundamental fermions.

- Among the predictions of the Standard Theory, some of those related to QED are confirmed with the highest precision.

The most precise test of QED is currently given by the experimental value of the anomalous magnetic moment of the electron. The Dirac equation assigns to the electron a magnetic moment equal to one Bohr magneton[2], $\frac{e}{2m}$, but this result must be corrected because of the interaction with the radiation field by a factor $(1 + a_e)$, where a_e is the *magnetic anomaly*, which can be expressed as a power series in α, the fine structure constant:

$$a_e = \frac{\alpha}{2\pi} + \cdots .$$

The experimental values for the electron and positron are known with

---

[2]Units and other conventions are discussed at the end of this chapter.

errors of the order of 0.004 parts per million:

$$a_{e-}^{\text{Exp}} = (1159652.1884 \pm 0.0043) \times 10^{-9} \ ,$$
$$a_{e+}^{\text{Exp}} = (1159652.1879 \pm 0.0043) \times 10^{-9} \ , \qquad (1.1)$$

to be compared with the theoretical prediction:

$$a_{e-}^{\text{Th}} = (1159652.1535 \pm 0.0240) \times 10^{-9} \ \ (0.02 \text{ parts per million}). (1.2)$$

The error quoted for the theoretical prediction is mainly due to the uncertainty of the value of α, obtained from a measurement of the quantum Hall effect. The theoretical prediction is based on calculations of a_e up to terms in α^4.

In this course we will discuss quantum electrodynamics using the method of Feynman path integrals, which has been shown to be definitely superior to the more traditional method of canonical quantisation [1] to handle quantum theories characterised by gauge symmetries. The concepts developed here provide a useful foundation to discuss the Standard Theory [2].

We recall that electromagnetism is characterised by the invariance of all the observable quantities, for example the electric and magnetic fields $\mathbf{E}$ and $\mathbf{B}$, with respect to transformation of the vector potentials:

$$A_\mu \rightarrow A'_\mu = A_\mu + \partial_\mu f \ , \qquad (1.3)$$

where f is an arbitrary function. We can consider this transformation as due to a transformation operator U_f, and write

$$U_f A_\mu U_f^{-1} = A_\mu + \partial_\mu f \ . \qquad (1.4)$$

If we carry out two successive transformations, U_f, U_g, we will have

$$U_g U_f A_\mu U_f^{-1} U_g^{-1} = A_\mu + \partial_\mu f + \partial_\mu g \ , \qquad (1.5)$$

and we can easily verify that the set of U_f form a commutative (Abelian) group:

$$U_g U_f = U_f U_g = U_{g+f} \ . \qquad (1.6)$$

In the case of electromagnetism, we are therefore confronted with a particularly simple type of gauge invariance, while in the Standard Theory we must deal with non-commutative (non-Abelian) gauge symmetries. Even if the Feynman path integral method is not indispensable for the quantum description of electromagnetism, it is preferred in the case of the Standard Theory.

1.2   UNITS AND OTHER CONVENTIONS.

Heisenberg and Schrödinger representations.   In what follows, we will principally use the Heisenberg representation, where the operators have a dependence on time:

$$O(t) = e^{itH} O e^{-itH} \ \ , \ \ \ O = O(0) \ . \qquad (1.7)$$

We will note in passing when it will be preferable to use the Schrödinger representation. In each case, operators and states which will not be shown with any explicit dependence on time, for example $O|m\rangle$, are meant to be operators and fixed states, equivalent to $O(t = 0)|m\rangle$ in the Heisenberg representation and to $O|m, t = 0\rangle$ in the Schrödinger representation.

Units. We will use the system of units in which $\hbar = 1$ and $c = 1$.

Other conventions. We will follow the conventions used in [1, 3] for 4-vectors and the Dirac matrices. In particular, the scalar product between two 4-vectors $p = \{p^0, \mathbf{p}\}$ and $q = \{q^0, \mathbf{q}\}$ will be denoted as $pq \equiv p^\mu q_\mu = p^0 q^0 - \mathbf{p} \cdot \mathbf{q}$.

THE FEYNMAN PATH INTEGRAL

CONTENTS

In this chapter we will derive the expression for the sum over paths from the usual formulation of non-relativistic quantum mechanics, considering the particularly simple case of a quantum system with only one degree of freedom. After deriving the expression for the transition amplitudes by means of the sum over paths, we will show how this method also allows calculations of the Green's functions in quantum mechanics. We will conclude the chapter by showing how these results can be extended to the case of systems with more discrete degrees of freedom and to field theory.

As a final demonstration of the equivalence between the different formulations of quantum mechanics, in a later chapter we will use the formulation based on the sum over paths to deduce the canonical commutation rules,

$$\left[p^m(t_0) , q^k(t_0) \right] = -i\hbar \, \delta^{m\,k} \, .$$

2.1   CALCULATION OF THE TRANSITION AMPLITUDE

To introduce the method of Feynman integrals, we consider first the simplest case: a one-dimensional quantum system described by the dynamic variable q and its conjugate momentum p. The Hamiltonian is therefore

$$H = K(p) + V(q) = \frac{p^2}{2m} + V(q) \, , \tag{2.1}$$

where we have denoted the kinetic energy as K and the potential energy as V.

We would like to calculate the transition amplitude from a state $|q_1\rangle$ at time $t = t_1$ to a state $|q_2\rangle$ at time $t = t_2 = t_1 + T$. Using the Schrödinger representation [1] and the system of units in which $\hbar = 1$ we can write

$$|q_1\rangle = \text{state at time } t = t_1 \,,$$

$$e^{-iHT}|q_1\rangle = \text{state at time } t = t_1 + T \,,$$

$$\langle q_2|e^{-iHT}|q_1\rangle = \text{transition amplitude to } |q_2\rangle \,. \qquad (2.2)$$

To know the transition amplitude as a function of q_1, q_2 and T is equivalent to having a complete description of our quantum system. We will see some examples of this assertion, and direct the reader to the work of Feynman and Hibbs [4] for full details. We recall that up to now in elementary particle physics we have actually been interested in the calculation of transition amplitudes and especially in the elements of the S-matrix.

We consider first the case $V(q) = 0$ (for the details of the calculation see Appendix A), which leads directly to:

$$\langle q_2|e^{-iKT}|q_1\rangle = \langle q_2|e^{-i\frac{p^2}{2m}T}|q_1\rangle = \sqrt{\frac{m}{2\pi i T}}e^{i\frac{m(q_2-q_1)^2}{2T}} \,, \quad V(q) = 0 \,. \quad (2.3)$$

We note that the result can be rewritten in terms of the average speed $v = (q_2 - q_1)/T$

$$\langle q_2|e^{-i\frac{p^2}{2m}T}|q_1\rangle = \sqrt{\frac{m}{2\pi i T}}e^{i\frac{mv^2}{2}T} \,. \qquad (2.4)$$

This result has a very simple interpretation: in the classical limit the particle moves with a constant speed v. The phase of the transition amplitude is therefore given by the *action along the classical trajectory*:

$$\langle q_2|e^{-i\frac{T p^2}{2m}}|q_1\rangle \propto e^{iS_{\text{cl}}} \,,$$

with

$$S_{\text{cl}} = \int_{t_1}^{t_2} dt\, L(q,\dot{q}) = \int_{t_1}^{t_2} dt\, \frac{m\dot{q}^2}{2} = \frac{mv^2}{2}T \,.$$

The correspondence between the quantum transition amplitude and the classical action was shown for the first time by Dirac [5].

In the general case, with an arbitrary potential $V(q) \neq 0$, we can calculate the transition amplitude through a limiting process which leads to the definition of the Feynman path integral, or the sum over paths. Subdividing the time interval T into N intervals $\epsilon = T/N$ we can write

$$\langle q_N|e^{-iHT}|q_0\rangle = \langle q_N|\left(e^{-iH\epsilon}\right)^N|q_0\rangle$$

$$= \int dq_1\ldots dq_{N-1}\langle q_N|e^{-iH\epsilon}|q_{N-1}\rangle\langle q_{N-1}|e^{-iH\epsilon}|q_{N-2}\rangle \cdots \langle q_1|e^{-iH\epsilon}|q_0\rangle \,.$$

$$(2.5)$$

We note that, since K and V do not commute,

$$e^{-i(K+V)\epsilon} = 1 - i(K+V)\epsilon - \frac{\epsilon^2}{2}(K^2 + V^2 + KV + VK) \, , \qquad (2.6)$$

while

$$e^{-iV\epsilon}e^{-iK\epsilon} = 1 - i(K+V)\epsilon - \frac{\epsilon^2}{2}(K^2 + V^2 + 2VK) \, . \qquad (2.7)$$

We can therefore write

$$e^{-iH\epsilon} = e^{-iV\epsilon}e^{-iK\epsilon} + \mathcal{O}(\epsilon^2) \, , \qquad (2.8)$$

observing that an error $\mathcal{O}(\epsilon^2)$ repeated N times is equivalent to a global error $\mathcal{O}(\epsilon)$, which will become negligible in the limit $\epsilon \to 0$. Each of the factors in (2.5) can therefore be approximated as

$$\langle q_k | e^{-iH\epsilon} | q_{k-1} \rangle \approx \langle q_k | e^{-iV\epsilon}e^{-iK\epsilon} | q_{k-1} \rangle = e^{-iV(q_k)\epsilon} \langle q_k | e^{-iK\epsilon} | q_{k-1} \rangle \, , \quad (2.9)$$

and, using (2.3), we obtain

$$\langle q_k | e^{-i\epsilon H} | q_{k-1} \rangle = \sqrt{\frac{m}{2\pi i\epsilon}} e^{i\left(\frac{m(q_k - q_{k-1})^2}{2\epsilon^2} - V(q_k)\right)\epsilon}$$

$$= \sqrt{\frac{m}{2\pi i\epsilon}} e^{i\left(\frac{mv_k^2}{2} - V(q_k)\right)\epsilon} \, . \qquad (2.10)$$

In the last step we have defined the velocity in the kth interval as

$$v_k = \frac{(q_k - q_{k-1})}{\epsilon} \, , \qquad (2.11)$$

so that we can recognise the phase factor of the Lagrangian $L = mv^2/2 - V(q)$. If we now substitute into (2.5) we obtain

$$\langle q_N | e^{-iHT} | q_0 \rangle \approx \left(\sqrt{\frac{m}{2\pi i\epsilon}}\right)^N \cdot$$

$$\cdot \int \prod_{k=1}^{N-1} dq_k \exp\left(i \sum_k \left[\frac{mv_k^2}{2} - V(q_k)\right]\epsilon\right) + \mathcal{O}(\epsilon) \, . \quad (2.12)$$

In the limit $\epsilon \to 0$, the set of points $\{q_N, q_{N-1}, \dots q_1, q_0\}$ form a trajectory $q(t)$ from the starting point q_0 to the final point q_N. The phase in (2.12) simply becomes the classical action along this trajectory, q_N

$$\sum_k \left[\frac{mv_k^2}{2} - V(q_k)\right]\epsilon \to \int dt L(q(t), \dot{q}(t)) = S(q(t)) \, ,$$

while the integration of (2.12) can be interpreted as a *sum over the trajectories*. In the limit $\epsilon \to 0$, the sum becomes an integral in the space of the trajectories $q(t)$. The integration measure is *defined* by the equation.[1] For a detailed discussion of the integration measure in path integrals, we direct the interested reader to Ref. [6].

$$\left(\sqrt{\frac{m}{2\pi i \epsilon}}\right)^N \int \prod_{k=1}^{N-1} dq_k \to \int d[q(t)] .$$

Taking the limit $\epsilon \to 0$, we finally obtain

$$\langle q_N | e^{-iTH} | q_0 \rangle = \int d[q(t)] e^{iS(q(t))} , \qquad (2.13)$$

where the integral is extended over all trajectories $q(t)$ such that $q(t_0) = q_0$, and $q(t_0 + T) = q_N$.

2.2   THE LATTICE APPROXIMATION

The Feynman integral is a *functional integral*, which is an integral carried out over all the functions $q(t)$ defined in the interval $[t_0, t_0+T]$. It is interesting to consider (2.12) as an approximate expression which, at least in principle, could be subject to an explicit calculation. This type of approximation, which we call the lattice approximation, is widely used in field theory, since it lends itself to numerical calculations in situations where it is not possible to obtain exact results, and where perturbative methods—that we will illustrate in the case of quantum electrodynamics—fail. The lattice approximation is shown to be of particular utility in the study of the fundamental theory of the strong interaction, quantum chromodynamics (QCD), which is not suitable for perturbative calculations except in some special cases. For example, in (2.12) the time t is represented by a network of points, $t_k = t_0 + k\epsilon$, and the function $q(t)$ through the value that it takes at times t_k, $q_k = q(t_k)$.

The approximation to the transition amplitude provided by (2.12) has errors $\mathcal{O}(\epsilon)$, which was sufficient in the preceding discussion to demonstrate convergence to the result of (2.13). If interest is focused on the method of numerical calculation, however, the *speed* with which the approximate results converge to the exact value becomes of great practical importance. For example, as the reader can easily show, the simple modification of equation (2.8)

$$e^{-iH\epsilon} = e^{-i(V/2)\epsilon} e^{-iK\epsilon} e^{-i(V/2)\epsilon} + \mathcal{O}(\epsilon^3) , \qquad (2.14)$$

allows the construction of a method of calculation which converges much more quickly to the exact result.

---

[1] In the general theory of integration, the *measure* is the weight one gives to a particular volume around a point x. Integrating a function in Cartesian coordinates, the measure is simply the product $dq_1 \cdots dq_n$. Integrating over a space of functions requires defining the weight to give to a particular set of functions which are infinitesimally close to each other, as does the limit of equation (2.12).

2.3 THE CLASSICAL LIMIT

The formulation of quantum mechanics by means of the sum over paths of equation (2.13) is particularly well adapted to the discussion of the classical limit of a quantum theory. To be clear, we held that, in every case, the "true" theory is the quantum theory and that classical theory is a special limiting case. Since the classical limit is obtained when $\hbar \to 0$, it is convenient to rewrite (2.13) making the dependence on $\hbar$ explicit,

$$\langle q_N | e^{-iTH/\hbar} | q_0 \rangle = \int d[q(t)] e^{iS(q(t))/\hbar} \ . \tag{2.15}$$

The situation becomes approximately classical if the value of the action is very large compared to $\hbar$. Let us suppose that there exists a trajectory $q_c(t)$, such that $q_c(t_0) = q_0$, and $q_c(t_0 + T) = q_N$, which minimises the action. The condition $\delta S = 0$ implies that trajectories close to $q_c(t)$ contribute to the integral (2.15) with the same phase (or very similar phases), and therefore they interfere constructively. Conversely, near every trajectory which *is not* a minimum of the action there are others with different phases which interfere destructively. Therefore the integral will be dominated by the contributions from trajectories close to $q_c(t)$, trajectories for which the action differs from $S(q_c(t))$ by less than $\hbar$.

In the limit $\hbar \to 0$ the motion of the quantum system will be described by the "classical" trajectory $q_c(t)$.

It is certainly interesting to note that this argument allows the explanation of an otherwise rather mysterious fact. The action principle $\delta S = 0$ is normally demonstrated starting from Newton's equation of motion, but this derivation does not explain the reason for its existence. Instead, the origin of the action principle is clear if we recall that classical mechanics is nothing more than the particular limit of quantum mechanics when $\hbar \to 0$.

2.4 TIME AS A COMPLEX VARIABLE

Up to now we have discussed the sum over paths without worrying excessively about the convergence of the integrals, for example those which appear in equation (2.13). In reality, by examining (2.13) it is immediately clear that there is a problem: the integrand $\exp[iS(q(t))]$ is of unit modulus, and therefore the definition of the integral requires some care. In fact, we have already encountered the same problem in the calculation of the transition amplitude in the absence of forces, equation (2.3) (see Appendix A). In that case, we saw that it is necessary to define the transition amplitude relative to a time T as a limit, for $\eta \to 0^+$, of the amplitude relative to a time interval $T - i\eta$. We will apply the same method to define the integral over paths from (2.13), which we rewrite again in a more explicit form,

$$\langle q_2 | e^{-iHT} | q_1 \rangle = \int d[q(t)] \exp(iS)$$

$$= \int d[q(t)] \exp\left(i \int_{t_1}^{t_2} dt \left[\frac{m\dot{q}^2}{2} - V(q) \right] \right) . \qquad (2.16)$$

To give a negative imaginary part to the time, we write $t = (1 - i\chi)\tau$, with τ real and χ constant, small and positive, so that $(1 - i\chi)^{-1}$ can be approximated by $(1 + i\chi)$. Therefore we have

$$t = (1 - i\chi)\tau, \qquad \text{such that} \qquad \begin{cases} dt = (1 - i\chi)\tau \\[2mm] \dot{q} = \frac{dq}{dt} = (1 + i\chi)\frac{dq}{d\tau} \end{cases} . \qquad (2.17)$$

and the integrand of (2.16) becomes $\exp(iS_\chi)$, where S_χ is the action calculated with the modified time:

$$\exp(iS_\chi) = \exp\left(i \int d\tau \left[\frac{m}{2} \left(\frac{dq}{d\tau} \right)^2 - V(q) \right] \right)$$

$$\cdot \exp\left(-\chi \int d\tau \left[\frac{m}{2} \left(\frac{dq}{d\tau} \right)^2 + V(q) \right] \right) . \qquad (2.18)$$

At this point, the integrand $\exp(iS_\chi)$ has a modulus equal to $\exp(-\chi I)$ where I is the integral

$$I = \int d\tau \left[\frac{m}{2} \left(\frac{dq}{d\tau} \right)^2 + V(q) \right] = \int d\tau \mathcal{H}(q, \dot{q}) ,$$

and $\mathcal{H}(q, \dot{q})$ is the energy of the particle. For clarity we note that, since $q(t)$ is an arbitrary trajectory, in general $\mathcal{H}(q, \dot{q})$ is not independent of time. Regarding the convergence of the functional integral in equation (2.16), we can distinguish several cases according to the behaviour of the potential energy $V(q)$:

$V(\mathbf{q}) = \mathbf{0}$. This is the case of a free particle, where the functional integral can be explicitly calculated with the limiting procedure outlined in Section 2.1 and the basic methods discussed in Appendix A. The functional integral converges.

$V(\mathbf{q})$ positive definite. In this case $I > I_0$, where I_0 is the value of I calculated along the trajectory for which $V(q) = 0$. Therefore the integral has at least the same convergence as in the preceding case.

V(q) limited from below. In this case, if $V(q) > V_0$, $I > I_0 + V_0 T$. The addition of a constant $V_0 T$ does not change the convergence compared to the two preceding cases.

V(q) not limited from below. It is necessary to evaluate case by case. If, for example, $V(q) = -q^n$, the convergence of the functional integral depends on the value of the exponent n. It can be shown that the functional integral converges if $0 \geq n \geq -1$, and does not converge if $n > 0$ or $n < -1$. Therefore the Coulomb potential is a limiting case.

We note that the cases excluded are those in which the alternative formulations of quantum mechanics—for example those based on wave mechanics—also fail.

2.5   STATISTICAL MECHANICS

In the preceding considerations, we substituted the integration along the real time axis with an integration in the complex plane of t, along the line identified by

$$t = (1 - i\chi)\tau \simeq e^{-i\chi}\tau = (\cos\chi - i\sin\chi)\tau . \qquad (2.19)$$

The result also converges for non-infinitesimal values of χ and we can take this to the extreme case in which $\chi = \pi/2$, or $t = -i\tau$. We then find the following expression for the transition amplitude between imaginary (!) times $t_1 = 0$, $t_2 = -i\beta$

$$\langle q_2 | e^{-\beta H} | q_1 \rangle = \int d[q(\tau)] \exp\left(-\int_0^\beta d\tau \left[\frac{m}{2}\left(\frac{dq}{d\tau}\right)^2 + V(q)\right]\right)$$

$$= \int d[q(\tau)] \exp\left(-\int_0^\beta d\tau \, \mathcal{H}(q, \dot{q})\right) , \qquad (2.20)$$

where the integral is along paths which go from q_1 for $\tau = 0$ to q_2 for $\tau = \beta$. This expression curiously resembles the partition function of statistical mechanics. To understand this relation we restrict the functional integral to *periodic paths*, i.e. such that $q_1 = q(0) = q_2 = q(-i\beta)$.

We assume that the eigenfunctions of the Hamiltonian of our quantum particle are given by $|m\rangle$ with corresponding eigenvalues E_m. With the help of (2.20) the partition function of a particle in thermal equilibrium at an inverse temperature[2] β can be expressed as a path integral,

$$
\begin{aligned}
Z(\beta) &= \sum_m e^{-\beta E_m} \\
&= \sum_m \langle m | e^{-\beta H} | m \rangle \qquad \text{(a trace...)} \\
&= \text{Tr } e^{-\beta H} = \int dq \langle q | e^{-\beta H} | q \rangle \\
&= \int d[q(\tau)] \exp \left(- \int_0^\beta d\tau \, \mathcal{H}(q, \dot{q}) \right) \qquad \text{(...a path integral)} , \quad (2.21)
\end{aligned}
$$

where the integral is over all the cyclic paths, which start from an arbitrary value of q at $t = 0$ and return to the same point for $t = -i\beta$. Thus defined, the path integral absorbs the integration which derives from the trace (the penultimate step of equation (2.21)).

We note also that in the limit $\beta \to \infty$, the partition function is dominated by the ground state:

$$
Z(\beta) \xrightarrow[\beta \to \infty]{} \exp(-\beta E_0)(1 + \text{terms exponentially small in } \beta) . \qquad (2.22)
$$

2.6   GREEN'S FUNCTIONS

We define as Green's functions the expectation values of the product of operators in the ground state; for example, in the case of a particle in one-dimensional motion, the products of the variable $q(t)$ taken at different times, $t_1, t_2, \ldots t_N$ *in order of decreasing time,*

$$
\langle 0 | q(t_1) q(t_2) \ldots q(t_N) | 0 \rangle \qquad (t_1 \geq t_2 \geq \ldots \geq t_N) .
$$

We have adopted the Heisenberg representation, or

$$
q(t_1) = e^{iHt} q e^{-iHt} .
$$

As we will see, an expression of this type can be simply written as a sum over paths.

To extend the definition to the case of arbitrary times $t_1, t_2, \ldots t_N$, we introduce the concept of the time-ordered product, $T(q(t_1) q(t_2) \ldots q(t_N))$, which is simply the product of the same operators ordered *according to decreasing time.* For example, in the case of two operators:

$$
T(q(t_1) q(t_2)) = \begin{cases} q(t_1) q(t_2) & \text{if } t_1 \geq t_2 \\ q(t_2) q(t_1) & \text{if } t_2 \geq t_1 \end{cases} . \qquad (2.23)
$$

---

[2]We recall that $\beta = 1/k_B T$, where T is the absolute temperature and k_B is Boltzmann's constant. So as not to cause confusion with time, in the text we use β instead of temperature.

We can then define the Green's function at N times as

$$G_N(t_1, t_2, \ldots t_N) = \langle 0|T\left(q(t_1)\, q(t_2)\, \ldots q(t_N)\right)|0\rangle . \qquad (2.24)$$

The motivation leading to the above definition needs to be clarified. Why should special importance be given to expectation values in the ground state $|0\rangle$? Why introduce time ordering?

The answer to these questions are to be found in field theory, for which we are preparing. In simple terms, the ground state of field theory is the "vacuum" state, which is devoid of particles. The vacuum has several properties which make it uniquely worthy of interest: from the vacuum, all other states can be created by means of creation operators; the vacuum is the only state of a field theory which is invariant under translations in space and time, and under Lorentz transformations, rotations, and more besides. For the time ordering, we will see that there is a direct relationship between the Green's functions defined as in equation (2.24) and the elements of the S-matrix. On the other hand, we already know (cf. Dyson's formula, in [1]) that the T-ordered product assumes a central role in perturbation theory.

We now show that equation (2.24) can be transformed into the following sum over paths

$$G_N(t_1, \ldots, t_N) = \frac{\int d[q(t)]\exp(iS)\, q(t_1)\, q(t_2)\, \ldots q(t_N)}{\int d[q(t)]\exp(iS)} . \qquad (2.25)$$

In the numerator, and in the denominator, the integral extends over all paths between $t = -\infty$ and $t = +\infty$, such that $q(+\infty) = q(-\infty)$. Moreover, it is understood that times are obtained as limits of complex times, in accordance with what was discussed in Section 2.4. More precisely

$$G_N(t_1, \ldots, t_N) = \lim_{\chi \to 0^+} \lim_{T \to \infty} \left[\frac{\int d[q(t')]\exp(iS)q(t'_1)\, q(t'_2)\, \ldots q(t'_N)}{\int d[q(t')]\exp(iS)}\right], \qquad (2.26)$$

and the integrals extend over all *periodic* paths between $t = -T' = -T(1-i\chi)$ and $t = T' = T(1-i\chi)$, such that $q(T') = q(-T')$, and $t'_{1,2,\ldots} = (1-i\chi)\, t_{1,2,\ldots}$.

To prove the equivalence of (2.26) and (2.24) we consider for simplicity the case of two operators, and make explicit the time dependence of the operators in (2.24). We consider the case in which $t_1 \geq t_2$

$$G_2(t_1, t_2) = \langle 0|q(t_1)\, q(t_2)|0\rangle = \langle 0|e^{iHt_1}q\, e^{-iH(t_1-t_2)}q\, e^{-iHt_2}|0\rangle \quad (t_1 \geq t_2)$$

$$= \frac{\langle 0|e^{-iH(T-t_1)}q\, e^{-iH(t_1-t_2)}q\, e^{-iH(t_2+T)}|0\rangle}{\langle 0|e^{-2iHT}|0\rangle} . \qquad (2.27)$$

In the second step, T is an arbitrary time, but such that $T > t_1 > t_2 > -T$. The denominator, $\langle 0|e^{-2iHT}|0\rangle = \exp(-2iE_0T)$, balances the introduction of the two factors e^{-iHT} in the numerator, so that the result does not depend on T. At this point we introduce the complex time, and write

$$G_2(t_1, t_2) = \lim_{\chi \to 0^+} \lim_{T \to \infty} \left[\frac{\langle 0|e^{-i(T'-t_1')H} q \, e^{-i(t_1'-t_2')H} q \, e^{-i(t_2'+T')H}|0\rangle}{\langle 0|e^{-2iT'H}|0\rangle} \right],$$

where $T' = (1 - i\chi)T$, $t_{1,2}' = (1 - i\chi) t_{1,2}$. The limit $T \to \infty$ is apparently useless, given our observation that the term in square brackets is independent of T. However this allows us to go from the expectation value in $|0\rangle$ to the trace of the operators in both the numerator and denominator

$$G_2(t_1, t_2) = \lim_{\chi \to 0^+} \lim_{T \to \infty} \left[\frac{\sum_m \langle m|e^{-i(T'-t_1')H} q \, e^{-i(t_1'-t_2')H} q \, e^{-i(t_2'+T')H}|m\rangle}{\sum_m \langle m|e^{-2iT'H}|m\rangle} \right].$$

For every $\chi > 0$ and for "large" T, the contribution of the excited states $|m\rangle \neq |0\rangle$ is in fact reduced, both in the numerator and denominator, by a factor $\exp\left[-2\chi(E_m - E_0)T\right]$ with respect to the ground state, and disappears in the limit $T \to \infty$, which is carried out *before* taking $\chi \to 0^+$. We note that the order of the limits is important!

Given that the trace is independent of the basis chosen to describe the states, we can use the basis of the position eigenstates, $|q\rangle$, to obtain

$$G_2(t_1, t_2) = \lim_{\chi \to 0^+} \lim_{T \to \infty} \left[\frac{\int d\tilde{q} \langle \tilde{q}|e^{-i(T'-t_1')H} q \, e^{-i(t_1'-t_2')H} q \, e^{-i(t_2'+T')H}|\tilde{q}\rangle}{\int d\tilde{q} \langle \tilde{q}|e^{-2iT'H}|\tilde{q}\rangle} \right].$$

To complete the proof it is sufficient to show that the expression in square brackets is identical to what we would obtain from (2.26) in the case of two operators,

$$\frac{\int d\tilde{q} \langle \tilde{q}|e^{-iH(T'-t_1')} q \, e^{-iH(t_1'-t_2')} q \, e^{-iH(t_2'+T')}|\tilde{q}\rangle}{\int d\tilde{q} \langle \tilde{q}|e^{-2iHT'}|\tilde{q}\rangle}$$

$$= \frac{\int d[q(t')] \exp(iS) q(t_1') \, q(t_2')}{\int d[q(t')] \exp(iS)} . \qquad (2.28)$$

Actually, the numerator and the denominator in the two expressions are separately equal. The denominator, $\langle \tilde{q}|e^{-2iT'H}|\tilde{q}\rangle$, is a transition amplitude, which we can express as $\int d[q(t')] \exp(iS)$, where the integration is over all the paths which originate from $\tilde{q}$ at $t = -T'$ and return to the same point for $t = T'$. To carry out the integration over $\tilde{q}$ it is enough to extend the path integral to all those paths which, starting from *any* point for $t = -T'$, return to the starting point for $t = T'$. The two denominators are therefore equal.

For the numerator we assume that $T > t_1 > t_2 > -T$, and we introduce alongside the two q operators, two sums over the complete set of states,

$\int dq |q\rangle\langle q| = 1$

$$\int d\tilde{q}\langle\tilde{q}|e^{-iH(T'-t'_1)}q\ e^{-iH(t'_1-t'_2)}q\ e^{-iH(t'_2+T')}|\tilde{q}\rangle$$

$$= \iiint d\tilde{q}dq_1dq_2\langle\tilde{q}|e^{-iH(T'-t'_1)}|q_1\rangle\langle q_1|q\ e^{-iH(t'_1-t'_2)}|q_2\rangle\langle q_2|q\ e^{-iH(t'_2+T')}|\tilde{q}\rangle$$

$$= \iiint d\tilde{q}dq_1dq_2\ q_1q_2\langle\tilde{q}|e^{-iH(T'-t'_1)}|q_1\rangle\langle q_1|\ e^{-iH(t'_1-t'_2)}|q_2\rangle\langle q_2|\ e^{-iH(t'_2+T')}|\tilde{q}\rangle\ .$$

The last expression contains the product of three transition amplitudes, which we can write as a single sum over paths which pass respectively through q_1 at time t'_1 and q_2 at t'_2:

$$[q(-T') = \tilde{q}] \rightarrow [q(t'_2) = q_2] \rightarrow [q(t'_1) = q_1] \rightarrow [q(T') = \tilde{q}]\ .$$

The integration over $\tilde{q}$, q_1, q_2 implies extending the functional integral over all periodic paths for which $[q(T') = q(-T')]$, substituting the factor q_1q_2 with $q(t'_1)q(t'_2)$, and so obtaining the numerator of the right-hand side of the equality (2.28).

We leave as an exercise the proof that the result is correct also for the other possible time ordering, $t_2 > t_1$. The proof is easily extended, following the same steps, to the case in which more than two operators are present.

TOWARDS A FIELD THEORY

CONTENTS

The procedure we described for the case of a single degree of freedom can be extended immediately to the case of a finite number of degrees of freedom. All the results obtained are directly applicable to a system with n degrees of freedom, provided it is understood that the symbol q should be interpreted as a vector with n components, $q = \{q_1 \ldots q_n\}$. For example, a Green's function can be defined as

$$G_{k_1, k_2 \ldots, k_N}(t_1, t_2, \ldots t_N) = \langle 0 | T \left[q_{k_1}(t_1) \, q_{k_2}(t_2) \, \ldots q_{k_N}(t_N) \right] | 0 \rangle \ . \qquad (3.1)$$

By "path" is meant the trajectory of the vector $q_k(t)$ between the initial time t_1 and the final time t_2, or the set of functions $q_k(t)$ in the interval $t_1 \geq t \geq t_2$.

These ideas are easily extended, at least from a formal viewpoint, to a field theory. For example, considering the case of a real scalar field $\phi(\mathbf{x}, t)$, we can define the Lagrangian L as the integral of a Lagrangian density and the action as the time integral of the Lagrangian:

$$\mathcal{L} = \mathcal{L}(\phi, \partial_\mu \phi) \ , \quad L = \int d^3x \; \mathcal{L}(\phi, \partial_\mu \phi) \ ,$$

$$S = \int dt \, L = \int d^4x \; \mathcal{L}(\phi, \partial_\mu \phi) \ . \qquad (3.2)$$

We can think of this system as the limit of a field defined on a lattice of points $\mathbf{x}_k$ with separation a, the *lattice spacing*, which covers a cube of side L. A double limit is taken, of $a \rightarrow 0$, $L \rightarrow \infty$. For each value of a, L has a finite number $n = (L/a)^3$ of points, and the field is described by n dynamic variables $\phi_k(t) = \phi(\mathbf{x}_k, t)$. In the discretised version, we will write the action as

$$S = \int dt \sum_k a^3 \, \mathcal{L}[\phi(\mathbf{x}_k, t), \partial_\mu \phi(\mathbf{x}_k, t)] = \int dt \sum_k a^3 \mathcal{L}_k \,,$$

where $\mathcal{L}_k$ is the Lagrangian density at point $\mathbf{x}_k$, calculated by approximating the field derivatives by differences. For example

$$\frac{\partial \phi(\mathbf{x}_k, t)}{\partial x} \approx \frac{\phi(x_k + a, y_k, z_k, t) - \phi(x_k, y_k, z_k, t)}{a} \,.$$

The description of a field by means of a lattice of points, and a transition towards the continuum limit, can be used to formally define a field theory but also to carry out numerical calculations of interesting quantities, for example the Green's functions. In this second case it is usually useful also to represent time as a discrete variable, as we mentioned in Section 2.2.

In conclusion, a field theory can be considered as a limiting case of a system with many degrees of freedom, and the corresponding quantum theory can be defined by means of the sum over paths. By "path", with fixed values of a and L, is meant the set of functions $\phi_k(t)$ in the relevant time interval. In the limit $a \to 0$, $L \to \infty$ this becomes the value of the function $\phi(\mathbf{x}, t)$ for all values of $\mathbf{x}$ and t in the given interval. The convergence of the procedure remains to be proven, something that cannot be done in general, except in the simplest case, that of free fields, and a few other examples. For interacting fields it is necessary to turn to perturbation theory, which will be discussed later. We will then see that an essential step is constituted by the procedure of renormalisation, which is necessary to eliminate—or better, interpret—the divergences which emerge in the limit of small distances, as $a \to 0$.

Supposing the problems of taking the limit to have been resolved, what we have said in the previous section is directly applicable to a field theory. For example, we can define the N-point Green's function as

$$G(x_1, x_2, \ldots x_N) = \langle 0 | T \left[\phi(x_1) \phi(x_2) \ldots \phi(x_N) \right] | 0 \rangle \,, \tag{3.3}$$

where $|0\rangle$ denotes the vacuum state, and x_k the 4-vector $(\mathbf{x}_k, t_k)$. As in the finite-dimensional case, G can be expressed as a functional integral:

$$G(x_1, \ldots x_N) = \frac{\int d[\phi(x)] \exp(iS) \, \phi(x_1) \phi(x_2) \ldots \phi(x_N)}{\int d[\phi(x)] \exp(iS)} \,, \tag{3.4}$$

where $d[\phi(x)]$ represents the measure in the space of the functions $\phi(x)$. In this case also, as in Section 2.6, this expression should be understood as a limit,

$$G(x_1, \ldots x_N) =$$

$$\lim_{\substack{\chi \to 0^+ \\ \substack{T \to \infty \\ L \to \infty}}} \left[\frac{\int d[\phi(\mathbf{x}, t')] \exp(iS) \, \phi(\mathbf{x}_1, t_1') \phi(\mathbf{x}_2, t_2') \ldots \phi(\mathbf{x}_N, t_N')}{\int d[\phi(\mathbf{x}, t')] \exp(iS)} \right] \,, \tag{3.5}$$

and the integral in square brackets extends over all functions $\phi(\mathbf{x}, t')$ with $t' = (1 - i\chi)t$ included between $\pm(1 - i\chi)T$, and periodic in both time and space, such that

$$\phi[\mathbf{x}, (1 - i\chi)T] = \phi[\mathbf{x}, -(1 - i\chi)T] ,$$
$$\phi(x, y, z, t) = \phi(x + L, y, z, t) = \phi(x, y + L, z, t) = \phi(x, y, z + L, t) . \tag{3.6}$$

The periodicity both in the spatial and time direction allows integration by parts to be freely carried out. For example

$$\int d^4x \, (\partial_\mu \phi \, \partial^\mu \phi) = -\int d^4x \, (\phi \, \partial_\mu \partial^\mu \phi) . \tag{3.7}$$

These expressions can be generalised to the case of many scalar fields, $\phi_k(x), \; (k = 1 \ldots n)$

$$
\begin{aligned}
G_{k_1 \ldots, k_N}(x_1, \ldots x_N) &= \langle 0| T \left[\phi_{k_1}(x_1) \ldots \phi_{k_N}(x_N) \right] |0\rangle \\
&= \frac{\int \prod_{k=1}^{n} d[\phi_k(x)] \exp(iS) \, \phi_{k_1}(x_1) \ldots \phi_{k_N}(x_N)}{\int \prod_{k=1}^{n} d[\phi_k(x)] \exp(iS)} .
\end{aligned} \tag{3.8}
$$

The case of a complex field can be treated as two real fields,[1]

$$\phi(x) = \frac{1}{\sqrt{2}}(\phi_1 + i\phi_2); \qquad \phi^\dagger(x) = \frac{1}{\sqrt{2}}(\phi_1 - i\phi_2) . \tag{3.9}$$

The factor $1/\sqrt{2}$ is chosen so that $d[\phi(x)]d[\phi^\dagger(x)] = d[\phi_1(x)]d[\phi_2(x)]$ as verified by the Jacobian. Using this relation, the calculation of Green's function leads to (3.8) and the general result is obtained as

$$
\begin{aligned}
&\langle 0| T \left[\phi(x_1) \ldots \phi(x_N) \, \phi^\dagger(y_1) \ldots \phi^\dagger(y_M) \right] |0\rangle \\
&= \frac{\int d[\phi(x)]d[\phi^\dagger(x)] \exp(iS) \, \phi(x_1) \ldots \phi(x_N) \, \phi^\dagger(y_1) \ldots \phi^\dagger(y_M)}{\int d[\phi(x)]d[\phi^\dagger(x)] \exp(iS)} .
\end{aligned} \tag{3.10}
$$

In the simple case of a two-point function

$$
\begin{aligned}
&\langle 0| T \left[\phi(x) \, \phi^\dagger(y) \right] |0\rangle \\
&= \frac{\int d[\phi(x)]d[\phi^\dagger(x)] \exp(iS) \, \phi(x) \, \phi^\dagger(y)}{\int d[\phi(x)]d[\phi^\dagger(x)] \exp(iS)} .
\end{aligned} \tag{3.11}
$$

The proof of equations (3.10) and (3.11) is elementary, it being sufficient to note that the Green's functions defined via the expectation values are linear

---

[1] We will use the † symbol to denote both Hermitian conjugation, in the case of operators, and complex conjugation of numerical functions which represent the trajectories of the same quantities.

in the fields at each point, and that this is obviously true from their definition in terms of a sum over paths, for which

$$
\begin{aligned}
\langle 0|T\left[\cdots(\phi_1(x) \pm i\phi_2(x))\cdots\right]|0\rangle &= \\
&= \langle 0|T\left[\cdots\phi_1(x)\cdots\right]|0\rangle \pm i\langle 0|T\left[\cdots\phi_2(x)\cdots\right]|0\rangle \\
&= \frac{\int d[\phi(x)]d[\phi^\dagger(x)]\exp(iS)\ \cdots(\phi_1(x) \pm i\phi_2(x))\cdots}{\int d[\phi(x)]d[\phi^\dagger(x)]\exp(iS)} .
\end{aligned}
$$

3.1   THE GENERATING FUNCTIONAL

In this section we introduce the formalism based on the use of the *generating functional*, which allows expression of the set of Green's functions in a compact form. We consider the case of a system with n degrees of freedom $q_k(t)$, $(k = 1 \ldots n)$. We define the generating functional Z as $[\mathbf{q} \equiv (q_1, \ldots, q_n)]$

$$
\begin{aligned}
Z[J] &\equiv Z\left[J_1(t), \ldots J_n(t)\right] = \\
&= \int d[\mathbf{q}(t)]\exp\left(iS\left(\mathbf{q}(t)\right) - i\int dt \sum_k q_k(t)J_k(t)\right) , \quad (3.12)
\end{aligned}
$$

where the $J_k(t)$ are n functions of time, and the functional integral is defined as the limit

$$
\begin{aligned}
Z[J] &\equiv Z\left[J_1(t), \ldots J_n(t)\right] = \\
&= \lim_{\chi\to 0^+} \lim_{T\to\infty} \int d[\mathbf{q}(t')]\exp\left(iS\left(\mathbf{q}(t')\right) - i\int dt' \sum_k q_k(t')J_k(t')\right) ,
\end{aligned}
$$

$$(3.13)$$

and the integrals in brackets extend over all *periodic* paths between $t = -T' = -T(1 - i\chi)$ and $t = T' = T(1 - i\chi)$, such that $q(T') = q(-T')$, and $t' = (1 - i\chi)t$.

From the definition, it follows immediately that $Z[0]$ is precisely the denominator which appears in the expression for the Green's functions in the path integral formalism. We will now prove that the Green's functions are obtained as functional derivatives of $Z[J]$.

The concept of "functional derivative" is very simple. We consider a function $f(x)$, and a functional $X[f(x)]$. We then consider a variation of the function $f(x)$ at the point $x = y$, i.e. $f(x) \to f(x) + \epsilon\delta^{(4)}(x - y)$. The functional derivative $\delta X[f(x)]/\delta f(y)$ is defined, in analogy with the derivative of an ordinary function of one variable, to be

$$
\frac{\delta X[f(x)]}{\delta f(y)} = \lim_{\epsilon\to 0} \frac{1}{\epsilon}\left\{X[f(x) + \epsilon\delta^{(4)}(x - y)] - X[f(x)]\right\} . \quad (3.14)
$$

Using (3.14) we now see how to obtain the numerator which appears in the expression for the Green's function in the case of the scalar free field, in which the generating functional is written

$$Z[J] = \int d[\phi] \; e^{S - i \int d^4x \phi(x) J(x)} \; . \tag{3.15}$$

We consider the functional derivative

$$
\begin{aligned}
i \frac{\delta Z[J]}{\delta J(x_2)} &= i \lim_{\epsilon \to 0} \left\{ \int d[\phi] \; e^{\left\{ iS - i \int d^4x \phi(x)[J(x) + \epsilon \delta^{(4)}(x - x_2))\right\}} \right. \\
&\qquad \left. - \int d[\phi] \; e^{\left\{ iS - i \int d^4x \phi(x) J(x)\right\}} \right\} \\
&= i \lim_{\epsilon \to 0} \left\{ \int d[\phi] \; e^{\left\{ iS - i \int d^4x \phi(x) J(x)\right\}} e^{-i\epsilon \int d^{(4)} \phi(x) \delta^{(4)}(x - x_2)} \right. \\
&\qquad \left. - \int d[\phi] \; e^{\left\{ iS - i \int d^4x \phi(x) J(x)\right\}} \right\} \; . \tag{3.16}
\end{aligned}
$$

Using the δ-function to carry out the integration, and expanding the exponential in a power series in ϵ we obtain

$$i \frac{\delta Z[J]}{\delta J(x_2)} = \int d[\phi] \; \phi(x_2) \; e^{\left\{ iS - i \int d^4x \phi(x) J(x)\right\}} \; , \tag{3.17}$$

and differentiating a second time

$$\frac{i\delta}{\delta J(x_1)} \frac{i\delta}{\delta J(x_2)} Z[J] = \int d[\phi] \; \phi(x_1) \; \phi(x_2) \; e^{\left\{ iS - i \int d^4x \phi(x) J(x)\right\}} \; . \tag{3.18}$$

We have therefore proven that the two-point Green's function can be written in the form

$$G(x_1, x_2) = \frac{1}{Z[0]} \left[\frac{i\delta}{\delta J(x_1)} \frac{i\delta}{\delta J(x_2)} Z[J] \right]_{J=0} \; . \tag{3.19}$$

The result obtained can be extended easily to the case of many-point Green's functions and to more complex field theories. In the case of N fields $\phi_1(x) \ldots \phi_N(x)$, the generating functional will depend on N functions of x, $J_k(x)$, $(k = 1 \ldots N)$ according to

$$Z[J] = \int d[\phi(x)] \exp \left(iS\left(\phi(x)\right) - i \int d^4x \sum_k \phi_k(x) J_k(x) \right) \; . \tag{3.20}$$

The definition of this integral requires the prescription of limits and periodicity specified in the previous section, equations (3.5) and (3.6). The N-point

Green's functions are then

$$
\begin{aligned}
G_{k_1,\ldots k_N}(x_1, \cdots x_N) &= \langle 0|T\left[\phi_1(x_1)\cdots\phi_N(x_N)\right]|0\rangle \\
&= \frac{\int d[\phi(x)]\phi(x_1)\ldots\phi(x_N)\exp\left(iS\left(\phi(x)\right)\right)}{\int d[\phi(x)]\exp\left(iS\left(\phi(x)\right)\right)} = \\
&= (i)^N \frac{1}{Z[0]} \left[\frac{\delta^N Z[J]}{\delta J_1(x_1)\cdots\delta J_N(x_N)}\right]_{J=0}
\end{aligned} \tag{3.21}
$$

We note that equation (3.21) is obtained by the simple substitution rule

$$
\phi_k(x) \to i\frac{\delta}{\delta J_k(x)} \ , \tag{3.22}
$$

which we can use, for example, in the following way: if $F[\phi]$ is an arbitrary functional of fields, we obtain

$$
\int d[q(t)]\, F[\phi]\exp\left(iS\right) = \left[F\left[i\frac{\delta}{\delta J}\right]Z[J]\right]_{J=0} , \tag{3.23}
$$

a result which we will use to derive the perturbative expansion of a field theory in the presence of interactions.

For a complex field we can define a generating functional by means of two real functions J_1, J_2 or, better, in terms of a complex function J and its complex conjugate $J^\dagger$,

$$
Z(J, J^\dagger) = \int d[\phi(x)]d[\phi^\dagger(x)]\times
$$

$$
\times \exp\left(iS\left(\phi(x),\phi^\dagger(x)\right) - i\int d^4x\left(\phi^\dagger(x)J(x) + J^\dagger(x)\phi(x)\right)\right) . \tag{3.24}
$$

The Green's functions are then obtained by the substitution rule [see equation (3.22)]

$$
\phi(x) \to i\frac{\delta}{\delta J^\dagger(x)} \quad , \quad \phi^\dagger(x) \to -i\frac{\delta}{\delta J(x)} \ , \tag{3.25}
$$

which can be extended in an obvious way to the case of further real or complex fields.

Invariance under translations and conservation of four-momentum. The form of the Green's function given in the second line of (3.21) allows us to discuss invariance under translation in a very simple way. If we consider a translation by a^μ, we can write:

$$
\begin{aligned}
G_{k_1,\ldots k_N}&(x_1 + a, \cdots x_N + a) \\
&= \frac{\int d[\phi(x)]\phi(x_1 + a)\ldots\phi(x_N + a)\exp\left(iS\left(\phi(x)\right)\right)}{\int d[\phi(x)]\exp\left(iS\left(\phi(x)\right)\right)} \\
&= \frac{\int d[\phi(x)]\phi(x_1 + a)\ldots\phi(x_N + a)\exp\left(iS\left(\phi(x + a)\right)\right)}{\int d[\phi(x)]\exp\left(iS\left(\phi(x)\right)\right)} \\
&= \frac{\int d[\phi(x)]\phi(x_1)\ldots\phi(x_N)\exp\left(iS\left(\phi(x)\right)\right)}{\int d[\phi(x)]\exp\left(iS\left(\phi(x)\right)\right)} \\
&= G_{k_1,\ldots k_N}(x_1, \cdots x_N) , \quad (3.26)
\end{aligned}
$$

where we have used the translational invariance of the action, and the last step follows from invariance of the measure of the functional integration following the substitution $\phi(x) \to \phi(x + a)$.

To conclude, we note that, while the Green's functions depend only on the differences between x_k, their Fourier transforms must contain a δ-function which expresses the conservation of four-momentum

$$
\begin{aligned}
\int \cdots \int d^4x_1 \ldots d^4x_N \; e^{i(p_1 x_1 + \cdots + p_N x_N)} G_{k_1,\ldots k_N}(x_1, \cdots x_N) \\
= (2\pi)^4 \delta^{(4)}(p_1 + p_2 + \ldots p_N) \; \widetilde{G}(p_1, \ldots, p_N) , \quad (3.27)
\end{aligned}
$$

where $\widetilde{G}$ is a regular function of four-momentum. We leave it to the reader to prove (3.27) starting from (3.26) (see, for example, [1], Chapter 12).

3.2   THE HARMONIC OSCILLATOR

In this section we will apply the concepts that we have illustrated to a particularly simple case, that of a harmonic oscillator of mass $m = 1$. The Hamiltonian and the Lagrangian of this system are, respectively

$$
H = \frac{1}{2}(p^2 + \omega^2 q^2) , \quad L = \frac{1}{2}(\dot{q}^2 - \omega^2 q^2) . \tag{3.28}
$$

We wish to use the method of the generating functional, Z, to calculate the Green's functions. We will describe the calculations in detail, because the same procedure is applied in interesting contexts, for example to field theory. Before proceeding with the functional Z, we will calculate the two-point Green's function starting from the usual formulation of quantum mechanics, to be able to confirm the equivalence of the two approaches.

We recall that, in terms of the creation and destruction operators (see, for example, [7]), we can write the relations

$$q = \frac{1}{\sqrt{2\omega}}(a + a^\dagger) ,$$

$$a^\dagger|0\rangle = |1\rangle \quad , \quad a^\dagger|1\rangle = \sqrt{2}|2\rangle \quad , \quad a|1\rangle = |0\rangle ,$$

$$H|0\rangle = \frac{\omega}{2}|0\rangle , \quad H|1\rangle = \frac{3\omega}{2}|1\rangle , \tag{3.29}$$

from which it follows that

$$q|0\rangle = \frac{1}{\sqrt{2\omega}}|1\rangle \quad , \quad q|1\rangle = \frac{1}{\sqrt{2\omega}}\left(|0\rangle + \sqrt{2}|2\rangle\right) . \tag{3.30}$$

We first consider the time ordering $t_1 > t_2$. In this case

$$\begin{aligned}
G(t_1, t_2) &= \langle 0|q(t_1)\, q(t_2)\, |0\rangle \qquad (t_1 > t_2) \\
&= \langle 0|e^{iHt_1} q\, e^{-iH(t_1-t_2)} q\, e^{-iHt_2}|0\rangle \\
&= \frac{1}{2\omega} e^{-i\omega(t_1-t_2)} ,
\end{aligned}$$

and combining the result with that obtained for $t_2 > t_1$, we find

$$G(t_1, t_2) = \frac{1}{2\omega}\left(e^{-i\omega(t_1-t_2)}\theta(t_1 - t_2) + e^{-i\omega(t_2-t_1)}\theta(t_2 - t_1)\right) . \tag{3.31}$$

For the method of the sum over paths, we note first of all that with an integration by parts

$$S = \frac{1}{2}\int dt \left(\dot{q}^2 - \omega^2 q^2\right) = -\frac{1}{2}\int dt \left(q\ddot{q} + \omega^2 q^2\right) = -\frac{1}{2}\int dt \left(q\mathcal{O}q\right) , \tag{3.32}$$

where $\mathcal{O}$ is the differential operator:

$$\mathcal{O} = \partial_t^2 + \omega^2 . \tag{3.33}$$

The generating functional will therefore be

$$Z[J] = \int d[q(t)] \exp\left[-\frac{i}{2}\int dt \left(q\,\mathcal{O}q + 2qJ\right)\right] . \tag{3.34}$$

The strategy to carry out the integral is the standard one for integrating functions of the type $\exp(-a\, x^2 + bx)$, used, for example, in Appendix A. It consists in rewriting the exponent as a perfect square, to obtain a Gaussian integral. We proceed formally, rewriting equation (3.34) as[2]

$$
Z[J] = \int d[q(t)]
$$

$$
\times \exp \frac{-i}{2} \left[(q + \mathcal{O}^{-1} J) \mathcal{O} (q + \mathcal{O}^{-1} J) - (\mathcal{O}^{-1} J) \mathcal{O} (\mathcal{O}^{-1} J) \right]
$$

$$
= \exp \left[\frac{i}{2} \int dt \, (\mathcal{O}^{-1} J) \mathcal{O} (\mathcal{O}^{-1} J) \right]
$$

$$
\times \int d[q(t)] \exp \left[\frac{-i}{2} \int dt \, (q + \mathcal{O}^{-1} J) \mathcal{O} (q + \mathcal{O}^{-1} J) \right] . \qquad (3.35)
$$

The functional integral in the last expression is in reality a constant independent of J, as can be seen with a change of variables[3] $q \to q' = q + \mathcal{O}^{-1} J$, and in conclusion

$$
Z[J] = K \exp \left[\frac{i}{2} \int dt \, (\mathcal{O}^{-1} J) \mathcal{O} (\mathcal{O}^{-1} J) \right] . \qquad (3.36)
$$

A multiplicative constant has no effect on the value of the Green's functions, for which reason we can simply write

$$
Z[J] = \exp \left[\frac{i}{2} \int dt \, J \mathcal{O}^{-1} J \right] . \qquad (3.37)
$$

The inverse of the differential operator $\mathcal{O}$ will be an integral operator defined through the equation

$$
(\mathcal{O}^{-1} J)(t) = - \int dt' G(t - t') J(t') , \qquad (3.38)
$$

where the function G is the so-called "propagator". The sign has been chosen in view of subsequent developments in field theory and of existing conventions. Obviously the following relation must be proven

$$
\mathcal{O}(\mathcal{O}^{-1} J)(t) = J(t) ,
$$

which implies that the function $G(t)$ must be a solution of the differential equation

$$
\mathcal{O} \, G(t) = -\delta(t) . \qquad (3.39)
$$

---

[2] To confirm this result, note that, with two integrations by parts,

$$
\int dt \, (\mathcal{O}^{-1} J) \, \mathcal{O}\phi = \int dt \, (\mathcal{O}(\mathcal{O}^{-1} J)) \, \phi = \int dt \, J\phi .
$$

[3] We recall that the functional integral extends over all periodic orbits, $q(+\infty) = q(-\infty)$, therefore we would also like q' to be periodic. It is therefore necessary to impose some restrictions on $J(t)$, for example that the allowable functions $J(t)$ tend fast enough to zero for $t \to \pm\infty$. An analogous restriction should be applied in the case of field theory.

However, we must remember that in equation (3.13) $Z[J]$ is defined as a limit starting from values of time with a small[4] imaginary part $t' = (1 - i\chi)t$. For small but not zero χ, the differential operator $\mathcal{O}$ becomes

$$\mathcal{O} = \partial_{t'}^2 + \omega^2 = (1 + 2i\chi)\partial_t^2 + \omega^2 , \qquad (3.40)$$

and the propagator $G(t)$ will be the limit for $\chi \to 0$ of a function $G(t, \chi)$ which satisfies the equation[5] [see equation (2.17)]

$$\left[(1 + 2i\chi)\partial_t^2 + \omega^2\right] G(t, \chi) = -\delta(t) . \qquad (3.41)$$

The general solution to this equation is the sum of a particular solution and the general solution of the homogeneous equation, which we can write, again neglecting terms $\mathcal{O}(\chi^2)$,

$$\partial_t^2 G = -\left[(1 - i\chi)\omega\right]^2 G \quad , \quad G = a\, e^{-i\omega(1-i\chi)t} + b\, e^{i\omega(1-i\chi)t} .$$

To find a solution of (3.41) we use the Fourier transforms

$$G(t, \chi) = \frac{1}{2\pi} \int dE\, G(E, \chi)e^{-iEt} , \qquad \delta(t) = \frac{1}{2\pi} \int dE e^{-iEt} . \qquad (3.42)$$

In this way we find

$$G(E, \chi) = \frac{-1}{\omega^2 - E^2(1 + 2i\chi)} = \frac{(1 - 2i\chi)}{E^2 - \omega^2(1 - 2i\chi)} , \qquad (3.43)$$

and neglecting the multiplicative factor $(1 - 2i\chi)$,

$$G(E, \chi) = \frac{1}{E^2 - \omega^2 + i\epsilon} \quad \text{or} \quad = \frac{1}{E^2 - (\omega - i\eta)^2} , \qquad (3.44)$$

where $\epsilon = 2\chi\omega^2$ and $\eta = \chi\omega$.

We can now calculate $G(t, \chi)$, given by

$$G(t, \chi) = \frac{1}{2\pi} \int dE\, \frac{e^{-iEt}}{E^2 - (\omega - i\eta)^2} . \qquad (3.45)$$

For $t > 0$ the integration path can be completed in the lower half-plane (see Figure 3.2) and includes the pole on the right, while for $t < 0$ the integration path is closed in the upper half-plane and includes the pole on the left. We note that adding an imaginary part to the time results in selection of the pole, which contributes to the integration. Applying the theorem of residues in each

---

[4]In the following manipulations we will neglect terms $\propto \chi^2$.

[5]We are neglecting multiplicative factors which tend to unity when $\chi \to 0$. For example we should have written $\delta(t') = \delta(t)/(1 - i\chi)$, but the additional factor is irrelevant in the limit.

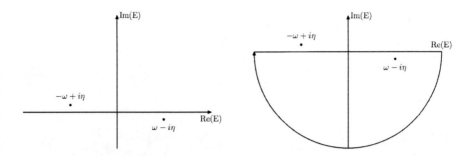

Figure 3.1 Position of the poles in the function $G(x, \chi)$ and integration paths for positive times.

of the two cases, $t > 0$ and $t < 0$, we therefore obtain a particular solution of (3.41)

$$G(t, \chi) = -\frac{i}{2\omega} \left(e^{-i\omega(1-i\chi)t} \theta(t) + e^{i\omega(1-i\chi)t} \theta(-t) \right) . \qquad (3.46)$$

The general solution is

$$G(t, \chi) = \frac{-i}{2\omega} \left(e^{-i\omega(1-i\chi)t} \theta(t) + e^{i\omega(1-i\chi)t} \theta(-t) \right)$$
$$+ a\, e^{-i\omega(1-i\chi)t} + b\, e^{i\omega(1-i\chi)t} . \qquad (3.47)$$

To determine the constants a and b we recall that, to carry out the calculation which leads to (3.37), we used a change of variable in the path integration

$$q(t) \rightarrow q'(t) = q(t) - \delta q(t) \quad , \quad \delta q(t) = \int dt' G(t - t')\, J(t') = q(t) .$$

Given that the integral extends over all the periodic paths between $t = \pm\infty$, the transformed paths must also be periodic, whatever the function $J(t)$. Therefore $\delta q(t)$ must be periodic. The particular solution satisfies this condition, since it tends to zero both for $x \rightarrow \infty$ and for $x \rightarrow -\infty$ (we recall that this limit is taken before that of $\chi \rightarrow 0$) thanks to the presence of the functions $\theta(\pm t)$. Therefore we must consider the effect of the additive terms. If we take $J(t) = \delta(t - t_1)$, with t_1 an arbitrary time, for the additive terms with a and b we obtain:

$$\delta q(t) = a\, e^{-i\omega(1-i\chi)(t-t_1)} + b\, e^{i\omega(1-i\chi)(t-t_1)} .$$

Since the first term diverges for $t \rightarrow -\infty$ and the second for $t \rightarrow +\infty$, the only acceptable solution for (3.41) is that with $a = b = 0$, i.e. solution (3.46), which, taking the limit, $\chi \rightarrow 0$ becomes

$$G(t) = -\frac{i}{2\omega} \left(e^{-i\omega t} \theta(t) + e^{i\omega t} \theta(-t) \right) . \qquad (3.48)$$

The generating functional, equations (3.37) and (3.38), can be written in the form

$$Z[J] = \exp\left[\frac{-i}{2} \iint dt\, dt'\, J(t)\, G(t - t')\, J(t')\right] , \qquad (3.49)$$

which implies

$$G(t, t') = \langle 0|T\left[q(t)\, q(t')\right]|0\rangle = -\frac{1}{Z(0)}\left[\frac{\delta^2 Z[J]}{\delta J(t)\delta J(t')}\right]_{J=0} . \qquad (3.50)$$

We have apparently done much more work than in the usual formulation—although in this case we have taken as known the properties of the creation and destruction operators, so the comparison is not entirely fair—but we have obtained more, since the generating functional of equation (3.49) contains all possible Green's functions of the harmonic oscillator. For example, we could calculate in two lines—we leave it as an exercise—the value of the four-point Green's function $\langle 0|T\left[q(t_1)\, q(t_2)\, q(t_3)\, q(t_4)\right]|0\rangle$.

It is interesting at this point to pretend to know nothing about the structure of the excited states of the harmonic oscillator and to see what we can learn directly from the knowledge of the Green's functions, i.e. from (3.31), which we rewrite for ease of use, in the case $t_1 > t_2$

$$\langle 0|q(t_1)\, q(t_2)|0\rangle = \frac{1}{2\omega}e^{-i\omega(t_1-t_2)} \qquad (t_1 > t_2) . \qquad (3.51)$$

Inserting a complete set of states

$$\langle 0|q(t_1)\, q(t_2)|0\rangle = \sum_X \langle 0|q(t_1)|X\rangle\langle X|q(t_2)|0\rangle \qquad (t_1 > t_2) , \qquad (3.52)$$

and comparing with the previous result we conclude that a state $|1\rangle$ must exist such that

$$\langle 0|q(t_1)|1\rangle = \frac{1}{\sqrt{2\omega}}e^{-i\omega t_1} , \qquad (3.53)$$

and therefore $E_1 = E_0 + \omega$. From the four-point Green's function we can learn something about the second excited state, and so on. The functional $Z[J]$ therefore contains information on the entire spectrum of states.

3.3 FREE SCALAR FIELDS: PROPAGATOR AND GENERATING FUNCTIONAL

The work developed in the previous section allows us to pass on directly to field theory, without introducing conceptual novelties. We will consider here the simplest case, that of a real scalar field $\phi(x)$, where x denotes the 4-vector $\{\mathbf{x}, t\}$. We recall also, that in the system of measurement units which we choose, the speed of light is $c = 1$. The Lagrangian density is, for a field in the absence of interactions,

$$\mathcal{L} = \frac{1}{2} \left[\partial_\mu \phi(x) \, \partial^\mu \phi(x) - m^2 \phi^2 \right] . \qquad (3.54)$$

Analogous with what was done for the harmonic oscillator, we would like to calculate the vacuum expectation value of the time-ordered product of two field operators $\langle 0 | T [\phi(x) \phi(y)] | 0 \rangle$. Following the steps carried out in the case of the harmonic oscillator, we calculate the generating functional [equation (3.20)]. With an integration by parts[6] we can write the action as

$$S = \int d^4x \, \frac{1}{2} \left[\partial_\mu \phi(x) \, \partial^\mu \phi(x) - m^2 \phi^2 \right]$$
$$= - \int d^4x \, \frac{1}{2} \left[\phi(x) \, \partial_\mu \partial^\mu \phi(x) + m^2 \phi^2 \right] = - \int d^4x \, \frac{1}{2} \, (\phi K \phi) , \qquad (3.55)$$

where K is the differential Klein–Gordon operator,

$$K = \partial_\mu \partial^\mu + m^2 = \Box + m^2 . \qquad (3.56)$$

We can then calculate the generating functional (3.20), noting the analogy with the method used in the previous section

$$Z[J] = \int d[\phi(x)] \, \exp \left(iS \, (\phi(x)) - \int d^4x \phi(x) J(x) \right)$$
$$= \int d[\phi(x)] \, \exp \left(-\frac{i}{2} \int d^4x \, [\phi K \, \phi + 2 \, \phi(x) J(x)] \right)$$
$$= \int d[\phi(x)] \exp \left(-\frac{i}{2} \int d^4x \, [(\phi + K^{-1}J) K (\phi + K^{-1}J) - (J \, K^{-1} \, J)] \right)$$
$$= \exp \left(\frac{i}{2} \int d^4x (J \, K^{-1} \, J) \right) \qquad (3.57)$$
$$\times \int d[\phi(x)] \, \exp \left(-\frac{i}{2} \int d^4x \, [(\phi + K^{-1}J) K (\phi + K^{-1}J)] \right) .$$

The residual functional integral can be carried out with a change of variables, $\phi \to \phi' = \phi + K^{-1}J$ and contributes to the result with a constant which can be omitted. We can therefore write simply

$$Z[J] = \exp \left(\frac{i}{2} \int d^4x (J \, K^{-1} \, J) \right) .$$

The inverse of the differential operator K will be an integral operator,

---

[6]In doing the integrations by parts we assume a periodicity for $\phi(x)$ in both time and space, as described in detail in Section 3.

which (to reproduce the usual sign convention for the function Δ_F) we can define through the relation

$$K^{-1}J(x) = -\int dy\, \Delta_F(x-y)J(y) \;, \tag{3.58}$$

where Δ_F must satisfy the equation

$$(\Box + m^2)\Delta_F(x) = -\delta^4(x) \;. \tag{3.59}$$

Again this time, as in the case of the harmonic oscillator, we must consider complex time values, for which we rewrite the operator K as [see equation (3.40)]

$$K = (1+2i\chi)\partial_t^2 - \nabla^2 + m^2 \;. \tag{3.60}$$

Taking the Fourier transforms we therefore find

$$\Delta_F(x) = \frac{1}{(2\pi)^4}\int d^4p\, \Delta_F(p)e^{-ipx} \;,\quad \delta^4(x) = \frac{1}{(2\pi)^4}\int d^4p\, e^{-ipx} \;, \tag{3.61}$$

where p is the 4-vector $\{E, \mathbf{p}\}$, and (see the discussion in the previous section)

$$\Delta_F(p) = \frac{1}{E^2 - \mathbf{p}^2 - m^2 + i\epsilon} \;. \tag{3.62}$$

As we saw, the term $i\epsilon$ in the denominator specifies the integration path.

Naturally, the same result for the two-point Green's function, including the $i\epsilon$ prescription, is obtained starting from the canonical quantisation of the scalar field, discussed in [1].

In conclusion, we have proven that the generating functional for a scalar field has the form

$$Z[J] = \exp\left(\frac{-i}{2}\iint d^4x\, d^4y\, J(x)\,\Delta_F(x-y)\,J(y)\right) \;, \tag{3.63}$$

from which we can derive the different Green's functions. In particular, the two-point function

$$\langle 0|\, T\left[\phi(x)\,\phi(y)\right]|0\rangle = i\Delta_F(x-y) \;. \tag{3.64}$$

The function $\Delta_F(x)$ is called the propagator of the field ϕ.

The result we have obtained is extended easily to the case of a complex field (or of two real fields) of mass m, defined by equation (3.9), in the absence of interactions. The corresponding Lagrangian density can be written

$$\mathcal{L} = \sum_{k=1}^{2}\frac{1}{2}\left[\partial_\mu\phi_k(x)\,\partial^\mu\phi_k(x) - m^2\phi_k^2\right] = \partial_\mu\phi^\dagger(x)\,\partial^\mu\phi(x) - m^2\phi^\dagger\phi \;. \tag{3.65}$$

Again in this case the generating functional $Z[J]$, defined by equation (3.24), is calculated explicitly using the procedure described. In this way we find the result

$$Z[J, J^\dagger] = \exp\left(-i \iint d^4x \, d^4y \, J^\dagger(x) \, \Delta_F(x-y) \, J(y)\right) \, . \qquad (3.66)$$

The two-point function, which is obtained using the substitution rule (3.25), has the form

$$\langle 0| \, T\left[\phi(x) \, \phi^\dagger(y)\right] |0\rangle = i\Delta_F(x-y) \, , \qquad (3.67)$$

while

$$\langle 0|T\left[\phi(x) \, \phi(y)\right] |0\rangle = \langle 0|T\left[\phi^\dagger(x) \, \phi^\dagger(y)\right] |0\rangle = 0 \, . \qquad (3.68)$$

We leave the derivation of this last result as an exercise.

3.4   FREE SCALAR FIELD: ONE-PARTICLE STATES

We can use equations (3.67) and (3.68) to study the spectrum of states in a scalar theory, as we did previously in the case of the harmonic oscillator. We will find that a complex scalar field describes two particles of the same mass, which we can call the *particle P* and *antiparticle A*. Each one of the two can exist in a state characterised by a momentum $\mathbf{p}$ and an energy $\omega_p = \sqrt{(\mathbf{p}^2 + m^2)}$.

We rewrite (3.67) using equations (3.61) and (3.62):

$$\langle 0| \, T\left[\phi(x) \, \phi^\dagger(y)\right] |0\rangle = \frac{i}{(2\pi)^3} \int d^3p \, e^{i\mathbf{p}(\mathbf{x}-\mathbf{y})} \left[\frac{1}{(2\pi)} \int dE \, \frac{e^{-iE(t_x-t_y)}}{E^2 - \omega_p^2 + i\epsilon}\right] \, .$$

The integral in square brackets is that from equation (3.45), which we already know [see equation (3.48)]. We can therefore write

$$\langle 0| \, T\left(\phi(x) \, \phi^\dagger(y)\right) |0\rangle = i\Delta_F(x-y)$$
$$= \frac{1}{(2\pi)^3} \int d^3p \, \frac{e^{i\mathbf{p}(\mathbf{x}-\mathbf{y})}}{2\omega_p} \left(e^{-i\omega_p(t_x-t_y)}\theta(t_x-t_y) + e^{-i\omega_p(t_y-t_x)}\theta(t_y-t_x)\right) \, .$$
$$(3.69)$$

If we consider the Fourier transform of the variable $\mathbf{y}$, defining

$$\phi(\mathbf{q}, t_y) = \int d^3y \, e^{-i\mathbf{q}\mathbf{y}} \phi(\mathbf{y}, t_y) \, , \quad \phi^\dagger(\mathbf{q}, t_y) = \int d^3y \, e^{i\mathbf{q}\mathbf{y}} \phi^\dagger(\mathbf{y}, t_y) \, , \quad (3.70)$$

we obtain

$$\langle 0|T\left[\phi(\mathbf{x}, t_x) \, \phi^\dagger(\mathbf{q}, t_y)\right] |0\rangle =$$
$$= \frac{e^{i\mathbf{q}\mathbf{x}}}{2\omega_q} \left(e^{-i\omega_q(t_x-t_y)}\theta(t_x-t_y) + e^{-i\omega_q(t_y-t_x)}\theta(t_y-t_x)\right) \, .$$

We analyse separately the two cases $t_x > t_y$ and $t_y > t_x$,

$$\langle 0| \, \phi(\mathbf{x}, t_x) \, \phi^\dagger(\mathbf{q}, t_y) \, |0\rangle = \frac{e^{i\mathbf{q}\mathbf{x}}}{2\omega_q} e^{-i\omega_q(t_x - t_y)} \qquad (t_x > t_y) \,, \qquad (3.71)$$

$$\langle 0| \, \phi^\dagger(\mathbf{q}, t_y) \, \phi(\mathbf{x}, t_x) \, |0\rangle = \frac{e^{i\mathbf{q}\mathbf{x}}}{2\omega_q} e^{-i\omega_q(t_y - t_x)} \qquad (t_y > t_x) \,. \qquad (3.72)$$

Defining two families of states $|P, \mathbf{q}\rangle$ and $|A, \mathbf{q}\rangle$,

$$|P, \mathbf{q}\rangle = \frac{\sqrt{2\omega_q}}{(2\pi)^{3/2}} \phi^\dagger(\mathbf{q}, 0)|0\rangle \,, \qquad (3.73)$$

$$|A, \mathbf{q}\rangle = \frac{\sqrt{2\omega_q}}{(2\pi)^{3/2}} \phi(-\mathbf{q}, 0)|0\rangle \,, \qquad (3.74)$$

from (3.71) we obtain

$$\langle 0| \, \phi(\mathbf{x}, t_x) \, |P, \mathbf{q}\rangle = \frac{1}{(2\pi)^{3/2} \sqrt{2\omega_q}} e^{i(\mathbf{q}\mathbf{x} - \omega_q t_x)} \,, \qquad (3.75)$$

and from the complex conjugate of (3.72),

$$\langle 0| \, \phi^\dagger(\mathbf{x}, t_x) \, |A, \mathbf{q}\rangle = \frac{1}{(2\pi)^{3/2} \sqrt{2\omega_q}} e^{i(\mathbf{q}\mathbf{x} - \omega_q t_x)} \,. \qquad (3.76)$$

This already tells us[7] that the states $|P, \mathbf{q}\rangle$ and $|A, \mathbf{q}\rangle$ have energy $\omega_q = \sqrt{m^2 + \mathbf{q}^2}$ and momentum $\mathbf{q}$, and are therefore interpretable as states[8] of a free particle of mass m.

With a second Fourier transform of $\mathbf{x}$ we can also obtain the scalar products of two P states, or two A states:

$$\langle P, \mathbf{p}|P, \mathbf{q}\rangle = \langle A, \mathbf{p}|A, \mathbf{q}\rangle = \delta^3(\mathbf{p} - \mathbf{q}) \,. \qquad (3.77)$$

If we now repeat this exercise starting from the Green's function with two fields ϕ, (3.68), we can prove that the states $|P\rangle$ and $|A\rangle$ are orthogonal to each other, i.e. that

$$\langle P, \mathbf{p}|A, \mathbf{q}\rangle = 0 \,. \qquad (3.78)$$

We are therefore discussing the states of two different particles, both of mass m.

That the two types of particle are in a particle–antiparticle relationship

---

[7] We can write $\phi(\mathbf{x}, t) = e^{i(Ht - \mathbf{p}\mathbf{x})} \phi(0, 0) e^{-i(Ht - \mathbf{p}\mathbf{x})}$ where H is the Hamiltonian and $\mathbf{p}$ the momentum operator.

[8] The normalisation of the kets which appear in equations (3.73) and (3.74) corresponds to the limit of infinite volume, and will be discussed in detail in the next section.

cannot be seen from such a simple theory, which describes non-interacting particles. However, by introducing the interaction with an electromagnetic field, we can easily prove that the two particles have opposite electric charge. Which should be called the particle and which the antiparticle remains an arbitrary choice. Obviously, in the case of a real scalar field, the situation is simpler, and we have just one type of particle.

3.5 CREATION AND DESTRUCTION OPERATORS

It is useful to connect the discussion of the previous section to the formulation of the theory of the free scalar field in terms of creation and destruction operators, discussed in [1]. We start in a finite volume, to then introduce a normalisation of the states and of the creation and destruction operators appropriate in the limit of infinite volume.

We will consider fields in a cubic volume V with periodic boundary conditions, and define a set of solutions of the Klein–Gordon (K–G) equation with positive frequencies

$$f_q(x) = \frac{1}{\sqrt{2\omega(\mathbf{q})V}} e^{-iqx} , \quad (\Box + m^2) f_q = 0 ,$$

$$\mathbf{q} = \frac{2\pi}{L}(n_1, n_2, n_3) , \quad \omega(\mathbf{q}) = +\sqrt{\mathbf{q}^2 + m^2} , \tag{3.79}$$

where $n_{1,2,3}$ are integers and L is the side of the cube, so $V = L^3$. The functions (3.79) are normalised in V according to

$$\int_V d^3x \, f_q^*(x) i \overleftrightarrow{\partial_t} f_{q'}(x) = \delta_{q,q'} , \tag{3.80}$$

where we have introduced the abbreviated notation

$$\overleftrightarrow{\partial_t} fg = f(\partial_t \, g) - (\partial_t \, f)g . \tag{3.81}$$

We can construct the creation and destruction operators starting from ϕ and f_q in the form

$$a_q = \int d^3x \left[f_q(x)^* i \overleftrightarrow{\partial_t} \phi(x)) \right] \qquad \text{(destruction)} ,$$

$$a_q^\dagger = (a_q)^\dagger \qquad \text{(creation)} ,$$

with

$$\left[a_q, a_{q'}^\dagger \right] = i\delta_{q,q'} . \tag{3.82}$$

After the spatial integration, the operators a_q could depend on time but it is quickly shown that they are in fact constant, as follows from

$$
\begin{aligned}
\partial_t a_q &= i \int d^3x \; [f_q(x)^*(\partial_t^2 \phi(x)) - (\partial_t^2 f_q(x)^*)\phi(x)] \\
&= i \int d^3x \; [f_q(x)^*(\Box \phi(x)) - (\Box f_q(x)^*)\phi(x)] \\
&= i \int d^3x \; [f_q(x)^*(\Box + m^2)\phi(x)] = 0 \; ,
\end{aligned}
\tag{3.83}
$$

since both f_q and ϕ satisfy the K–G equation.

To take the continuum limit, we introduce the projection operator onto the momentum states included in a three-dimensional interval, $\Delta^3 n$:

$$
P = \sum_{\Delta^3 n} |p\rangle\langle p| \; ,
\tag{3.84}
$$

with $P^2 = P$ because of the orthonormality of the states $|p\rangle$. If we now take the limit, we obtain

$$
P = \int_{\Delta^3 p} |p\rangle \frac{V \, d^3p}{(2\pi)^3} \langle p| \; ,
\tag{3.85}
$$

which suggests to define the normalised kets

$$
| \, \widetilde{p} \, \rangle = \sqrt{\frac{V}{(2\pi)^3}} |p\rangle \; ,
\tag{3.86}
$$

for which

$$
P = \int_{\Delta^3 p} | \, \widetilde{p} \, \rangle \, d^3p \, \langle \, \widetilde{p} \, | \; .
\tag{3.87}
$$

We note that the condition $P^2 = P$ requires, as the normalisation condition of the new kets,

$$
\langle \, \widetilde{p}\,' \, | \widetilde{p} \, \rangle = \delta^{(3)}(p' - p) \; .
\tag{3.88}
$$

The destruction and creation operators which correspond to the new states are evidently,

$$
\widetilde{a}_p = \sqrt{\frac{V}{(2\pi)^3}} a_p \; , \quad \widetilde{a}_p^\dagger = (\widetilde{a}_p)^\dagger \; ,
\tag{3.89}
$$

and the new commutation rules are obtained from (3.88)

$$
\delta^{(3)}(p' - p) = \langle 0|\widetilde{a}_{p'}\widetilde{a}_p^\dagger|0\rangle = \langle 0| \left[\widetilde{a}_{p'}, \widetilde{a}_p^\dagger \right] |0\rangle \; ,
\tag{3.90}
$$

which implies

$$[\tilde{a}_{p'}, \tilde{a}_p^\dagger] = \delta^{(3)}(p' - p) \ . \tag{3.91}$$

The expansion of the field is written now in the form

$$\phi(x) = \sum_p \frac{1}{\sqrt{2\omega(\mathbf{p})V}} \left[a_p \, e^{-ipx} + a_p^\dagger \, e^{ipx} \right] =$$

$$= \int \frac{d^3p}{(2\pi)^{3/2}} \frac{1}{\sqrt{2\omega(\mathbf{p})}} \left[\tilde{a}_p \, e^{-ipx} + \tilde{a}_p^\dagger \, e^{ipx} \right] \ . \tag{3.92}$$

The new states are normalised so as to have, instead of one particle in the reference volume, a constant density of particles. This property is shown by calculating the energy of the field, which is equal to (we use the normal product of the operators, defined in [1])

$$H = \int d^3x \, \frac{1}{2} \left[: (\partial_t \phi)^2 : + : (\nabla \phi)^2 : + m^2 : \phi^2 : \right]$$

$$= \int d^3p \, \omega(\mathbf{p}) \, \tilde{a}_p^\dagger \tilde{a}_p \ ,$$

with

$$E_p = \langle \, \tilde{p} \, |H| \, \tilde{p} \, \rangle = \delta^{(3)}(0)\omega(\mathbf{p}) = \frac{V}{(2\pi)^3}\omega(\mathbf{p}) \ . \tag{3.93}$$

In this way it is evident that the particle density in the state $| \, \tilde{p} \, \rangle$ is equal to $\rho = 1/(2\pi)^3$.

To finish, we note that the relation which links the fields to the destruction operators is written

$$\tilde{a}_q = \int d^3x \left[\tilde{f}_q(x)^* i \overset{\leftrightarrow}{\partial}_t \phi(x)) \right] \ , \tag{3.94}$$

with

$$\tilde{f}_q(x) = \frac{1}{(2\pi)^{3/2}\sqrt{2\omega(\mathbf{p})}} e^{-ipx} \ . \tag{3.95}$$

In what follows we will adopt the continuum normalisation, omitting for brevity the tilde on both the operators and the functions f_q.

CHAPTER 4

EQUATIONS OF MOTION, SYMMETRIES AND WARD'S IDENTITY

CONTENTS

In this chapter, we want to construct, starting from the Feynman method, some general properties of quantum mechanics (QM) and of field theory. In particular:

- Equations of motion in operator form

- Canonical commutation rules in QM: $[q_i, p_k] = i\hbar\delta_{ik}$

- Symmetries and conserved quantities – Noether's theorem

- The Ward identity

4.1   SUM OVER PATHS AND OPERATORS

We begin with a very general consideration, which shows the way to obtain relations between operators in the language of the sum over paths. A relation between operators can always be stated in the form $E = 0$. For example, the canonical commutation rules can be written in terms of $E = [q^i, p^k] - i\hbar\delta^{ik}$, and $E = 0$ will hold if $\langle B|E|A\rangle = 0$ for any pair of states $|A\rangle$, $|B\rangle$.

Apparently the Feynman method does not allow us direct access to the matrix elements of operators between arbitrary states, but only to "Green's functions", i.e. to expectation values of the T-ordered product of operators in the ground state, which in field theory coincides with the vacuum, $|0\rangle$. However, we can use the Green's functions to prove relations between operators.

To fix the ideas, we consider the case of a QM system with a finite number of degrees of freedom,[1] which we will identify with N variables collectively denoted q

$$q = \{q_1 \ldots q_N\} \ .$$

If

$$\int [\Pi dq^i(t)] e^{iS(q)/\hbar} E(t)\, q^{i_1}(t_1)\, q^{i_2}(t_2) \ldots q^{i_n}(t_n)$$

$$= \langle 0|T\left[E(t)\, q^{i_1}(t_1)\, q^{i_2}(t_2) \ldots q^{i_n}(t_n)\right]|0\rangle = 0 \ , \qquad (4.1)$$

for any value of n, and for arbitrary values of $t_1, t_2 \ldots t_n$ different from t, it follows that $E(t) = 0$. Actually, if we choose $t_1 > t_2 \ldots > t_k > t > t_{k+1} \ldots > t_n$, equation (4.1) is equivalent to

$$\langle 0|q^{i_1}(t_1)q^{i_2}(t_2) \ldots q^{i_k}(t_k)E(t)q^{i_{k+1}}(t_{k+1}) \ldots q^{i_n}(t_n)|0\rangle = 0 \ ,$$

which we can rewrite as

$$\langle B|E(t)|A\rangle = 0 \ , \qquad \text{with} \quad \begin{cases} |A\rangle = q^{i_{k+1}}(t_{k+1}) \ldots q^{i_n}(t_n)|0\rangle \ , \\ \langle B| = \langle 0|q^{i_1}(t_1)q^{i_2}(t_2) \ldots q^{i_k}(t_k) \ . \end{cases}$$

Since the $q^i(t)$ are a complete set of operators, and the number of q^i appearing in $|A\rangle$ and $|B\rangle$, as well as their arguments, are arbitrary, $|A\rangle$ and $|B\rangle$ are arbitrary vectors.[2] Therefore it follows from (4.1) that $E(t) = 0$.

---

[1] In this case also we will denote the ground state of the system with $|0\rangle$. The case of a QM system with finite number of degrees of freedom is used here to illustrate methods which can be applied later to field theory. Normally we use units in which $\hbar = 1$, but it is useful to make this explicit, at least in the case of QM, to obtain the canonical commutation rules in the standard form.

[2] In simple terms we can think of the formulation in terms of wave functions. If $\Psi_0(q)$ is the ground state wave function, any other wave function can be obtained multiplying $\Psi_0(q)$ by a function $f(q)$ and, if $f(q)$ is regular, it can be approximated by a series of powers in q.

4.1.1 Derivatives

For what follows, in particular to derive the "fundamental identity" which we will use widely in this chapter, we will need to establish the relation between a Green's function in which a general operator $O(t)$ appears, for example

$$\int [\Pi dq^i(t)] e^{iS(q)/\hbar} O(t_0)\, q^{i_1}(t_1) \ldots q^{i_n}(t_n) \tag{4.2}$$

$$= \langle 0 | T \left[O(t_0)\, q^{i_1}(t_1) \ldots q^{i_n}(t_n) \right] | 0 \rangle \ ,$$

and one in which its derivative $dO(t)/dt$ appears. We will assume that the path integral is regular enough to be able to exchange the integral with the derivative, from which we obtain

$$\int [\Pi dq^i(t)] e^{iS(q)/\hbar}\, \frac{d}{dt_0} O(t_0)\, q^{i_1}(t_1) \ldots q^{i_n}(t_n)$$

$$= \frac{d}{dt_0} \int [\Pi dq^i(t)] e^{iS(q)/\hbar}\, O(t_0)\, q^{i_1}(t_1) \ldots q^{i_n}(t_n)$$

$$= \frac{d}{dt_0} \langle 0 | T \left[O(t_0)\, q^{i_1}(t_1) \ldots q^{i_n}(t_n) \right] | 0 \rangle \ . \tag{4.3}$$

4.2 THE FUNDAMENTAL IDENTITY

The various results that we wish to derive, which range from the commutation rules between quantum operators, to relationships between symmetries and conservation laws, to the Ward identity, can be derived from an identity which we will obtain in this section. We remain with the case of the QM of a system with a finite number of degrees of freedom, and we will now consider an infinitesimal variation of $q(t)$,

$$q \to q' = q + \delta q \ , \ \text{i.e.} \ q^i \to q^i + \delta q^i \quad (i = 1 \ldots N) \ .$$

Examples are:

- A translation in q: $\delta q^i(t) = \eta^i(t)$.

- A linear transformation: $q^{i'} = a_{ik}(t) q^k(t)$. For an infinitesimal transformation, the matrix a_{ik} can be written as $\delta_{ik} + \epsilon_{ik}$, with ϵ_{ik} infinitesimal, hence $\delta q^i(t) = \epsilon_{ik}(t) q^k(t)$.

In the case of a field theory, the variables are the fields themselves, $\phi^i(\mathbf{x}, t)$, and also in this case we can consider:

- Translations, $\delta \phi^i(\mathbf{x}, t) = \eta^i(\mathbf{x}, t)$.

- Linear transformations, $\delta \phi^i(\mathbf{x}, t) = \epsilon_{ik}(\mathbf{x}, t) \phi^k(\mathbf{x}, t)$.

Both in the case of QM and of a field theory, we will limit ourselves to the consideration of transformations which leave invariant the *measure* of the path integral, $[dq] = [dq']$. This is automatically true in the case of translations, but in the case of linear transformations implies that the transformation matrices are limited to those with unit determinant. Actually, the determinant of a_{ik} is the Jacobian of the transformation

$$dq^{1'}(t) \ldots dq^{N'} N(t) = \left| \frac{\partial q^{i'}}{\partial q^k} \right| dq^1(t) \ldots dq^N(t) = \det(a_{ik}) dq^1(t) \ldots dq^N(t) .$$

In the infinitesimal form, since $\det(1 + \epsilon) = 1 + \mathrm{Tr}(\epsilon) + O(\epsilon^2)$, this implies a restriction to infinitesimal transformations with $\mathrm{Tr}(\epsilon) = 0$.

Exercise 1. *Prove that* $\det(1 + \epsilon) = 1 + \mathrm{Tr}(\epsilon) + O(\epsilon^2)$.

Given a n-point Green's function we have

$$\int [dq^i] e^{iS(q+\delta q)} q^{i_1}(t_1) \ldots q^{i_n}(t_n)$$

$$= \int [d(q^i + \delta q^i)] e^{iS(q+\delta q)} q^{i_1}(t_1) \ldots q^{i_n}(t_n)$$

$$= \int [dq^i] e^{iS(q)} \left(q^{i_1}(t_1) - \delta q^{i_1}(t_1) \right) \ldots \left(q^{i_n}(t_n) - \delta q^{i_n}(t_n) \right) ,$$

where we have used $[dq^{i'}(t)] = [dq^i(t)]$, and the last line is obtained with the change of variables $q^i \to q^i - \delta q^i$.

The above identity is clearly satisfied if $\delta q = 0$. Expanding the first and last lines to first order in δq, we obtain

$$i \int [dq^i] e^{iS(q)} \delta S(q) q^{i_1}(t_1) \ldots q^{i_n}(t_n)$$

$$= - \int [dq^i] e^{iS(q)} \delta q^{i_1}(t_1) \ldots q^{i_n}(t_n)$$

$$- \ldots \tag{4.4}$$

$$- \int [dq^i] e^{iS(q)} q^{i_1}(t_1) \ldots \delta q^{i_n}(t_n) .$$

On the left-hand side of the equality, the variation of S appears, which we can express in terms of a variation of the Lagrangian,

$$\delta S(q) = \int dt \, \delta L(q(t), \dot{q}(t)) ,$$

to rewrite the identity (4.4) in the form (for convenience we make the dependence on Planck's constant explicit)

$$\frac{i}{\hbar} \int dt \int [dq^i] e^{iS(q)/\hbar} \, \delta L \left(q(t), \dot{q}(t) \right) \, q^{i_1}(t_1) \dots q^{i_n}(t_n)$$

$$= \; - \int [dq^i] e^{iS(q)/\hbar} \, \delta q^{i_1}(t_1) \dots q^{i_n}(t_n)$$

$$- \; \dots \tag{4.5}$$

$$- \int [dq^i] e^{iS(q)/\hbar} \, q^{i_1}(t_1) \dots \delta q^{i_n}(t_n) \; .$$

The transformation of this identity into relationships between Green's functions requires care, taking account of the rules established in Section 4.1.1 for the treatment of the time derivatives. Identities of this type, called "Ward identities", connect a Green's function with $n+1$ operators (the n q operators and δL of the left-hand side) to Green's functions with n operators. As we will see, the original Ward identity of quantum electrodynamics, which connects the vertex function, a three-point Green's function, to a two-point Green's function, the self-energy of the electron, is obtainable directly from (4.5).

4.3   QUANTUM MECHANICS

In this section we study the consequences of the Ward identity (4.5) in QM. We will give three examples: the equations of motion in operator form, the canonical commutation rules, and the connection between symmetries and observed quantities in the case of translations in time. The steps may seem complicated on a first reading, but in reality are elementary.

4.3.1   Equations of motion and commutation rules

For the first two examples we will consider an infinitesimal translation of the variables $q^i(t)$, $\delta q^i(t) = \eta^i(t)$. We will then have

$$\delta L \left(q(t), \dot{q}(t) \right) = \eta^i(t) \frac{\partial L}{\partial q^i} + \dot{\eta}^i(t) \frac{\partial L}{\partial \dot{q}^i} = \eta^i(t) F^i(t) + \dot{\eta}^i(t) p^i(t) \; , \tag{4.6}$$

where we have defined $p^i(t)$, the conjugate momentum at $q^i(t)$, and $F^i(t)$, the i-th component of the force. The identity (4.5) becomes:

$$\frac{i}{\hbar} \int dt \; \dot{\eta}^i(t) \langle 0 | T \left[p^i(t) \, q^{i_1}(t_1) \dots q^{i_n}(t_n) \right] | 0 \rangle$$

$$+ \frac{i}{\hbar} \int dt \; \eta^i(t) \langle 0 | T \left[F^i(t) \, q^{i_1}(t_1) \dots q^{i_n}(t_n) \right] | 0 \rangle$$

$$= - \eta(t_1) \langle 0 | T \left[q^{i_2}(t_2) \dots q^{i_n}(t_n) \right] | 0 \rangle \dots$$

$$- \eta(t_n) \langle 0 | T \left[q^{i_1}(t_1) \dots \delta q^{i_{n-1}}(t_{n-1}) \right] | 0 \rangle \; .$$

If we now assume that

$$\lim_{t \to \pm\infty} \eta^i(t) = 0 \ , \tag{4.7}$$

we can integrate the left-hand side by parts and, reorganising the terms, we obtain

$$\frac{i}{\hbar} \int dt \ \eta^i(t) \frac{d}{dt} \langle 0|T \left[p^i(t) \, q^{i_1}(t_1) \ldots q^{i_n}(t_n) \right] |0\rangle$$

$$= \frac{i}{\hbar} \int dt \ \eta^i(t) \langle 0|T \left[F^i(t) \, q^{i_1}(t_1) \ldots q^{i_n}(t_n) \right] |0\rangle$$

$$+ \eta(t_1) \langle 0|T \left[q^{i_2}(t_2) \ldots q^{i_n}(t_n) \right] |0\rangle \ldots$$

$$+ \eta(t_n) \langle 0|T \left[q^{i_1}(t_1) \ldots \delta q^{i_{n-1}}(t_{n-1}) \right] |0\rangle \ .$$

Finally, we set $\eta^i(t) = \delta^{im} \delta(t - t_0)$ and carry out the integrals

$$\frac{i}{\hbar} \frac{d}{dt_0} \langle 0|T \left[p^m(t_0) \, q^{i_1}(t_1) \ldots q^{i_n}(t_n) \right] |0\rangle$$

$$= \frac{i}{\hbar} \langle 0|T \left[F^m(t_0) \, q^{i_1}(t_1) \ldots q^{i_n}(t_n) \right] |0\rangle \tag{4.8}$$

$$+ \delta^{i_1 m} \delta(t_1 - t_0) \langle 0|T \left[q^{i_2}(t_2) \ldots q^{i_n}(t_n) \right] |0\rangle \ldots$$

$$+ \delta^{i_n m} \delta(t_n - t_0) \langle 0|T \left[q^{i_1}(t_1) \ldots \delta q^{i_{n-1}}(t_{n-1}) \right] |0\rangle \ .$$

The derivative of the left-hand side requires a certain care, because the time-ordered product has singularities when two of the times are equal, in particular if $t_0 = t_1$, or $t_0 = t_2$, and so on. To simplify the discussion, we assume that $t_1, t_2 \ldots t_n$ are all different, with $t_1 > t_2 \ldots > t_n$. Thus we have

$$T \left[p^m(t_0) \ q^{i_1}(t_1) q^{i_2}(t_2) \ldots q^{i_n}(t_n) \right] \ , \quad \text{for } t_1 > t_2 \ldots > t_n$$

$$= p^m(t_0) \, q^{i_1}(t_1) q^{i_2}(t_2) \ldots q^{i_n}(t_n) \theta(t_0 - t_1) \tag{4.9}$$

$$+ q^{i_1}(t_1) p^m(t_0) q^{i_2}(t_2) \ldots q^{i_n}(t_n) \theta(t_1 - t_0) \theta(t_0 - t_2)$$

$$+ q^{i_1}(t_1) q^{i_2}(t_2) p^m(t_0) \ldots q^{i_n}(t_n) \theta(t_2 - t_0) \theta(t_0 - t_3) + \ldots \ .$$

We can now take the derivative of the left-hand side of (4.8), which involves the derivative of $p^m(t_0)$ as well as the derivatives of the θ function. We note that on the right-hand side of (4.9) both $\theta(t_0 - t_1)$ and $\theta(t_1 - t_0)$ appear, whose derivatives are $\pm \delta(t_0 - t_1)$, and since $t_1 > t_2$, $\delta(t_0 - t_1)\theta(t_0 - t_2) = \delta(t_0 - t_1)$, and so on. We therefore obtain

$$\frac{i}{\hbar} \frac{d}{dt_0} \langle 0|T \left[p^m(t_0) \, q^{i_1}(t_1) q^{i_2}(t_2) \ldots q^{i_n}(t_n) \right] |0\rangle \ , \quad \text{for } t_1 > t_2 \ldots > t_n$$

$$= \frac{i}{\hbar} \langle 0|T \left[\dot{p}^m(t_0) \, q^{i_1}(t_1) q^{i_2}(t_2) \ldots q^{i_n}(t_n) \right] |0\rangle$$

$$+ \frac{i}{\hbar} \delta(t_0 - t_1) \langle 0| \left[p^m(t_0), q^{i_1}(t_1) \right] q^{i_2}(t_2) \ldots q^{i_n}(t_n) |0\rangle$$

$$+ \frac{i}{\hbar} \delta(t_0 - t_2) \langle 0|q^{i_1}(t_1) \left[p^m(t_0), q^{i_2}(t_2) \right] \ldots q^{i_n}(t_n) |0\rangle + \ldots \ .$$

If we now substitute this expression into (4.8), we find in this identity some non-singular terms (without δ-functions), terms proportional to $\delta(t_0 - t_1)$, $\delta(t_0 - t_2)$, and so on up to terms $\propto \delta(t_0 - t_n)$. Because equation (4.8) must hold, on the left and right-hand side both the non-singular terms and the coefficients of each δ-function should be equal to one another. From the non-singular terms we find

$$\frac{i}{\hbar}\langle 0|T\left(\dot{p}^m(t_0)\, q^{i_1}(t_1)q^{i_2}(t_2)\ldots q^{i_n}(t_n)\right)|0\rangle$$
$$= \frac{i}{\hbar}\langle 0|T\left(F^m(t_0)\, q^{i_1}(t_1)q^{i_2}(t_2)\ldots q^{i_n}(t_n)\right)|0\rangle , \quad (4.10)$$

which, on the basis of the discussion in Section 4.1, given that n and the times $t_1\ldots t_n$ are arbitrary, expresses the validity of the equations of motion in operator form,

$$\dot{p}^m(t) = F^m(t) . \quad (4.11)$$

The above result justifies the definition of $F^m = \frac{\partial L}{\partial q^m}$, given by equation (4.6), as "force components". Equating the coefficients of the generic δ-function $\delta(t_0 - t_k)$ we obtain the relation[3]

$$\frac{i}{\hbar}\langle 0|q^{i_1}(t_1)\ldots q^{i_{k-1}}(t_{k-1})\left[p^m(t_0),\, q^{i_k}(t_0)\right]q^{i_{k+1}}(t_{k+1})\ldots q^{i_n}(t_n)|0\rangle$$
$$= \delta^{m\,i_k}\langle 0|q^{i_1}(t_1)\ldots q^{i_{k-1}}(t_{k-1})\, q^{i_{k+1}}(t_{k+1})\ldots q^{i_n}(t_n)|0\rangle , \quad (4.12)$$

which, according to the argument discussed in Section 4.1, is equivalent to the canonical commutation rules

$$\left[p^m(t_0),\, q^{i_k}(t_0)\right] = -i\hbar\,\delta^{m\,i_k} . \quad (4.13)$$

4.3.2 Symmetries

We now consider transformations of the dynamic variables, $q^i \to q^i + \delta_g q^i$ which leave the Lagrangian—and therefore the action—invariant, i.e. the symmetries of our physical system

$$\delta_g L = \frac{\partial L}{\partial q^i}\delta_g q^i + \frac{\partial L}{\partial \dot{q}^i}\frac{d}{dt}\delta_g q^i = 0 . \quad (4.14)$$

However, it is useful, also in the case of symmetry, to consider some more general variations, obtained by multiplying $\delta_g q$ by an arbitrary function of time[4]: $\delta q^i = f(t)\delta_g q^i$. We will also assume that $f(t)$ tends to zero at infinity,

$$\lim_{t\to\pm\infty} f(t) = 0 .$$

---

[3]Since these terms are multiplied by $\delta(t_0 - t_k)$, we can put $t_k = t_0$.
[4]In field theory, as we will see, an arbitrary function of $\mathbf{x}$ and t is used.

This more general variation is not itself a symmetry, and we will have

$$\delta L = \frac{\partial L}{\partial q^i} f \delta_g q^i + \frac{\partial L}{\partial \dot{q}^i} \left(f \frac{d}{dt} \delta_g q^i + \dot{f} \delta_g q^i \right)$$

$$= \dot{f} \frac{\partial L}{\partial \dot{q}^i} \delta_g q^i \tag{4.15}$$

$$= \dot{f}(t) J_g(t) \,,$$

where

$$J_g = \frac{\partial L}{\partial \dot{q}^i} \delta_g q^i \,. \tag{4.16}$$

Substituting into the "fundamental identity" (4.5), and with an integration by parts, we find

$$\frac{i}{\hbar} \int dt \, f(t) \frac{d}{dt} \langle 0| T \left[J_g(t) \, q^{i_1}(t_1) \ldots q^{i_n}(t_n) \right] |0\rangle$$

$$= f(t_1) \langle 0| T \left[\delta_g q^{i_1}(t_1) \ldots q^{i_n}(t_n) \right] |0\rangle \tag{4.17}$$

$$+ \ldots$$

$$+ f(t_n) \langle 0| T \left[q^{i_1}(t_1) \ldots \delta_g q^{i_n}(t_n) \right] |0\rangle \,.$$

At this point we can follow the steps of the previous section: we set $f(t) = \delta(t - t_0)$ and carry out the integral on the left-hand side

$$\frac{i}{\hbar} \frac{d}{dt_0} \langle 0| T \left[J_g(t_0) \, q^{i_1}(t_1) \ldots q^{i_n}(t_n) \right] |0\rangle$$

$$= \delta(t_1 - t_0) \langle 0| T \left[\delta_g q^{i_1}(t_1) \ldots q^{i_n}(t_n) \right] |0\rangle \tag{4.18}$$

$$+ \ldots$$

$$+ \delta(t_n - t_0) \langle 0| T \left[q^{i_1}(t_1) \ldots \delta_g q^{i_n}(t_n) \right] |0\rangle \,.$$

As we did previously, we assume that $t_1, t_2 \ldots t_n$ are all distinct, with $t_1 > t_2 \ldots > t_n$, we make the time ordering explicit as done in equation (4.9), and rewrite the left-hand side of (4.18) as

$$\frac{i}{\hbar} \frac{d}{dt_0} \langle 0| T \left[J_g(t_0) \, q^{i_1}(t_1) q^{i_2}(t_2) \ldots q^{i_n}(t_n) \right] |0\rangle \,, \quad \text{for } t_1 > t_2 \ldots > t_n$$

$$= \frac{i}{\hbar} \langle 0| T \left[\dot{J}_g(t_0) \, q^{i_1}(t_1) q^{i_2}(t_2) \ldots q^{i_n}(t_n) \right] |0\rangle$$

$$+ \frac{i}{\hbar} \delta(t_1 - t_0) \langle 0| \left[J_g(t_0), \, q^{i_1}(t_1) \right] q^{i_2}(t_2) \ldots q^{i_n}(t_n) |0\rangle$$

$$+ \frac{i}{\hbar} \delta(t_2 - t_0) \langle 0| q^{i_1}(t_1) \left[J_g(t_0), \, q^{i_2}(t_2) \right] \ldots q^{i_n}(t_n) |0\rangle + \ldots \,.$$

The first term, which is not singular (it does not contain δ-functions), does not have a counterpart on the right-hand side of (4.18) and hence, by the arguments of Section 4.1, $\dot{J}_g(t) = 0$ must hold, i.e. the quantity $J_g(t)$ is conserved. This is essentially Noether's theorem.

Now equating the coefficients of the general δ-function, $\delta(t_k - t_0)$, on the left- and right-hand sides, we obtain the relation

$$\frac{i}{\hbar} \langle 0|q^{i_1}(t_1) \dots q^{i_{k-1}}(t_{k-1}) \left[J_g(t_0), \, q^{i_k}(t_0) \right] q^{i_{k+1}}(t_{k+1}) \dots q^{i_n}(t_n)|0\rangle$$
$$= \langle 0|q^{i_1}(t_1) \dots q^{i_{k-1}}(t_{k-1}) \, \delta_g q^{i_k}(t_0) \, q^{i_{k+1}}(t_{k+1}) \dots q^{i_n}(t_n)|0\rangle \,, \quad (4.19)$$

which, again following the discussion in Section 4.1, turns out to be equivalent to the commutation rules

$$\left[J_g(t), \, q^{i_k}(t) \right] = -i\hbar \, \delta_g \, q^{i_k}(t_0) \,. \quad (4.20)$$

These commutation rules confirm that $J_g(t)$ is the generator of the transformation $q \to q + \delta_g q$. Because an infinitesimal $J_g(t)$ will correspond to an infinitesimal transformation δ_g, equation (4.20) is equivalent to

$$\left(1 + \frac{i}{\hbar} J_g(t_0) \right) q^i(t) \left(1 - \frac{i}{\hbar} J_g(t_0) \right) = q^i(t) + \delta_g \, q^i(t) \,. \quad (4.21)$$

We can continue from here with the construction of finite transformations, and the entire group to which the infinitesimal transformations δ_g belong. But remaining on the subject of infinitesimal transformations, we note that a particular case is provided by translational , i.e. the situation in which the transformation is $\delta_g q^i(t) = \epsilon \delta^{im}$, with ϵ an infinitesimal constant independent of time. If this is a symmetry—of the system along the direction m—we can apply the methods developed in this section, by defining a time-dependent transformation $\delta q^i = f(t)\delta_g q^i(t) = f(t)\epsilon \delta^{im}$. But this is exactly the transformation which we considered in the previous section, with the only difference that, δ_g being a symmetry, the "force" F is zero. The conserved quantity associated with this symmetry is the m-th component of the momentum, p^m, and its commutation rules are given by equation (4.13).

4.3.3 The Hamiltonian function

A system is invariant under translations in time, $q^i \to q^i + \epsilon \dot{q}^i$, if the Lagrangian depends on time only implicitly through the time dependence of q and $\dot{q}$. In this case, setting $\delta_g q = \epsilon \dot{q}$, we find

$$\delta_g L = \epsilon \left(\frac{\partial L}{\partial q^i} \dot{q}^i + \frac{\partial L}{\partial \dot{q}^i} \ddot{q}^i \right) = \epsilon \frac{d}{dt} L \,. \quad (4.22)$$

Since the variation of the Lagrangian is a total derivative, the action is invariant. Again in this case it is useful to use a variation which depends on time, by multiplying δ_g by a function $f(t)$ which is zero at infinity. We obtain

$$\delta q^i = f(t)\delta_g q^i = f(t)\epsilon \dot{q}^i$$

$$\delta L = f(t)\delta_g L + \dot{f}\frac{\partial L}{\partial \dot{q}^i}\delta_g q^i = \epsilon\left(f(t)\frac{d}{dt}L + \dot{f}(t)\frac{\partial L}{\partial \dot{q}^i}\dot{q}^i\right) . \tag{4.23}$$

If we now define the Hamiltonian function

$$H = \frac{\partial L}{\partial \dot{q}^i}\dot{q}^i - L = p^i\dot{q}^i - L , \tag{4.24}$$

by substituting into (4.5) and with simple manipulations—integration by parts, change of sign, omission of the common factor ϵ—we obtain

$$\frac{i}{\hbar}\int dt f(t)\int [dq^i]e^{iS(q)/\hbar}\frac{d}{dt}H(t)\, q^{i_1}(t_1)\ldots q^{i_n}(t_n) ,$$

$$= \int [dq^i]e^{iS(q)/\hbar}\, f(t_1)\dot{q}^{i_1}(t_1)\ldots q^{i_n}(t_n) \tag{4.25}$$

$$+ \ldots$$

$$+ \int [dq^i]e^{iS(q)/\hbar}\, q^{i_1}(t_1)\ldots f(t_n)\dot{q}^{i_n}(t_n) ,$$

which, using (4.3), becomes

$$\frac{i}{\hbar}\int dt\, f(t)\frac{d}{dt}\langle 0|T\left(H(t)\, q^{i_1}(t_1)\ldots q^{i_n}(t_n)\right)|0\rangle$$

$$= f(t_1)\langle 0|T\left(\dot{q}^{i_1}(t_1)\ldots q^{i_n}(t_n)\right)|0\rangle \tag{4.26}$$

$$+ \ldots$$

$$+ f(t_n)\langle 0|T\left(q^{i_1}(t_1)\ldots \dot{q}^{i_n}(t_n)\right)|0\rangle .$$

From the above relation, following the same steps as in the previous section, we obtain two results

1. The Hamiltonian is independent of time.

2. The commutation rules $[H, q^i] = -i\hbar\dot{q}^i$ apply.

4.4   FIELD THEORY

In this section we give an example of application of the ideas we have developed to a field theory for systems with a finite number of degrees of freedom. The relation between sums over paths and Green's functions is given by

$$\int [d\phi^i(x)]e^{iS(\phi)}\, \phi^{i_1}(x_1)\ldots\phi^{i_n}(x_n) = \langle 0|T\left[\phi^{i_1}(x_1)\ldots\phi^{i_n}(x_n)\right]|0\rangle , \tag{4.27}$$

where the fields ϕ^i can be scalar or vector, or fermion, fields. The action is written in terms of the Lagrangian density $\mathcal{L}(\phi^i(x), \phi^i_\mu(x))$ as

$$S(\phi) = \int d^4x \mathcal{L}(\phi^i(x), \phi^i_\mu(x)) \tag{4.28}$$

where we adopt the notation $\phi_\mu(x) = \partial_\mu \phi(x)$.

The identity in (4.5) becomes (with $\hbar = 1$)

$$i \int d^4x \int [d\phi^i] e^{iS(\phi)} \, \delta\mathcal{L}\left(\phi^i(x), \phi^i_\mu(x)\right) \, \phi^{i_1}(x_1) \ldots \phi^{i_n}(x_n)$$

$$= -\int [d\phi^i] e^{iS(\phi)} \, \delta\phi^{i_1}(x_1) \ldots \phi^{i_n}(x_n)$$

$$- \ldots$$

$$- \int [d\phi^i] e^{iS(\phi)} \, \phi^{i_1}(x_1) \ldots \delta\phi^{i_n}(x_n) \, . \tag{4.29}$$

Also in the case of field theories it is necessary to treat with care the case in which the derivative of an operator must be calculated, which we discussed in Section 4.1.1 in the case of QM. Equation (4.3) becomes

$$\int [\Pi d\phi^i(x)] e^{iS(\phi)} \, \frac{\partial}{\partial x^\mu} O(x) \, \phi^{i_1}(x_1) \ldots \phi^{i_n}(x_n)$$

$$= \frac{\partial}{\partial x^\mu} \int [\Pi d\phi^i(x)] e^{iS(\phi)} \, O(t) \, \phi^{i_1}(x_1) \ldots \phi^{i_n}(x_n)$$

$$= \frac{\partial}{\partial x^\mu} \langle 0| T\left[O(x) \phi^{i_1}(x_1) \ldots \phi^{i_n}(x_n)\right] |0\rangle \, . \tag{4.30}$$

4.4.1 Symmetries in field theory

We recall initially that here we will be interested in groups of continuous transformations, which include infinitesimal transformations

$$\phi^i(x) \rightarrow \phi^i(x) + \delta_g \phi^i(x) \, , \tag{4.31}$$

and that discrete symmetries like parity or charge conjugation will be excluded. A continuous transformation that leaves the action invariant is a symmetry of the system. We can subdivide continuous symmetries into two main classes, according to whether or not the invariance of the action is realised through the invariance of the Lagrangian density. To the second class belong the coordinate transformations—translations, rotations, Lorentz transformations—which require a separate treatment, as we saw in the case of QM for time translations. We will now examine symmetries of the first type, i.e. transformations $\delta_g \phi(x)$ which leave the Lagrangian density invariant, such that

$$\delta_g \mathcal{L} = \frac{\partial \mathcal{L}}{\partial \phi^i} \delta_g \phi^i + \frac{\partial \mathcal{L}}{\partial \phi^i_\mu} \partial_\mu \delta_g \phi^i = 0 \, . \tag{4.32}$$

As we did in the case of QM, it is useful to consider a transformation $\delta\phi^i$ obtained from $\delta_g\phi^i$ by multiplying by a function $f(x)$ which vanishes in the limit $|\mathbf{x}| \to \infty$ or $|t| \to \infty$,

$$\delta\phi^i(x) = f(x)\delta_g\phi \ . \tag{4.33}$$

As a consequence, taking account of (4.32), we find

$$\delta\mathcal{L} = \partial_\mu f(x)\, J_g^\mu(x) \ , \tag{4.34}$$

where

$$J_g^\mu = \frac{\partial\mathcal{L}}{\partial\phi_\mu^i}\delta_g\phi^i \ . \tag{4.35}$$

Substituting into (4.29), and with an integration by parts and a change of sign, we find the relation

$$
\begin{aligned}
i\int d^4x\, f(x) &\int [d\phi^i]e^{iS(\phi)}\, \partial_\mu J_g^\mu(x)\, \phi^{i_1}(x_1)\ldots\phi^{i_n}(x_n) \\
&= f(x^1)\int [d\phi^i]e^{iS(\phi)}\, \delta_g\phi^{i_1}(x_1)\ldots\phi^{i_n}(x_n) \\
&\quad + \ldots \\
&\quad + f(x^n)\int [d\phi^i]e^{iS(\phi)}\, \phi^{i_1}(x_1)\ldots\delta_g\phi^{i_n}(x_n) \ ,
\end{aligned}
\tag{4.36}
$$

which, owing to equation (4.30), translates into

$$
\begin{aligned}
i\int d^4x\, f(x)\frac{\partial}{\partial x^\mu} &\langle 0|T\left(J_g^\mu(x)\,\phi^{i_1}(x_1)\ldots\phi^{i_n}(x_n)\right)|0\rangle \\
&= f(x^1)\langle 0|T\left(\delta_g\phi^{i_1}(x_1)\ldots\phi^{i_n}(x_n)\right)|0\rangle \\
&\quad + \ldots \\
&\quad + f(x^n)\langle 0|T\left(\phi^{i_1}(x_1)\ldots\delta_g\phi^{i_n}(x_n)\right)|0\rangle \ .
\end{aligned}
\tag{4.37}
$$

Setting $f(x) = \delta^4(x - x_0)^5$ we obtain the Ward identity in the form

$$
\begin{aligned}
i\frac{\partial}{\partial x_0^\mu} &\langle 0|T\left(J_g^\mu(x_0)\,\phi^{i_1}(x_1)\ldots\phi^{i_n}(x_n)\right)|0\rangle \\
&= \delta(x^1 - x_0)\langle 0|T\left(\delta_g\phi^{i_1}(x_1)\ldots\phi^{i_n}(x_n)\right)|0\rangle \\
&\quad + \ldots \\
&\quad + \delta(x^1 - x_0)\langle 0|T\left(\phi^{i_1}(x_1)\ldots\delta_g\phi^{i_n}(x_n)\right)|0\rangle \ ,
\end{aligned}
\tag{4.38}
$$

---

[5]We are following the procedure of Section 4.3.2 step by step, but adapting it to the space-time structure of field theory.

implying conservation of the current J_g^μ and the commutation rules

$$\partial_\mu J_g^\mu(x) = 0 \ , \tag{4.39}$$

$$i[J_g^0(\mathbf{x}_0, t) \, , \, \phi^i(\mathbf{x}_1, t)] = \delta^{(3)}(\mathbf{x}_0 - \mathbf{x}_1)\delta_g \phi^i(x_1) \ . \tag{4.40}$$

The details of the derivation of the above equations, which we leave for an exercise, follow the steps described in Section 4.3.2.

4.4.2 Ward's identity

In this section we want to obtain from (4.38) the relation between the vertex function and the electron self-energy in QED, generally referred to as the Ward identity (cf. also [3]).

The QED Lagrangian,[6]

$$\mathcal{L} = \bar{\psi} \left[i\gamma^\mu(\partial_\mu - ie_0 A_\mu) - m_0 \right] \psi - \frac{1}{2}\partial_\mu A^\nu \partial^\mu A_\nu \ ,$$

is invariant under global phase transformations $\psi(x) \rightarrow e^{i\alpha}\psi(x)$, and in particular under infinitesimal transformations

$$\delta_g \psi(x) = -i\epsilon\psi \ , \quad \delta_g \bar{\psi}(x) = i\epsilon\bar{\psi} \ . \tag{4.41}$$

To apply the methods of the previous section, we use a transformation, $\epsilon \rightarrow f(x)$, dependent on position. The Lagrangian is no longer invariant, and we obtain

$$\delta\mathcal{L} = \partial_\mu f(x)\bar{\psi}\gamma^\mu\psi = \partial_\mu f(x)j^\mu(x) \ .$$

We can therefore directly apply the considerations of the previous section and we obtain (4.38) where we should identify J_g^μ with the Dirac current $j^\mu = \bar{\psi}\gamma^\mu\psi$, while each one of the fields ϕ^i which appear in (4.38) can be identified with one of the three fields ψ, $\bar{\psi}$, A^μ. To obtain the usual Ward identity, we consider the case with two fields and (4.38) becomes:

$$i\frac{\partial}{\partial x^\mu}\langle 0|T\left[j^\mu(x)\psi(x_1)\bar{\psi}(x_2)\right]|0\rangle =$$
$$i[\delta(x - x_1) - \delta(x - x_2)]\langle 0|T\left[\psi(x_1)\bar{\psi}(x_2)\right]|0\rangle \ . \tag{4.42}$$

To rewrite equation (4.42) in a more familiar form, we must make a Fourier transform with respect to the variables x, x^1, x^2, and use the full definition of the propagator

$$\langle 0|T\left[\psi(x_1)\bar{\psi}(x_2)\right]|0\rangle = iS_F(x_1 - x_2) \ . \tag{4.43}$$

---

[6]We have chosen the Feynman gauge. We recall that all conclusions regarding gauge-invariant quantities remain invariant, regardless of the choice of gauge.

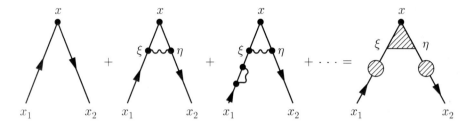

Figure 4.1   Feynman diagrams for the Green's function (4.45).

In this way, from the two terms on the right-hand side, we obtain

$$i \int d^4 x d^4 x_1 d^4 x_2 e^{-ipx} e^{-ikx_1} e^{iqx_2} [\delta(x - x_1) - \delta(x - x_2)] i S_F(x_1 - x_2)$$
$$= -(2\pi)^4 \delta(p + k - q)[S_F(q) - S_F(k)] . \tag{4.44}$$

Figure 4.1 shows some diagrams which contribute to the three-point Green's function

$$\langle 0|T \left[j^\mu(x)\psi(x_1)\bar\psi(x_2) \right] |0\rangle . \tag{4.45}$$

The first diagram corresponds to the theory without interactions: a vertex j^μ from which two lines emerge which represent the free propagator of the field ψ. Introducing an interaction, the lines are subject to corrections, as for example in the third diagram. For each line, the set of corrections reconstruct the complete propagator. In the same time the corrections to the vertex (second and third diagram) are summed to form a one-particle irreducible, or 1PI, vertex[7] denoted Γ^μ.

The result can be expressed (diagram on the right) as[8]

$$\langle 0|T \left[j^\mu(x)\psi(x_1)\bar\psi(x_2) \right] |0\rangle$$
$$= - \int d^4 \xi d^4 \eta S_F(x_1 - \xi)\Gamma^\mu(\xi - x, x - \eta)S_F(\eta - x_2) . \tag{4.46}$$

Carrying out the Fourier transform we obtain

$$\int d^4 x \, d^4 x_1 \, d^4 x_2 \, e^{-ipx} e^{-ikx_1} e^{iqx_2} \langle 0|T \left[j^\mu(x)\psi(x_1)\bar\psi(x_2) \right] |0\rangle$$
$$= -S_F(k)\Gamma^\mu(k, q)S_F(q) \, (2\pi)^4 \delta(p + k - q) , \tag{4.47}$$

---

[7]As we will see Chapter 9, a diagram is one-particle irreducible if it cannot be split into two parts involving non-trivial interactions by cutting a single line.

[8]In this equation we define the sign and normalisation of Γ. Invariance under translation implies that Γ depends only on the difference between the coordinates of its three points x, ξ, η.

From (4.47), it follows immediately that

$$\int d^4x \, d^4x_1 \, d^4x_2 \, e^{-ipx} e^{-ikx_1} e^{iqx_2} i \frac{\partial}{\partial x^\mu} \langle 0|T\,[\ldots]\,|0\rangle$$
$$= (q-k)_\mu S_F(k) \Gamma^\mu(k,q) S_F(q) \, (2\pi)^4 \delta(p+k-q) \, . \, (4.48)$$

Hence, equating (4.48) to (4.44) we obtain

$$(q-k)_\mu S_F(k) \Gamma^\mu(k,q) S_F(q) = S_F(k) - S_F(q) \, , \qquad (4.49)$$

or

$$(q-k)_\mu \Gamma^\mu(k,q) = S_F^{-1}(q) - S_F^{-1}(k) \, . \qquad (4.50)$$

We now consider the case $q \to k$, in which we can set $q = k + \delta$, and

$$S_F^{-1}(q) = S_F^{-1}(k) + \left(\frac{\partial S_F^{-1}(k)}{\partial k_\mu} \right) \delta_\mu \, . \qquad (4.51)$$

From (4.50) and (4.51), using $(q-k)_\mu = \delta_\mu$, and given the arbitrariness of the vector δ_μ, we finally obtain the relation

$$\Gamma^\mu(k,k) = \frac{\partial S_F^{-1}}{\partial k_\mu} \, . \qquad (4.52)$$

If we write Γ^μ in the form

$$\Gamma^\mu(k,q) = \gamma^\mu + \Lambda^\mu(k,q); \qquad (4.53)$$

and for the inverse of the propagator

$$S_F^{-1}(k) = \slashed{k} - m + \Sigma(k) \, , \qquad (4.54)$$

where $\Sigma(k)$ is the fermion self-energy, we finally obtain

$$\gamma^\mu + \Lambda^\mu(k,k) = \frac{\partial}{\partial k_\mu} \left(k_\mu \gamma^\mu - m + \Sigma(k) \right) \, , \qquad (4.55)$$

or

$$\Lambda^\mu(k,k) = \frac{\partial \Sigma}{\partial k_\mu} \, , \qquad (4.56)$$

which is the Ward identity in its usual form. We note that this result is valid to all orders of perturbation theory. However, the manipulations which we have carried out are naturally valid only in a properly regularised theory.

In Chapter 11 we will verify the identity (4.56) to second order of perturbation theory.

4.5   THE SYMMETRIES OF THE VACUUM

A celebrated theorem proved by Coleman shows the importance of the field theory ground state, the vacuum state, for the determination of the symmetries of a given theory.

As a preamble to Coleman's theorem, we prove an important property of local operators. By this we mean operators, $\mathcal{O}(x)$, which depend on the spacetime coordinate x and satisfy the *microcausality* condition, i.e. commute with the field operators calculated at points with spacelike separations from x[9]

$$[\mathcal{O}(x), \phi(y)] = 0 \quad , \quad \text{if } (x - y)^2 < 0 . \tag{4.57}$$

We will use a property of the Green's functions of field operators which is implicit in what was discussed in the previous section. We consider the general Green's function in a field theory, equation (4.27), for simplicity, restricting ourselves to a real scalar field with a single component

$$G(x_1, \cdots, x_n) = \int [d\phi(x)] e^{iS(\phi)} \, \phi(x_1) \ldots \phi(x_n)$$
$$= \langle 0 | T \left[\phi(x_1) \ldots \phi(x_n) \right] | 0 \rangle . \tag{4.58}$$

From the path integral, assuming appropriate conditions of regularity on the functions which appear in the integrand, it is clear that the Green's function is an *analytic function*[10] in the variables $x_1, \cdots, x_n$. If we consider a finite domain $\mathcal{D}$ of space-time and we know G for $x_1, \cdots, x_n$ in $\mathcal{D}$, we can then obtain it in other space-time regions by analytic continuation.

Therefore if we know *all* the Green's functions with restricted arguments in a finite region of space-time, $\mathcal{D}$, we can reconstruct the entire set of Green's functions in all of space-time. This is equivalent to a result demonstrable starting from the axioms of a local field theory, according to which the states obtained by applying to the vacuum all the possible polynomials of the fields at the points of $\mathcal{D}$

$$|x_1, \cdots, x_n> = \phi(x_1) \cdots \phi(x_n) |0> \quad , \quad x_1, \cdots, x_n \in \mathcal{D} , \tag{4.59}$$

form a dense set in the Hilbert space of states.

Theorem (P. Federbush, K. Johnson, 1960):   A local operator $\mathcal{O}(x)$ which annihilates the vacuum state, namely such that:

$$\mathcal{O}(x)|0\rangle = 0 \tag{4.60}$$

is the null operator.

---

[9]Here we consider boson operators. In the case of fermion operators, anticommutators should be considered.

[10]The Green's function can have singularities when two or more coordinates coincide.

Proof: We consider the Green's function with the insertion of $\mathcal{O}(x)$:

$$\int [d\phi(x)]e^{iS(\phi)} \, \mathcal{O}(x)\phi(x_1)\ldots\phi(x_n) =$$

$$\langle 0|T\left[\mathcal{O}(x)\phi(x_1)\ldots\phi(x_n)\right]|0\rangle \ , \quad x_1,\cdots,x_n \in \mathcal{D} \ , \quad (4.61)$$

and take x with a spacelike separation with respect to the points of $\mathcal{D}$

$$(x-y)^2 < 0 \ , \quad \text{for } y \in \mathcal{D} \ . \tag{4.62}$$

Evidently, given (4.57), we can deduce

$$\langle 0|T\left[\mathcal{O}(x)\phi(x_1)\ldots\phi(x_n)\right]|0\rangle = \langle 0|T\left[\phi(x_1)\ldots\phi(x_n)\right]\mathcal{O}(x)|0\rangle = 0 \ . \tag{4.63}$$

Let us now consider two sets of products of operators on $\mathcal{D}$, say $P(\phi(x_1),\phi(x_2),\cdots)$ and $Q(\phi(y_1),\phi(y_2),\cdots)$. Evidently,

$$\langle 0|P\mathcal{O}(x)Q|0\rangle = \langle 0|PQ\mathcal{O}(x)|0\rangle = 0 \ . \tag{4.64}$$

However, the first term on the left-hand side is the matrix element of $\mathcal{O}(x)$ between two arbitrary states of a set dense in Hilbert space and it vanishes, due to the right-hand side. Therefore $\mathcal{O}(x) = 0$.

We note that the locality of $\mathcal{O}$ is an essential condition for the validity of the theorem. An example of the opposite is given by the annihilation operators $a_{\mathbf{k}}$, which are not local operators and, although they annihilate the vacuum state

$$a_{\mathbf{k}}|0>= 0 \ , \tag{4.65}$$

are not null operators, as shown by

$$\langle 0|a_{\mathbf{k}} = \langle k| \neq 0 \ . \tag{4.66}$$

We now consider a situation in which a current $J_\mu(x)$ and the associated charge exist

$$Q = \int d^3x \, J_0(x) \ . \tag{4.67}$$

If we suppose that

$$Q|0>= 0 \ , \tag{4.68}$$

the transformations generated by Q are symmetries of the vacuum

$$e^{i\alpha Q}|0>= |0> \ . \tag{4.69}$$

We can prove the following theorem.

Theorem (S. Coleman, 1965): If the charge (4.67) annihilates the vacuum, according to equation (4.68), the corresponding current is conserved, and the transformations generated by the charge are exact symmetries for all states: *the symmetries of the vacuum are the symmetries of the world.*

Proof. We consider a general state with zero spatial momentum, $|n, \mathbf{p} = 0 >$. From (4.68) we obtain

$$0 = < n, \mathbf{p} = 0|Q|0 > = < n, \mathbf{p} = 0| \int d^3x \ J_0(x)|0 >,$$

$$0 = iE_n < n, \mathbf{p} = 0|Q|0 > = < n, \mathbf{p} = 0|\frac{dQ}{dt}|0 >$$

$$= < n, \mathbf{p} = 0| \int d^3x \ \partial^0 J_0(x)|0 > = < n, \mathbf{p} = 0| \int d^3x \ \partial^\mu J_\mu(x)|0 >$$

$$= V < n, \mathbf{p} = 0| \ \partial^\mu J_\mu(0)|0 > , \qquad (4.70)$$

where V is the volume in which we restrict our system and we have assumed fields and currents which vanish at infinity.

The operator $\partial^\mu J_\mu(x)$ is Lorentz invariant, and so is the vacuum state. If $U(\Lambda)$ is a general Lorentz transformation, we can also obtain

$$0 = < n, \mathbf{p} = 0|U(\Lambda)^\dagger U(\Lambda)(\partial^\mu J_\mu(0))U(\Lambda)^\dagger U(\Lambda)|0 >$$

$$= < n, \mathbf{p} = 0|U(\Lambda)^\dagger \partial^\mu J_\mu(0)|0 > .$$

As n and Λ vary, the state $U(\Lambda)|n, \mathbf{p} = 0 >$ can reach any state of the system, therefore the state $\partial^\mu J_\mu(0)|0 >$ is orthogonal to all the possible states, or

$$\partial^\mu J_\mu(x)|0 > = 0, \ \text{from which} \ \partial^\mu J_\mu(x) = 0 . \qquad (4.71)$$

It follows that the current is conserved as operator, so that Q generates a symmetry for all states, proving Coleman's statement.

THE ELECTROMAGNETIC FIELD

CONTENTS

In this chapter we will deal with quantisation of the free electromagnetic field. Once again we will adopt the conventions of [1] and [3].

The field tensor $F^{\mu\nu}$ can be expressed in terms of potentials A^μ by means of

$$F^{\mu\nu} = \partial^\nu A^\mu - \partial^\mu A^\nu , \qquad (5.1)$$

and Maxwell's equations for the free field

$$\partial_\nu F^{\mu\nu} = \Box A^\mu - \partial^\mu(\partial_\nu A^\nu) = 0 , \qquad (5.2)$$

can be derived from the Lagrangian density

$$\mathcal{L} = -\frac{1}{4}F_{\mu\nu}F^{\mu\nu} . \qquad (5.3)$$

A transformation of the potentials

$$A^\mu(x) \rightarrow A'^\mu(x) = A^\mu(x) + \partial^\mu \Lambda(x), \qquad (5.4)$$

where Λ is an arbitrary function, leaves invariant both the field tensor $F^{\mu\nu}$ and the Lagrangian density $\mathcal{L}$, and hence the action integral. More generally, a gauge transformation should not have an effect on any physical process, and in particular on the results of any measurement. A corollary of this statement is that the electromagnetic potentials A^μ are not themselves observables.

Gauge invariance is central to the theory of the electromagnetic field and its interactions. The requirement that even in the presence of interactions the theory should be invariant with respect to gauge transformations determines the possible type of interactions with other fields. The theory of the electromagnetic field is the prototype of modern theories of the fundamental interactions, all based on a particular gauge invariance.

5.1 THE CHOICE OF GAUGE

To describe the quantum theory of the electromagnetic field by means of the sum over paths we must overcome a problem linked to gauge invariance.

Where is the problem? Readers will remember that to define a quantum theory we must be able to guarantee the convergence of the functional integrals which define the sum over paths. To obtain this result, we considered an analytic continuation in the complex plane of the time variable, through the prescription $t \to t(1 - i\chi)$. In the case of the electromagnetic field, the gauge invariance introduces a new type of divergence which is immune to this remedy. We consider the functional integral

$$I = \int d[A^\mu] e^{iS[A^\mu]} \mathcal{O}[A^\mu] , \qquad (5.5)$$

where $\mathcal{O}[A^\mu]$ is a gauge-invariant functional of A^μ, which can therefore represent some physical quantity. As usual, the functional integral extends over periodic A^μ between $t = \pm\infty$.

According to the general principles of path integrals, the quantity I in (5.5) should represent the matrix element of $\mathcal{O}$ on the vacuum, at least to within a multiplicative constant independent of $\mathcal{O}$. However it is quickly seen that the integral I defined by $\mathcal{O}[A^\mu]$ is, in reality, infinite. Since $S[A^\mu]$ is gauge invariant, for every path $A^\mu(t, \mathbf{x})$ there exists an infinity of others, obtained by a gauge transformation, for which the integrand has the same value. Since the space of possible gauge transformations, the space of the functions $\Lambda(\mathbf{x})$, is infinite, the integral is necessarily divergent.

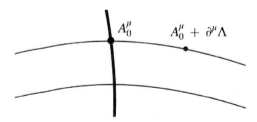

Figure 5.1 In the space of functions $A^\mu(x)$, trajectories composed of paths connected by gauge transformations can be identified (horizontal lines). Along the orthogonal trajectories (vertical lines) we find physically distinct paths.

To overcome this divergence it is necessary to find a way of factorising each integral of the type (5.5). The set of paths $A^\mu(x)$ connected by gauge transformations is called a *gauge trajectory*. What we would like to do is to establish a coordinate system in the space of paths such that a subset of such coordinates (the horizontal lines in Figure 5.1) correspond to the gauge trajectories and the remaining coordinates (the vertical line) serve to distinguish non-equivalent, hence physically distinct, paths under gauge transformations.

If this were possible, we could re-express the integral (5.5) as

$$\int d[\Lambda] \int d[A_0^\mu] e^{iS[A^\mu]} O[A^\mu],$$

but since the integrand is by hypothesis gauge invariant, this can be rewritten as

$$\left[\int d[\Lambda]\right] \int d[A_0^\mu] e^{iS[A_0^\mu]} O[A_0^\mu]$$

and the integral over the gauge transformations, while still divergent, becomes a multiplicative common factor in all the integrals of the type (5.5) and can be omitted in the calculation of the Green's functions which are [see equation (2.25)] *ratios* of integrals of this type.

Obviously things are not quite so simple since, dealing with a change of variables, we must also include the Jacobian of the transformations of A^μ to $\{A_0^\mu, \Lambda\}$. The calculation of the Jacobian determinant in the general case of a gauge theory will be illustrated in Chapter 14. The result is that the Jacobian determinant, for an Abelian gauge theory, contributes a field-independent factor, hence a contribution *irrelevant for theories based on Abelian gauge transformations*, as in the case of QED. We can therefore continue in a *naive* way, simply ignoring the problem and imposing a gauge condition, which restricts the integration to the vertical line of Figure 5.1.

To impose the *gauge condition* we employ the method of Lagrange multipliers, adding to the Maxwell Lagrangian a term which vanishes for configurations which satisfy the same condition (known as *gauge fixing*). Since we are interested in keeping the relativistic invariance visible, we consider the *Lorenz gauge*,[1] characterised by the condition

$$\partial_\nu A^\nu = 0 . \tag{5.6}$$

We therefore write the Lagrangian as

$$S_0 = \int d^4x \left[-\frac{1}{4}(\partial^\nu A^\mu - \partial^\mu A^\nu)(\partial_\nu A_\mu - \partial_\mu A_\nu) - \frac{1}{2\alpha}(\partial_\mu A^\mu)^2 \right] . \tag{5.7}$$

Integrating by parts inside the integral, we rewrite the action in the form

$$S_0 = \int d^4x \, \frac{1}{2} A_\mu \mathcal{O}^{\mu\nu} A_\nu , \tag{5.8}$$

---

[1]Introduced by the Danish theoretical physicist Ludwig Valentin Lorenz (1829 –1891), not to be confused with the more famous Hendrik Antoon Lorentz of the Lorentz transformations.

with

$$\mathcal{O}^{\mu\nu} = g^{\mu\nu}\Box + \frac{1-\alpha}{\alpha}\partial^\mu\partial^\nu \ . \tag{5.9}$$

For $\alpha = 1$, we obtain the Fermi action

$$S_F = \int d^4x \frac{1}{2}\partial_\mu A_\nu \partial^\mu A^\nu \ , \tag{5.10}$$

with the equation of motion of the form

$$\Box A^\mu = 0 \ . \tag{5.11}$$

5.2   GENERATING FUNCTIONAL AND PROPAGATOR

The Green's functions of the electromagnetic field,

$$\langle 0|T\left[A_{\mu_1}(x_1)\cdots A^{\mu_n}(x_n)\right]|0\rangle \ ,$$

can be deduced from a generating functional that is dependent on an auxiliary function $J_\mu(x)$,

$$Z[J_\mu] = \int d[A^\mu]\exp\left[\frac{-i}{2}\int d^4x \left(A_\mu\, \mathcal{O}^{\mu\nu}\, A_\nu + 2J_\mu A^\mu\right)\right] \ , \tag{5.12}$$

by means of the correspondence rule

$$A^\mu(x) \quad \rightarrow \quad i\frac{\delta}{\delta J_\mu(x)} \ . \tag{5.13}$$

To calculate $Z[J]$ we must make the exponent into a perfect square by defining

$$\left[\mathcal{O}^{-1}\right]^{\mu\nu} J_\nu(x) = -\int d^4y\, \Delta_F^{\mu\nu}(x-y;0)\, J_\nu(y) \ ,$$

so that

$$Z[J_\mu] = \exp\left[\frac{-i}{2}\int d^4x \left(J_\mu\left[\mathcal{O}^{-1}\right]^{\mu\nu} J_\nu\right)\right]$$
$$\times \int d[A^\mu]\exp\left[\frac{i}{2}\int d^4x \left(A_\mu - \left[\mathcal{O}^{-1}\right]_{\mu\sigma} J^\sigma\right)\mathcal{O}^{\mu\nu}\left(A_\nu - \left[\mathcal{O}^{-1}\right]_{\nu\rho} J^\rho\right)\right].$$

In this case the residual functional integral is also a constant factor which can be omitted. We obtain

$$Z[J_\mu] = \exp\left[\frac{i}{2}\int d^4x\, d^4y\, J_\mu(x)\Delta_F^{\mu\nu}(x-y;0)J_\nu(y)\right] \ , \tag{5.14}$$

and the two-point function, in the Feynman gauge, becomes

$$\langle 0|T\left(A^{\mu}(x)\,A^{\nu}(y)\right)|0\rangle = i\Delta_{F}^{\mu\nu}(x-y;0)\,. \tag{5.15}$$

We now explicitly calculate the photon propagator in the general gauge in which the Lagrangian takes the form corresponding to equation (5.9). We must compute the inverse of

$$g^{\mu\nu}\Box + (\frac{1}{\alpha}-1)\partial^{\mu}\partial^{\nu}\,,$$

or, by a Fourier transform, the inverse of

$$K^{\mu\nu} = g^{\mu\nu}p^{2} + (\frac{1}{\alpha}-1)p^{\mu}p^{\nu}\,. \tag{5.16}$$

If we write the required operator as

$$D^{\mu\nu} = A(p^{2})g^{\mu\nu} + B(p^{2})p^{\mu}p^{\nu}\,,$$

and impose the condition

$$D^{\mu\nu}K_{\nu\rho} = \delta^{\mu}_{\rho}\,, \tag{5.17}$$

we obtain the result

$$A = \frac{1}{p^{2}}\quad,\quad B = (\alpha-1)\frac{p^{\mu}p^{\nu}}{(p^{2})^{2}}\,,$$

implying

$$D^{\mu\nu} = \frac{1}{p^{2}}\left[g^{\mu\nu} + (\alpha-1)\frac{p^{\mu}p^{\nu}}{p^{2}}\right]\,. \tag{5.18}$$

In conclusion, we give the photon propagator in the general gauge

$$i[\Delta_{F}^{(\alpha)}]_{\mu\nu}(x-y) = \int \frac{d^{4}p}{(2\pi)^{4}} e^{-ip(x-y)} \frac{i}{p^{2}+i\epsilon}(-g^{\mu\nu} + (1-\alpha)\frac{p^{\mu}p^{\nu}}{p^{2}})\,. \tag{5.19}$$

For $\alpha = 1$, the propagator in the Feynman gauge is obtained. Instead, $\alpha = 0$ corresponds to the so-called *Landau gauge*, in which the propagator satisfies the Lorenz condition.

5.3   SINGLE PHOTON STATES

The Lorenz condition does not fix the gauge completely. In the absence of charges (free electromagnetic field) we may further impose the condition: $\phi(x) = 0$, which corresponds to the so-called Coulomb gauge[2]. In this case, the Lorenz condition becomes $\nabla \cdot \mathbf{A} = 0$. For a photon of

---

[2]For a discussion of the Coulomb gauge, we direct the reader to Mandl and Shaw [3], in particular the first chapter.

momentum $\mathbf{k}$, it follows from this that the polarisation vector $\boldsymbol{\epsilon}$ must be orthogonal to $\mathbf{k}$. The photon therefore has only *two polarisation states*, which correspond to two vectors $\boldsymbol{\epsilon}_r$ such that $\boldsymbol{\epsilon}_r \cdot \mathbf{k} = 0$ and $\boldsymbol{\epsilon}_r \cdot \boldsymbol{\epsilon}_s = \delta_{rs}$ $(r = 1, 2)$.

These conclusions, which have clear physical significance, should be independent of any choice of gauge, and of the quantisation method, and should therefore be valid if we proceed with a quantisation in the Feynman gauge, as we did in the previous section. From this point springs a rather subtle problem. The Fermi action (5.10) deals with the four components of the field A^μ in a symmetric way, and although the Fermi Lagrangian should be equivalent to the gauge-invariant one of (5.3) since $\partial_\mu A^\mu = 0$, this condition does not follow directly from the Fermi Lagrangian, or from the equations of motion (5.11). Actually in the Feynman gauge, four polarisation states are apparently present, and a surprise! Let us see what it is.

The two-point function in the case $t_x > t_y$ can be calculated directly from the expression for $\Delta_F^{\mu\nu}$ [see equation (3.69)],

$$\langle 0|A^\mu(x) A^\nu(y)|0\rangle = \frac{-g^{\mu\nu}}{(2\pi)^3} \int d^3p \frac{e^{i\mathbf{p}(\mathbf{x}-\mathbf{y})}}{2\omega_p} e^{-i\omega_p(t_x - t_y)} .$$

If we now define the spatial Fourier transforms (taking into account that $A^\mu(x)$ is real):

$$A^\mu(\mathbf{k}, t_x) = \frac{\sqrt{2\omega_k}}{(2\pi)^{3/2}} \int d^3x e^{-i\mathbf{k}\mathbf{x}} A^\mu(\mathbf{x}, t_x) ,$$

$$A^{\nu\,\dagger}(\mathbf{q}, t_\mathbf{y}) = \frac{\sqrt{2\omega_q}}{(2\pi)^{3/2}} \int d^3y e^{i\mathbf{q}\mathbf{y}} A^\nu(\mathbf{y}, t_y) = A^\nu(-\mathbf{q}, t_y) ,$$

we obtain from a Fourier transform in $\mathbf{y}$

$$\langle 0|A^\mu(x) A^{\nu\,\dagger}(\mathbf{q}, t_y)|0\rangle = \frac{-g^{\mu\nu}}{(2\pi)^{3/2}\sqrt{2\omega_q}} e^{i\mathbf{q}\mathbf{x}} e^{-i\omega_q(t_x - t_y)} , \qquad (5.20)$$

from which, setting $\mu = \nu$, we learn that the state

$$|\mu; \mathbf{q}\rangle = \mathbf{A}^{\mu\,\dagger}(\mathbf{q}, 0)|0\rangle ,$$

has momentum $\mathbf{q}$ and energy $\omega_q = \sqrt{\mathbf{q}^2}$.

This state therefore describes a zero mass particle. A second transformation in $\mathbf{x}$, in the limit $t_x = t_y = 0$, leads to

$$\langle \mu; \mathbf{k}|\nu; \mathbf{q}\rangle = \langle 0|A^\mu(\mathbf{k}, 0) A^{\nu\,\dagger}(\mathbf{q}, 0)|0\rangle = -g^{\mu\nu}\delta^{(3)}(\mathbf{k} - \mathbf{q}) . \qquad (5.21)$$

Here, therefore, is the surprise: not only for every value of $\mathbf{q}$ are there four states, and not the two which we expected, but while the states with $\mu = 1, 2, 3$ have positive modulus squared, since $g^{11} = g^{22} = g^{33} = -1$, implying

$$\langle \mu; \mathbf{k}|\nu; \mathbf{q}\rangle = \langle 0|A^\mu(\mathbf{k}, 0) A^{\nu\,\dagger}(\mathbf{q}, 0)|0\rangle = \delta_{\mu\nu}\delta^{(3)}(\mathbf{k} - \mathbf{q}) \quad (\mu, \nu = 1.2.3) ,$$

the state $|\mu = 0; \mathbf{q}\rangle$ has negative modulus squared

$$\langle 0; \mathbf{k} | 0; \mathbf{q} \rangle = \langle 0 | A^0(\mathbf{k}, 0)\, A^{0\,\dagger}(\mathbf{q}, 0) | 0 \rangle \, , = -\delta^{(3)}(\mathbf{k} - \mathbf{q}) \, .$$

A situation which seems in open conflict with the principles of quantum mechanics: we have a state with negative norm, leading to negative probabilities.[3]

The solution to this problem stems from the gauge invariance of electromagnetism, which is not completely lost because of the choice of the Feynman gauge. In fact, the Fermi action is invariant under a restricted class of gauge transformations, characterised by functions $f(x)$ such that $\Box f(x) = 0$. Actually (see equation (5.10)):

$$\int d^4x \, \partial^\nu (A^\mu + \partial^\mu f)\, \partial_\nu (A_\mu + \partial_\mu f) = \int d^4x \, (\partial^\nu A^\mu)\,(\partial_\nu A_\mu)$$

$$+ \int d^4x \, (\partial^\nu \partial^\mu f)\partial_\nu A_\mu + \int d^4x \, (\partial^\nu A^\mu)(\partial_\nu \partial_\mu f) + \int d^4x \, (\partial^\nu \partial^\mu f)(\partial_\nu \partial_\mu f) \, ,$$

and with an integration by parts it can be seen that all the terms in the second line vanish if $\Box f(x) = 0$.

In the next section we will see how the presence of this invariance resolves the problem of the additional states which appear in the Feynman gauge.[4]

For every value of $\mathbf{q}$ we can choose four polarisation vectors:

ϵ_1, ϵ_2 : Two *transverse* polarisations: $\epsilon^\mu_{1,2} = \{0, \epsilon_{1,2}\}$, with $(\epsilon_{1,2} \cdot \mathbf{q}) = 0$.

ϵ_3 : *Longitudinal* polarisation: a spatial vector parallel to $\mathbf{q}$: $\epsilon^\mu_3 = \{0, \hat{\mathbf{q}}\}$.

ϵ_0 : *Timelike* polarisation: a timelike vector, $\epsilon^\mu_0 = \eta^\mu = \{1, 0\}$.

Corresponding to the four vectors, we speak of transverse, longitudinal or timelike photons. If, for example, $\mathbf{q}$ is in direction 3, we can choose the four vectors as

$$\epsilon_1 = \{0, 1, 0, 0\} \quad , \quad \epsilon_2 = \{0, 0, 1, 0\} \quad , \quad \epsilon_L = \{0, 0, 0, 1\} \quad , \quad \epsilon_T = \{1, 0, 0, 0\} \, .$$

For the transverse states, for each value of $\mathbf{q}$, we can choose two purely spatial ($\epsilon^0_{1,2} = 0$) polarisation vectors $\epsilon^\nu_r(\mathbf{q})$ $(r = 1, 2)$ such that

$$\mathbf{q} \cdot \epsilon_\mathbf{r} = 0 \quad , \quad \epsilon_\mathbf{r} \cdot \epsilon_\mathbf{s} = \delta_{\mathbf{rs}} \, .$$

If we then define the photon states as

$$|\gamma; \mathbf{q}, r\rangle = \epsilon^\nu_r(\mathbf{q}) A^\dagger_\nu(\mathbf{q}, 0)|0\rangle \, ,$$

---

[3] A state of this kind is also dubbed a *ghost state*, or simply a *ghost*.

[4] A deeper discussion and citations from the original literature, in particular to the work of Gupta and Bleuler, can be found in Chapter 5 of Mandl and Shaw [3].

from (5.21) we obtain

$$\langle \gamma; \mathbf{k}, s | \gamma; \mathbf{q}, r \rangle = \delta_{rs} \delta^{(3)}(\mathbf{k} - \mathbf{q}) \ , \tag{5.22}$$

and, from (5.20),

$$\langle 0 | A^\mu(x) | \gamma; \mathbf{q}, r \rangle = \frac{\epsilon_r^\mu(\mathbf{q})}{(2\pi)^{3/2}\sqrt{2\omega_q}} e^{i\mathbf{q}\mathbf{x}} e^{-i\omega_q(t_x - t_y)} \ . \tag{5.23}$$

This last result is useful to establish the rules for calculating the S-matrix. In the present work we consider the ϵ_r^ν as vectors with *real* components which describe linearly polarised photons. We recall that to describe circularly polarised states, in particular photons of definite helicity, it is necessary to use vectors ϵ_r^ν with *complex* components.

5.4 VIRTUAL PHOTONS

The interaction between the electromagnetic field and matter occurs through the electromagnetic current. The corresponding Lagrangian density reads [1]

$$\mathcal{L}_I = eA_\mu(x)J^\mu(x) \ , \tag{5.24}$$

and the gauge invariance prescribes that the current is conserved, i.e. that

$$\partial_\mu J^\mu(x) = 0 \ . \tag{5.25}$$

In QED, for the electron

$$J^\mu = \bar{\psi}\gamma^\mu\psi \ , \tag{5.26}$$

and conservation of the current is guaranteed by Noether's theorem.

The simplest apparatus with which we can experiment on the electromagnetic field is composed of a transmitter and a receiver. In the transmitter, an electron undergoes a transition between a state A and a state B, while in the receiver another electron passes from A' to B'.

The overall transition amplitude, $A \rightarrow B, A' \rightarrow B'$, to second order of perturbation theory, is given by

$$
\begin{aligned}
S_{fi} &= \langle B, B' | \int d^4x d^4y \ T\left[\mathcal{L}_I(x)\mathcal{L}_I(y)\right] | A, A' \rangle \\
&= \frac{(ie)^2}{2} \int d^4x d^4y \ \langle B, B' | T\left[A_\mu(x)J^\mu(x)A_\nu(y)J^\nu(y)\right] | A, A' \rangle \ .
\end{aligned} \tag{5.27}
$$

Factorising the quantum states of the transmitter and receiver ($|A, A'\rangle =$

$|A\rangle|A'\rangle$, etc.) and taking into account that, to this order, the currents commute among themselves and with the vector potential, we can write

$$S_{fi} = \frac{(ie)^2}{2} \int d^4x d^4y \tag{5.28}$$
$$\times \left(\langle B|J^\mu(x)|A\rangle\langle B'|J^\nu(y)|A'\rangle + x \leftrightarrow y\right) \langle 0|T\left[A_\mu(x)A_\nu(y)\right]|0\rangle$$
$$= (ie)^2 \int d^4x d^4y \ \langle B|J^\mu(x)|A\rangle\langle 0|\ T\left[A_\mu(x)A_\nu(y)\right]|0\rangle \ \langle B'|J^\nu(y)|A'\rangle \ ,$$

where we have used the symmetry between the integration variables x and y to cancel the factor $\frac{1}{2}$. We can represent the amplitude (5.28) by the Feynman diagram in Figure 5.2 where the wavy line represents the photon propagator. In this case, this is described as the exchange of a *virtual photon* between the two currents.

Taking the Fourier transforms and using (5.19) we obtain

$$S_{fi} = (ie)^2 \int \frac{d^4p}{(2\pi)^4} \tilde{J}^\mu_{BA}(-p)\frac{i}{p^2 + i\epsilon}\left[-g_{\mu\nu} + (1-\alpha)\frac{p_\mu p_\nu}{p^2}\right]\tilde{J}^\nu_{B'A'}(p). \tag{5.29}$$

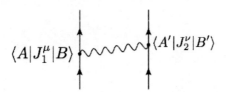

$$\langle A|J^\mu_1|B\rangle \qquad \langle A'|J^\nu_2|B'\rangle$$

Figure 5.2   Feynman diagram representing the amplitude of the transmitter–receiver interaction due to exchange of a virtual photon. The continuous lines represent the world lines in space-time of the radiator and the absorber. The wavy line between the vertices represents the propagation of the electromagnetic field.

A first consequence of (5.29) is that S_{fi} is independent of the gauge; the terms dependent on α in the propagator make no contribution thanks to conservation of the current (5.25), which implies that $p_\mu J^\mu(p) = 0$.

We can therefore write

$$S_{fi} = (ie)^2 \int \frac{d^4p}{(2\pi)^4}(\tilde{J}_{BA})^\mu(-p)\frac{i(-g_{\mu\nu})}{p^2 + i\epsilon}\ (\tilde{J}_{B'A'})^\nu(p) \ . \tag{5.30}$$

This equation draws our attention again to a problem: the residue of the pole at $p^2 = 0$ is proportional to $-g_{\mu\nu}$, which seems to indicate that all four types of photon can contribute; see (5.21). However, the pathology is only apparent and it is again cured by the conservation of current.[5] In connection

---

[5]We note that the amplitude can also be written as $\int d^4x \ J^\mu_{A'B'}(x)(A_{AB})_\mu(x)$, where

with the 4-momentum p, we introduce the four polarisation vectors $\epsilon^{\mu}(p)$, equation (5.3), which form an orthonormal basis in the space of p (naturally with reference to the metric $g^{\mu\nu}$). It is easy to confirm that the completeness condition for this basis is written:

$$-g^{\mu\nu} = \sum_{i=0,\ldots,3} x_i \, \epsilon_i^{\mu} \epsilon_i^{\nu} \ ,$$

with

$$x_0 = -1 \ , \qquad x_{1,2,3} = +1 \ , \tag{5.31}$$

or

$$-g^{\mu\nu} = \left(\sum_{i=1,2} \epsilon_i^{\mu} \epsilon_i^{\nu} \right) + (\epsilon_3^{\mu} \epsilon_3^{\nu} - \eta^{\mu} \eta^{\nu}) \ . \tag{5.32}$$

If we express ϵ_3 as a function of p and η^{μ}, we obtain:

$$\epsilon_3^{\mu} = \frac{1}{|\mathbf{p}|} \left(p^{\mu} - p^0 \eta^{\mu} \right) \ , \tag{5.33}$$

$$(\epsilon_3^{\mu} \epsilon_3^{\nu} - \eta^{\mu} \eta^{\nu}) = \frac{p^2}{|\mathbf{p}|^2} \eta^{\mu} \eta^{\nu} + \ldots \ , \tag{5.34}$$

where the dots denote terms proportional to p^{μ} and/or to p^{ν}, which give zero contributions when inserted into (5.29), again by virtue of the current conservation.

In conclusion, with this substitution, equation (5.29) becomes

$$S_{fi} = \int \frac{d^4 p}{(2\pi)^4} \, \tilde{J}_{BA}^{\mu}(-p) \frac{i \left(\sum_{i=1,2} (\epsilon_i)_{\mu} (\epsilon_i)_{\nu} \right)}{p^2 + i\epsilon} \tilde{J}_{B'A'}^{\nu}(p)$$

$$+ \int \frac{d^4 p}{(2\pi)^4} \, \tilde{J}_{BA}^0(-p) \frac{-i}{|\mathbf{p}|}^2 \tilde{J}_{B'A'}^0(p) \ . \tag{5.35}$$

The first term on the right-hand side of (5.35) shows that only the transverse photons contribute to the pole at $p^2 = 0$, which are therefore the only states present at large distance from the interaction. There are no photons of other types in the asymptotic "in" and "out" states which we will introduce later. The contribution of "longitudinal" and "timelike" photons combine, by virtue of the current conservation, in the second term which, as seen from the factor $1/|\mathbf{p}|$, is simply the instantaneous Coulomb interaction between the two currents.

$(A_{AB})_{\mu}(\mathbf{x})$ is the classical field generated by the current $J_{AB}^{\mu}(x)$. At this point, the discussion follows almost word for word that stated in [1] concerning the Green's function of the classical electromagnetic field.

In actual fact, turning to the space of coordinates, we have

$$\int \frac{d^4 p}{(2\pi)^4} \, \tilde{J}^0_{BA}(-p) \frac{-i}{|\mathbf{p}|}^2 \, \tilde{J}^0_{B'A'}(p) =$$

$$= \int d^4x d^4y \, J^0_{BA}(x) J^0_{B'A'}(y) \int \frac{d^4p}{(2\pi)^4} \, e^{-ip(x-y)} \frac{-i}{|\mathbf{p}|}^2$$

$$= -i \int d^4x d^4y \, J^0_{BA}(x) \, \delta(x^0 - y^0) \frac{1}{4\pi|\mathbf{x} - \mathbf{y}|} J^0_{B'A'}(y) \,, \qquad (5.36)$$

which is nothing more than multiplication by $-i$ of the Hamiltonian matrix element for the electrostatic interaction between the two currents.

If we had carried out the calculation in the Coulomb gauge, we would have written the Hamiltonian of the interaction as the sum of the interactions of the currents with the transverse photons (of order e) plus the Coulomb interaction between the charge density, of order e^2:

$$H_I = e \int d^3x \, \mathbf{A}(x,t)\mathbf{j}(x,t) + \frac{e^2}{2} \int d^3x d^3y \, j_0(x,t) \frac{1}{4\pi|\mathbf{x} - \mathbf{y}|} j_0(x,t) \,. \ (5.37)$$

The amplitude (5.35) is obtained by taking, in the S-matrix, the Coulomb interaction to first order and the interaction with the transverse photons to second order. The decisive advantage of the relativistic formulation is to calculate *everything together*, with the single diagram of Figure 5.2.

CHAPTER 6

FERMION FIELDS

CONTENTS

6.1 HARMONIC AND FERMI OSCILLATORS

In this chapter we move from the treatment of the electromagnetic field to that of the Dirac fields which describe spin $\frac{1}{2}$ particles. The problem posed is how to treat fields which must obey anticommutation rules and the Pauli exclusion principle. In the elementary treatments of field theory we saw that a free field which describes non-interacting bosons is equivalent to a set of harmonic oscillators, one for each state in which a particle can be found. Focusing on a single oscillator we can define the creation and destruction operators, $a^\dagger$, a which obey the commutation rules

$$[a, a^\dagger] = 1 \ . \tag{6.1}$$

If we denote with $|n\rangle$ the state in which the oscillator contains n particles, we have

$$a|n\rangle = \sqrt{n}|n-1\rangle \ , \quad a^\dagger|n\rangle = \sqrt{n+1}|n+1\rangle \qquad \text{(bosons)}. \tag{6.2}$$

A spinor field can also be expanded in "oscillators", but of a different type, the *Fermi oscillators*. Each one of them can contain a maximum of one particle, and the action of the corresponding creation and destruction operators on states of 0 or 1 particle is

$$a^\dagger|0\rangle = |1\rangle \ , \quad a^\dagger|1\rangle = 0 \ , \quad a|1\rangle = |0\rangle \ , \quad a|0\rangle = 0 \qquad \text{(fermions)}. \tag{6.3}$$

This gives rise to anticommutation rules

$$\{a,\, a^\dagger\} = 1 \,. \tag{6.4}$$

In both cases the Hamiltonian becomes[1]

$$H = \hbar\omega\, a^\dagger a \,, \tag{6.5}$$

with ω the particle energy. We can formally derive this Hamiltonian from a Lagrangian

$$L = i\hbar a^\dagger \dot{a} - \hbar\omega a^\dagger a \,, \tag{6.6}$$

yielding

$$\pi = \frac{\partial L}{\partial \dot{a}} = i\hbar a^\dagger \,, \quad H = \pi\dot{a} - L = \hbar\omega\, a^\dagger a \,. \tag{6.7}$$

We note that $\partial L/\partial \dot{a}^\dagger = 0$, since L does not depend on $\dot{a}^\dagger$. Actually the relation between a and $a^\dagger$ is symmetric, since with an integration by parts we can express the action in two equivalent ways in which the roles of a and $a^\dagger$ are exchanged,

$$S = \hbar \int dt \,(ia^\dagger \dot{a} - \omega a^\dagger a) = \hbar \int dt \,(-i\dot{a}^\dagger a - \omega a^\dagger a) \,.$$

If we apply the canonical commutation rules we obtain the "boson" result

$$[a,\, a^\dagger] = \frac{1}{i\hbar}[a,\, \pi] = \frac{1}{i\hbar}i\hbar = 1 \,. \tag{6.8}$$

We already know how to derive all the properties of the boson harmonic oscillator by means of the sum over paths of the variable $q(t)$. As we will soon see, we can directly use paths in the variables $a(t)$ and $a^\dagger(t)$. How should the sum over paths be modified to obtain the fermion result? The right idea emerges by considering, instead of a and $a^\dagger$, the variables $\tilde{a} = \sqrt{\hbar}a$ and $\tilde{a}^\dagger = \sqrt{\hbar}a^\dagger$, which do not depend on Planck's constant, and are therefore classical variables:

$$\tilde{a} = \sqrt{\frac{m\omega}{2}}\left(x + \frac{ip}{m\omega}\right) \,, \quad \tilde{a}^\dagger = \sqrt{\frac{m\omega}{2}}\left(x - \frac{ip}{m\omega}\right) \,. \tag{6.9}$$

In terms of them, the Lagrangian becomes

$$L = i\tilde{a}^\dagger \dot{\tilde{a}} - \omega\tilde{a}^\dagger \tilde{a} \,, \tag{6.10}$$

and the commutation rules are

$$[\tilde{a},\, \tilde{a}^\dagger] = \hbar \,. \tag{6.11}$$

---

[1] As will become clear shortly, it is useful to include the dependence on Planck's constant explicitly in this phase.

In the sum over paths, $\tilde{a}(t)$ and $\tilde{a}^\dagger(t)$ are treated as functions with numerical values, i.e. as quantities which commute. Therefore, it is as if a classical limit, $\hbar \to 0$, were taken in which $\tilde{a}(t)$ and $\tilde{a}^\dagger(t)$ become commuting quantities:

$$[\tilde{a}, \tilde{a}^\dagger] = 0 \ . \tag{6.12}$$

The fermion case starts from the anticommutation rules, which in the limit $\hbar \to 0$ become simply

$$\{\tilde{a}, \tilde{a}^\dagger\} = 0 \ . \tag{6.13}$$

Therefore, the paths in the fermion case should be described by a function whose value is not a normal number, but a quantity which anticommutes, a *Grassman variable*.

In what follows we return to using units in which $\hbar = 1$.

6.1.1 Anticommuting variables

The rules for calculating with anticommuting quantities are very simple, and the page which follows contains an entire manual for differential and integral calculations with Grassman variables. Suppose we have n variables of this type, $a_1 \cdots a_n$, such that

$$\{a_h, a_k\} = 0 \ , \tag{6.14}$$

that in particular imply that $(a_i)^2 = 0$. Then, the following rules hold:

Numerical coefficients, linear combinations: An anticommuting variable can be multiplied by an ordinary number c with which it commutes: $c\,a = a\,c$. Linear combinations can be constructed as $c_1 a_1 + c_2 a_2 + \cdots$.

Functions: Since $a_k^2 = 0$, the most general function is an n-th order polynomial

$$F = C_0 + \sum_k C_1(k)a_k + \sum_{h>k} C_2(h,k)\,a_h\,a_k \cdots + C_n\,a_1\,a_2\cdots a_n, \tag{6.15}$$

where C are numerical coefficients.

Differentials and derivatives: The differential da of an anticommuting quantity a is itself anticommuting: since $a_1\,a_2 = -a_2\,a_1$, it must be true that $a_1\,da_2 = -da_2\,a_1$. The derivative with respect to an anticommuting variable d/da_k is defined according to the following rules:

- $(d/da_k)\,1 = 0 \ , \quad (d/da_k)\,a_h = \delta_{hk}.$

- The operation d/da_k anticommutes with other Grassman variables. This can be understood by considering a product of anticommuting quantities $c\,d\cdots a\cdots$; d/da "extracts" a from the product and to do this must move a to the first position. Hence $d/da\,(b\cdot a) = d/da\,(-a\cdot b) = -b$ or $d/da\,b\cdot a = -b\cdot d/da\,a$, and so on.

- The derivatives anticommute with each other, for example $d/da\,d/db\,(ba) = d/da\,(b) = 1$, while $d/db\,d/da\,(ba) = d/db\,(-b) = -1$.

Integrals: The integral over anticommuting variables is defined with the following rules:

$$\int da = 0\ ,\quad \int da\,a = 1\ . \tag{6.16}$$

It follows from this that for anticommuting variables the integral and the derivative are the same operation.

$$\int da\,F = \frac{d}{da}\,F\ . \tag{6.17}$$

This definition is motivated in the following way: the integrals $\int da$ and $\int da\,a$ should be defined as constants which do not depend on any anticommuting quantity, hence as ordinary numbers. At the same time, since da is anticommuting, $\int da$ must be anticommuting, therefore the only possibility is that it must be zero, the only ordinary number which is also anticommuting: $0\,x = -x\,0$. The second integral, $\int da\,a$, can be any number; to set $\int da\,a = 1$ is equivalent to defining the normalisation of the a values. If, for example, we were to have $\int da\,a = X$, we could define a new variable, $a = a'X^{1/2}$, such that $\int da'\,a' = 1$.

We note that if P_i denotes products of an even number of anticommuting quantities and A_k products of an odd number, we have

$$P_iP_k = P_kP_i\ ,\quad P_iA_k = A_kP_i\ ,\quad A_iA_k = -A_kA_i\ . \tag{6.18}$$

For example, if a, b, c, d are anticommuting, $(ab)(cd) = (cd)(ab)$, $(ab)c = c(ab)$, while $(abc)d = -d(abc)$. Therefore the product of an even number of anticommuting quantities behaves like a commuting quantity.

6.1.2 Sum over paths for the two oscillators

In this section we develop the rules for calculating the sum over paths of anticommuting quantities by applying them to a concrete case: the Fermi oscillator defined in Section 6.1. As we will see, the sum over paths leads to results which are in full agreement with those obtained in the traditional way.

In Section 3.2 we calculated the generating functional of the harmonic oscillator in the $q(t)$ language. We would now like to do it in the language of $a(t)$ and $a^\dagger(t)$, being careful to carry out operations which are equally valid for commuting variables (boson case) and anticommuting (fermion case) and noting along the way the differences between the two cases. We therefore define $Z(J, J^\dagger)$ as

$$
Z(J, J^\dagger) =
\int d[a(t)]\, d[a^\dagger(t)] \, \exp\left[i \int dt \, \left(a^\dagger(t)\, D\, a(t) - J^\dagger(t)\, a(t) - a^\dagger(t)\, J(t)\right)\right] ,
$$
$$(6.19)$$

where (see equation(6.10)) the differential operator$^2 D$ is given by

$$
D = i\frac{d}{dt} - \omega \ . \tag{6.20}
$$

Since the action must in every case be a commuting quantity, in the fermion case both J and $J^\dagger$ must be anticommuting. As in Section 3.2 we introduce a function $S(t)$ such that

$$
D\,S(t) \;=\; \delta(t) \quad \rightarrow \quad \dot{S}(t) \;=\; -i\omega S(t) \;-\; i\,\delta(t) \ . \tag{6.21}
$$

We can rewrite the term in $(a^\dagger J)$ of (6.19) as

$$
\int dt \, a^\dagger(t)\, J(t) \;=\; \int dt \, a^\dagger(t)\, D\left(\int dt'\, S(t - t')\, J(t')\right) ,
$$

while the term in $(J^\dagger a)$ can be written as

$$
\int dt \, J^\dagger(t)\, a(t) \;=\; \int dt \left(\int dt'\, J^\dagger(t')\, S(t' - t)\right) D\, a(t) ,
$$

as can be seen with an integration by parts

$$
\int dt \left(\int dt'\, J^\dagger(t')\, S(t' - t)\right) \left(i\frac{d}{dt} - \omega\right) a(t) =
$$
$$
\int dt \left(-i\frac{d}{dt} - \omega\right) \int dt'\, J^\dagger(t')\, S(t' - t)\, a(t) , \quad (6.22)
$$

and noting that

$$
\left(-i\frac{d}{dt} - \omega\right) S(t' - t) = \left(i\frac{d}{d(t' - t)} - \omega\right) S(t' - t) = \delta(t' - t) \ .
$$

---

^{2}This is none other than the Dirac operator $(i\gamma^\mu \partial_\mu - m)$, but in a space of only one dimension.

We can then rewrite the generating functional as

$$Z(J, J^\dagger) = \exp\left[-i \int dt' \, dt \, J^\dagger(t') \, S(t' - t) \, J(t)\right]$$

$$\times \int d[a(t)] \, d[a^\dagger(t)]$$

$$\exp\left[i \int dt \left(a^\dagger(t) - \int dt' J^\dagger(t')S(t' - t)\right) D \left(a(t) - \int dt' S(t - t') J(t')\right)\right].$$

The functional integral can be carried out with a change of variables

$$a(t) \;\to\; a'(t) = a(t) - \int dt' S(t - t') \, J(t') \;,$$

and similarly for $a^\dagger$, and is reduced to a multiplicative constant which corresponds to the value of $Z[0]$ and can be omitted. This change of variables is, however, allowed only if (see Section 3.2)

$$\lim_{t \to \infty} S\left(t(1 - i\chi)\right) = \lim_{t \to -\infty} S\left(t(1 - i\chi)\right) = 0 \;. \tag{6.23}$$

The general solution of (6.21) is $S(t) = A e^{-i\omega t} - i\theta(t) e^{-i\omega t}$, but (6.23) imposes $A = 0$, hence

$$S(t) = -i\theta(t) e^{-i\omega t} \;, \tag{6.24}$$

and the generating functional becomes

$$Z(J, J^\dagger) = \exp\left[-i \int dt' \, dt \, J^\dagger(t') \, S(t' - t) \, J(t)\right] . \tag{6.25}$$

Since we have been careful not to change the order of the quantities which in the fermion case anticommute with each other, the procedure followed up to now is equally valid for both the boson and fermion case. Some differences appear in the calculation of the Green's functions, where it is necessary to take into account the anticommuting character of the operators. For example, if a is an anticommuting quantity, equation (2.23) is redefined as

$$T[a(t_1)\, a(t_2)] = \begin{cases} a(t_1)\, a(t_2) & \text{if } t_1 \geq t_2 \\ -a(t_2)\, a(t_1) & \text{if } t_2 \geq t_1 \end{cases} \qquad \text{fermions}, \tag{6.26}$$

and hence also

$$T[a(t_1)\, a(t_2)] = -T[a(t_2)\, a(t_1)] \;, \tag{6.27}$$

and this property extends to the time-ordered product of more operators, and therefore also to the Green's functions.

The rules for the use of the generating functional are slightly different in the two cases. While the correspondence rule

$$a(t) \to i \frac{\delta}{\delta J^\dagger(t)} \;, \qquad \text{bosons or fermions}, \tag{6.28}$$

is the same in both cases, we have

$$
a^\dagger(t) \quad \rightarrow \quad
\begin{cases}
i\frac{\delta}{\delta J(t)} & \text{bosons} , \\[2mm]
-i\frac{\delta}{\delta J(t)} & \text{fermions} ,
\end{cases}
\tag{6.29}
$$

as seen from (6.19) noting that in the fermion case $\frac{\delta}{\delta J(t)}$ anticommutes with $a^\dagger(t)$.

We will calculate some Green's functions; for the two-point function we obtain, in both the fermion and boson case (we leave the derivation to the reader),

$$
\langle 0|T\left[a(t)a^\dagger(\tau)\right]|0\rangle = \frac{\delta}{\delta J^\dagger(t)}\frac{\delta}{\delta J(\tau)}Z[J,J^\dagger]\Big|_{J=J^\dagger=0}
$$
$$
= iS(t-\tau) = \theta(t-\tau)e^{-i\omega(t-\tau)} . \tag{6.30}
$$

We first consider the case with $t > \tau$, for which we obtain, assigning an energy $E_0 = 0$ to the state $|0\rangle$,

$$
\langle 0|a(t)a^\dagger(\tau)|0\rangle \equiv \langle 0|a\, e^{-iH(t-\tau)}a^\dagger|0\rangle = e^{-i\omega(t-\tau)} .
$$

This result tells us that[3] a state $|1\rangle$ exists with energy $E_1 = \omega$, and that $|\langle 1|a^\dagger|0\rangle|^2 = 1$. Hence we can define the phase of $|1\rangle$ so that $a^\dagger|0\rangle = |1\rangle$. Conversely, if $\tau > t$ we obtain

$$
\langle 0|a^\dagger(\tau)\, a(t)|0\rangle = 0 ,
$$

and introducing a complete set of states $|A\rangle$ with energy E_A,

$$
\sum_A |\langle A|a|0\rangle|^2 e^{iE_A(\tau-t)} = 0 ,
$$

from which we obtain (considering the case with $\tau = t$) $a|0\rangle = 0$.

For the four-point function, there is a difference between the boson and fermion case. In the steps which follow, where the symbol $\pm$ appears it means that the $+$ sign applies to the boson case, $-$ to the fermion case, and where it does not appear, the result is the same in both cases.

$$
\langle 0|T\left(a(t_1)\,a(t_2)\,a^\dagger(\tau_1)\,a^\dagger(\tau_2)\right)|0\rangle
$$
$$
= \frac{\delta}{\delta J^\dagger(t_1)}\frac{\delta}{\delta J^\dagger(t_2)}\frac{\delta}{\delta J(\tau_1)}\frac{\delta}{\delta J(\tau_2)}\frac{1}{2}\left[-i\int dt'\, dt\, J^\dagger(t')\, S(t'-t)\, J(t)\right]^2
$$
$$
= -\frac{\delta}{\delta J^\dagger(t_1)}\frac{\delta}{\delta J^\dagger(t_2)}\left[\int dt' J^\dagger(t')S(t'-\tau_1)\right]\left[\int dt'' J^\dagger(t'')S(t''-\tau_2)\right]
$$
$$
= -\left[S(t_2-\tau_1)\,S(t_1-\tau_2) \pm S(t_1-\tau_1)\,S(t_2-\tau_2)\right]
$$
$$
= e^{i\omega(t_1+t_2-\tau_1-\tau_2)}\left[\theta(t_2-\tau_1)\,\theta(t_1-\tau_2) \pm \theta(t_1-\tau_1)\,\theta(t_2-\tau_2)\right] .
$$

---

[3] See the discussion at the end of Section 3.2.

The $-$ sign which appears in the fermion case reflects the Pauli exclusion principle. For example, in the case with $t_1 > t_2 > \tau_1 > \tau_2$ we obtain

$$\langle 0|(a(t_1)\,a(t_2)\,a^\dagger(\tau_1)\,a^\dagger(\tau_2)|0\rangle = \begin{cases} 2\,e^{-i\omega(t_1+t_2-\tau_1-\tau_2)} & \text{bosons,} \\ 0 & \text{fermions.} \end{cases} \tag{6.31}$$

In this expression, if we take the limit $t_1 \to t_2$ and $\tau_1 \to \tau_2$, in the boson case we obtain

$$\langle 0|[a(t_2)]^2\,[a^\dagger(\tau_2)]^2|0\rangle = 2e^{-i2\omega(t_2-\tau_2)} \ , \tag{6.32}$$

from which we learn that a state $|2\rangle$ with energy $E_2 = 2\omega$ exists, and that $[a^\dagger]^2|0\rangle = \sqrt{2}|2\rangle$. Since $a^\dagger|0\rangle = |1\rangle$, it must be true that $a^\dagger|1\rangle = \sqrt{2}|2\rangle$. Similarly we deduce that $a^2|2\rangle = \sqrt{2}|0\rangle$. Instead, in the fermion case we learn that, as follows from the Pauli principle, $[a^\dagger]^2|0\rangle = 0$, i.e. that a second excited state of the Fermi oscillator does not exist.

6.1.3 Gaussian integrals for commuting and anticommuting variables

For later use, we provide here the calculations of the Gaussian integrals over anticommuting variables, to compare with the result for the analogous integrals over commuting variables.

The simplest integral is

$$A(\lambda) = \int da^\dagger da \ e^{-\lambda a^\dagger a} \ , \tag{6.33}$$

which is carried out using the rules (6.16)

$$A(\lambda) = \int da^\dagger da\,(1 - \lambda a^\dagger a) = \lambda \int da^\dagger a^\dagger da\,a = \lambda \ . \tag{6.34}$$

The result (6.34) can immediately be generalised to the case of N pairs of variables

$$A(\lambda_1,\dots,\lambda_N) = \int \left(\prod_i da_i^\dagger da_i\right) e^{\sum_k \lambda_k a_k^\dagger a_k} = \prod_i \lambda_i \ , \tag{6.35}$$

and hence to the case of a bilinear form

$$A(\Lambda) = \int \left(\prod_i da_i^\dagger da_i\right) e^{(a^\dagger \Lambda a)} \ ,$$

$$(a^\dagger \Lambda a) = \sum_{ij} a_i^\dagger \Lambda_{ij} a_j \ . \tag{6.36}$$

In this case, we diagonalise the matrix Λ with a unitary transformation

$$\tilde{a}_i = \sum_k U_{ik} a_k; \quad U^\dagger U = \mathbf{1} \ , \tag{6.37}$$

so that

$$(a^\dagger \Lambda a) = \sum_k \lambda_k \tilde{a}_k^\dagger \tilde{a}_k \ , \tag{6.38}$$

where the λ_k are the eigenvalues of Λ. It should be noted that the Jacobian of the transformation between the variables $a^\dagger, a$ and $\tilde{a}^\dagger, \tilde{a}$ is equal to unity:

$$\begin{aligned}
J &= \det \left[\frac{\partial(\tilde{a}^\dagger, \tilde{a})}{\partial(a^\dagger, a)} \right] \\
&= \det \left[\frac{\partial(\tilde{a}^\dagger)}{\partial(a^\dagger)} \right] \det \left[\frac{\partial(\tilde{a})}{\partial(a)} \right] \\
&= |\det U|^2 = 1 \ ,
\end{aligned} \tag{6.39}$$

from which it follows that

$$A(\Lambda) = \int \left(\prod_i d\tilde{a}_i^\dagger d\tilde{a}_i \right) e^{\sum_k \lambda_k \tilde{a}_k^\dagger \tilde{a}_k} = \det \Lambda \ . \tag{6.40}$$

This result can be compared with that for the Gaussian integral over commuting complex variables

$$C(\Lambda) = \int \left(\prod_i dx_i^\dagger dx_i \right) e^{-(x^\dagger \Lambda x)} \ . \tag{6.41}$$

In the basis in which Λ is diagonal, we put

$$x_i = \frac{u_i + iv_i}{\sqrt{2}}; \qquad x_i^\dagger = \frac{u_i - iv_i}{i\sqrt{2}} \ , \tag{6.42}$$

from which we obtain

$$C(\Lambda) = \int \prod \left(du_i dv_i \ e^{-\frac{\lambda_i(u_i^2 + v_i^2)}{2}} \right) = \prod_i \left(\frac{2\pi}{\lambda_i} \right) = (2\pi)^n \, [\det \Lambda]^{-1} \ . \tag{6.43}$$

The step from anticommuting to commuting variables involves the change $\det \Lambda \rightarrow [\det \Lambda]^{-1}$, at least to within an inessential multiplicative constant.

We finally note that in the case of commuting variables we find

$$C(\Lambda) = \int \prod \left(du_i \ e^{-\frac{\lambda_i u_i^2}{2}} \right) = \prod_i \sqrt{\frac{2\pi}{\lambda_i}} = \frac{(2\pi)^{n/2}}{\sqrt{\det \Lambda}} \ . \tag{6.44}$$

With the help of the preceding formulae, we can consider the case of Gaussian functional integrals of the form

$$A(D) = \int d[\psi^\dagger(x)] d[\psi(x)] \ e^{-(\psi^\dagger D \psi)} \ ,$$

with

$$(\psi^\dagger D \psi) = \int dx dy \ \psi^\dagger(x) D(x,y) \psi(y) \ . \tag{6.45}$$

Once the set of eigenfunctions of the operator D is defined

$$\int dy D(x,y)\psi_i(y) = \lambda_i \psi_i(x) \ , \tag{6.46}$$

we can expand the anticommuting functions in the basis of the eigenfunctions:

$$\psi(x) = \sum_i a_i \psi_i(x) \ , \tag{6.47}$$

where a_i are anticommuting variables. In the new basis, we write the functional integral as

$$A(D) = \int \prod_i da_i^\dagger da_i \ e^{-(a^\dagger D a)} \ , \quad D_{ij} = \int dx dy \ \psi_i^\dagger(x) D(x,y) \psi_j(y) \ , \tag{6.48}$$

from which, as before, it follows that

$$A(D) = \det D \ , \tag{6.49}$$

with the determinant of D given by (6.40). The case of commuting variables is treated in an analogous way.

6.2 QUANTISATION OF THE DIRAC FIELD

In this section we explicitly calculate the generating functional for a free Dirac field and the two-point function. We give as a footnote the standard description of this system, based on the canonical quantisation rules, and the properties of the γ-matrices and of the plane wave solutions to the Dirac equation. We adopt the notation and conventions of [1].

We recall that two types of 4-spinors exist: the "normal" type, represented by the Dirac field $\psi(x)$ and the "adjoint" type represented by the field $\bar{\psi}(x) = \psi^\dagger \gamma^0$. In general we will omit the spinor indices, meaning in particular that two contiguous indices, one "normal" and one "adjoint", should be summed. In the Dirac matrices, the first index is considered "normal" and the second "adjoint".

The Dirac equation for a particle of mass m will be written as

$$(i\gamma^\mu \partial_\mu - m)\psi(x) = 0 \ , \tag{6.50}$$

and can be derived from a Lagrangian density

$$\mathcal{L} = \bar{\psi}(x)(i\gamma^\mu \partial_\mu - m)\psi(x) \ . \tag{6.51}$$

Equation (6.51) will be our starting point. Analogous with what we did for the oscillator, we define the generating functional by introducing two auxiliary functions $J_\rho(x)$, $\bar{J}_\rho(x)$.

$$Z(J, \bar{J}) = \int d[\psi]\, d[\bar{\psi}]$$
$$\times \exp\left[i \int d^4x \left(\bar{\psi}(x)\, D\, \psi(x) - \bar{J}(x)\, \psi(x) - \bar{\psi}(x)\, J(x) \right) \right] , \quad (6.52)$$

where D is the Dirac operator

$$D = i\gamma^\mu \partial_\mu - m ,$$

and we will consider ψ, $\bar{\psi}$, J, $\bar{J}$ as *anticommuting* quantities. To carry out the integral we follow the steps of the previous section. We introduce a function $S_F(x)$, the propagator, such that

$$D\, S_F(x) = \delta^4(x) , \quad (6.53)$$

so that we can write

$$Z(J, \bar{J}) = \exp\left[-i \int d^4x'\, d^4x\, \bar{J}(x')\, S_F(x' - x)\, J(x) \right]$$
$$\times \int d[\psi]\, d[\bar{\psi}] \exp\left[i \int d^4x \left(\bar{\psi}(x) - \int d^4x'\, \bar{J}(x')\, S_F(x' - x) \right) \right.$$
$$\left. \times D \left(\psi(x) - \int d^4x'\, S_F(x - x')\, J(x') \right) \right] .$$

To prove this transformation we begin with the relation

$$\int d^4x\, \bar{\psi}(x)\, J(x) = \int d^4x\, \bar{\psi}(x)\, (i\gamma^\mu \frac{\partial}{\partial x_\mu} - m) \int d^4x'\, S_F(x - x')\, J(x') ,$$

which can be confirmed directly from (6.53). Furthermore

$$\int d^4x'\, d^4x\, \bar{J}(x')\, S_F(x' - x)\, (i\gamma^\mu \frac{\partial}{\partial x_\mu} - m)\, \psi(x)$$
$$= \int d^4x'\, d^4x\, \bar{J}(x')\, S_F(x' - x)\, (-i\gamma^\mu \frac{\overleftarrow{\partial}}{\partial x_\mu} - m)\, \psi(x)$$
$$= \int d^4x'\, d^4x\, \bar{J}(x')\delta^4(x' - x)\psi(x) = \int d^4x\, \bar{J}(x)\psi(x) ,$$

where in the first step we carried out an integration by parts, and in the second we used equation (6.57), which we will demonstrate shortly. The arrow denotes that the derivative is taken on the function to the left.

The functional integral is carried out with a change of variables

$$\psi(x) \to \psi'(x) = \psi(x) - \int d^4x' S_F(x - x') J(x') ,$$

$$\bar{\psi}(x) \to \bar{\psi}'(x) = \bar{\psi}(x) - \int d^4x' \bar{J}(x') S_F(x' - x) ,$$

and results in a multiplicative constant (the value of $Z[0]$) which can be omitted. Thus we simply obtain

$$Z(J, \bar{J}) = \exp\left[-i \int d^4x' \, d^4x \, \bar{J}(x') S_F(x' - x) J(x)\right] . \tag{6.54}$$

As seen earlier, this procedure is allowed only if

$$\lim_{t \to \pm\infty} S_F \left((t(1 - i\chi), \mathbf{x})\right) = 0 . \tag{6.55}$$

A solution of (6.53) with the desired properties is obtained by setting

$$S_F(x) = (i\gamma^\mu \partial_\mu + m)\Delta_F(x) = \frac{1}{(2\pi)^4} \int d^4p \, e^{-ipx} \frac{\slashed{p} + m}{E^2 - \mathbf{p}^2 - m^2 + i\epsilon} . \tag{6.56}$$

Indeed, substituting into (6.53) we find[4]

$$(i\gamma^\mu \partial_\mu - m)(i\gamma^\mu \partial_\mu + m)\Delta_F(x) = -(\Box + m^2)\Delta_F(x) = \delta^4(x) .$$

From the expression (6.56) follows a relation that we used earlier,

$$S_F(y - x)(i\gamma^\mu \frac{\overleftarrow{\partial}}{\partial x^\mu} + m) = -\delta^4(x - y) . \tag{6.57}$$

Substituting into the left-hand side of (6.56) we obtain $(\partial/\partial x = -\partial/\partial y)$

$$S_F(y - x)(i\gamma^\mu \frac{\overleftarrow{\partial}}{\partial x^\mu} + m) = (i\gamma^\mu \frac{\partial}{\partial y^\mu} + m)(i\gamma^\mu \frac{\partial}{\partial x^\mu} + m)\Delta_F(y - x)$$

$$= (i\gamma^\mu \frac{\partial}{\partial y^\mu} + m)(-i\gamma^\mu \frac{\partial}{\partial y^\mu} + m)\Delta_F(y - x)$$

$$= -\delta^4(x - y) . \tag{6.58}$$

The correspondence rules to go from the generating functional to the Green's functions are deduced from (6.52)

$$\psi_\alpha(x) \to i\frac{\delta}{\delta \bar{J}_\alpha(x)} , \qquad \bar{\psi}_\beta(x) \to -i\frac{\delta}{\delta J_\alpha(x)} , \tag{6.59}$$

---

[4]See Section 3.3. We recall that $(\gamma^\mu \partial_\mu)^2 = \partial^\mu \partial_\mu = \Box$. It is easily confirmed that this is the only acceptable solution, since the homogeneous equation which corresponds to (6.53) is nothing other than the Dirac equation, whose solutions are superpositions of plane waves which fail the condition (6.55), for $t \to -\infty$ for positive frequencies if they are of the form $e^{-i\omega t}$, and for $t \to \infty$ for negative frequencies, if they are of the form $e^{i\omega t}$.

and the two-point Green's function becomes

$$\langle 0| \, T \left[\psi_\alpha(x) \, \bar{\psi}_\beta(y) \right] |0\rangle = i(S_F)_{\alpha\beta}(x-y) \,. \tag{6.60}$$

As we saw in the oscillator case, the minus sign in the second correspondence rule is balanced by a second minus sign which arises from the anticommuting character of the functional derivatives. We have therefore obtained exactly the same result as if we had treated the Dirac field as a commuting quantity.

We note also that the Green's functions with two ψ or $\bar{\psi}$ are equal to zero:

$$\langle 0|T \left[\psi_\alpha(x) \, \psi_\beta(y) \right] |0\rangle = \langle 0|T \left[\bar{\psi}_\alpha(x) \, \bar{\psi}_\beta(y) \right] |0\rangle = 0 \,. \tag{6.61}$$

6.2.1   Fermion propagator

For later applications, we record the relevant formulae for the fermion propagator.

$$
\begin{aligned}
i(S_F)_{\alpha\beta}(x-y) &= \int \frac{d^4p}{(2\pi)^4} \, e^{-ipx} \, (i\tilde{S}_F)_{\alpha\beta}(p) \\
&= \int \frac{d^4p}{(2\pi)^4} \, e^{-ipx} \, \frac{i \, (\slashed{p}+m)_{\alpha\beta}}{p^2 - m^2 + i\epsilon} \,.
\end{aligned}
\tag{6.62}
$$

We can simplify the expression for the Fourier transform of S_F by using the relation

$$(\slashed{p} + m)\,(\slashed{p} - m) = p^2 - m^2 \,, \tag{6.63}$$

to write

$$i(S_F)_{\alpha\beta}(x-y) = \int \frac{d^4p}{(2\pi)^4} \, e^{-ipx} \, \left(\frac{i}{\slashed{p} - m + i\epsilon} \right)_{\alpha\beta} \,. \tag{6.64}$$

6.2.2   The spin-statistics theorem

The spin-statistics theorem, according to which particles of integer spin are described by commuting fields while those of half-integer spin require anticommuting fields, is one of the few exact results in field theory. In this section we verify that the quantum theory of a free Dirac field necessarily requires that ψ and $\bar{\psi}$ should be anticommuting quantities, and that, conversely, in the theory of the free scalar field, ϕ is necessarily a commuting quantity. Clearly this confirmation is not a general proof of the theorem, which applies also to the case of interacting fields with arbitrary spin.

In the case of the Dirac theory, we write the right-hand side of (6.60) in a more explicit form using equation (3.69).

$$\langle 0| T\left(\psi_\alpha(x)\,\bar{\psi}_\beta(y)\right)|0\rangle$$

$$= \frac{(i\not\partial + m)_{\alpha\beta}}{(2\pi)^3} \int d^3p \frac{e^{i\mathbf{p}\,(\mathbf{x}-\mathbf{y})}}{2\omega_p} \left(e^{-i\omega_p\,(t_x-t_y)}\theta(t) + e^{i\omega_p\,(t_x-t_y)}\theta(-t)\right)$$

$$= \begin{cases} \frac{1}{(2\pi)^3} \int d^3p \frac{e^{-ip\,(x-y)}(\not p + m)_{\alpha\beta}}{2\omega_p} & (t_x > t_y) \\ \frac{1}{(2\pi)^3} \int d^3p \frac{e^{ip\,(x-y)}(-\not p + m)_{\alpha\beta}}{2\omega_p} & (t_x < t_y) \end{cases},$$

where we have changed the sign of the integration variable $\mathbf{p}$ in the term with negative frequencies $e^{i\omega_p(t_x-t_y)}$.

In the following steps we will use the properties of the projection operators (see Appendix A of [3])

$$\frac{(\not p + m)_{\alpha\beta}}{2m} = \sum_{r=1}^{2} u_{r\alpha}(\mathbf{p})\bar{u}_{r\beta}(\mathbf{p}) \;, \quad \frac{(-\not p + m)_{\alpha\beta}}{2m} = -\sum_{r=1}^{2} v_{r\alpha}(\mathbf{p})\bar{v}_{r\beta}(\mathbf{p}), \quad (6.65)$$

and the orthogonality properties

$$\left(u_r^\dagger(\mathbf{q})\,u_s(\mathbf{q})\right) = \left(v_r^\dagger(\mathbf{q})\,v_s(\mathbf{q})\right) = \frac{\omega_q}{m}\delta_{rs} \;, \quad \left(u_r^\dagger(\mathbf{q})\,v_s(-\mathbf{q})\right) = 0\;. \quad (6.66)$$

Hence, for $t_x > t_y$ we obtain

$$\langle 0|\,\psi_\alpha(x)\,\bar{\psi}_\beta(y)\,|0\rangle = \frac{1}{(2\pi)^3} \int d^3p\, e^{-ip\,(x-y)}\frac{m}{\omega_p}\sum_{r=1}^{2} u_{r\alpha}(\mathbf{p})\bar{u}_{r\beta}(\mathbf{p}) \;, \quad (6.67)$$

and, for $t_x < t_y$,

$$-\langle 0|\,\bar{\psi}_\beta(y)\,\psi_\alpha(x)\,|0\rangle = -\frac{1}{(2\pi)^3} \int d^3p\, e^{ip\,(x-y)}\frac{m}{\omega_p}\sum_{r=1}^{2} v_{r\alpha}(\mathbf{p})\bar{v}_{r\beta}(\mathbf{p}) \;, \quad (6.68)$$

where the minus sign on the right-hand side derives from the projection operator for negative energies, while that on the left-hand side derives from the anticommuting property of the fields.

Multiplying left and right-hand sides by γ^0, we transform $\bar{\psi} \to \psi^\dagger$, $\bar{u} \to u^\dagger$ and $\bar{v} \to v^\dagger$, and the two equations become

$$\langle 0|\,\psi_\alpha(x)\,\psi_\beta^\dagger(y)\,|0\rangle = \frac{1}{(2\pi)^3} \int d^3p\, e^{-ip\,(x-y)}\frac{m}{\omega_p}\sum_{r=1}^{2} u_{r\alpha}(\mathbf{p})u_{r\beta}^\dagger(\mathbf{p}) \;, \quad (6.69)$$

$$\langle 0|\,\psi_\beta^\dagger(y)\,\psi_\alpha(x)\,|0\rangle = \frac{1}{(2\pi)^3} \int d^3p\, e^{-ip\,(y-x)}\frac{m}{\omega_p}\sum_{r=1}^{2} v_{r\alpha}(\mathbf{p})v_{r\beta}^\dagger(\mathbf{p}) \;. \quad (6.70)$$

It is easy to see that the anticommutativity is indispensable. Let us consider equation (6.70), taking the limit $y \to x$, and with $\alpha = \beta$. Re-expressing the left-hand side by introducing a complete set of states $|X\rangle\langle X|$ we find

$$\langle 0|\, \psi_\alpha^\dagger(x)\, \psi_\alpha(x)\, |0\rangle = \sum_X |\langle X|\psi_\alpha(x)|0\rangle|^2 = \frac{1}{(2\pi)^3} \int d^3p\, \frac{m}{\omega_p} \sum_{r=1}^{2} |v_{r\alpha}(\mathbf{p})|^2\, .$$

Both the left- and right-hand sides, are positive-definite quantities. If we had considered ψ and $\bar\psi$ as *commuting* quantities, the right-hand side of this equation would have a negative sign (the minus sign on the left-hand side of equation (6.68) would be missing) and we would have obtained a nonsensical result. The Dirac fields must be *anticommuting*.

The opposite conclusion is reached in the case of a scalar field. We consider the case of a complex scalar field (see Section 3.4). It is easy to show for both the commuting and anticommuting cases that the two-point function should be given by equation (3.69); in the manipulations which led to this result starting from (3.24) we consistently maintained the ordering of the various quantities. If in (3.69) we consider the case $t_x < t_y$, we obtain, for commuting quantities,

$$\langle 0|\, \phi^\dagger(y)\, \phi(x)\, |0\rangle = \frac{1}{(2\pi)^3} \int d^3p\, \frac{e^{-ip(x-y)}}{2\omega_p}\, ,$$

and in the limit $y \to x$,

$$\langle 0|\, \phi^\dagger(x)\, \phi(x)\, |0\rangle = \sum_X |\langle X|\psi(x)|0\rangle|^2 = \frac{1}{(2\pi)^3} \int d^3p\, \frac{1}{2\omega_p}\, ,$$

an equality between positive-definite quantities. In the *anticommuting* case the change of sign on the left-hand side would have led to a nonsensical result.

6.2.3   One-particle states of the Dirac field

We would now like to show that the Dirac field describes two types of particle—particle and antiparticle—each with two polarisation states. To do this we define the following operators obtained from ψ and $\bar\psi$ with spatial Fourier transformations projected onto the spinors u and v:

$$c_r(\mathbf{q};t) = \left(\frac{m}{(2\pi)^3\omega_q}\right)^{1/2} \sum_\alpha \int d^3x\, e^{-i\mathbf{q}\cdot\mathbf{x}} u_{r\alpha}^\dagger(\mathbf{q})\psi_\alpha(\mathbf{x},t)\, , \tag{6.71}$$

$$c_r^\dagger(\mathbf{q};t) = \left(\frac{m}{(2\pi)^3\omega_q}\right)^{1/2} \sum_\alpha \int d^3y\, e^{i\mathbf{q}\cdot\mathbf{y}} \psi_\beta^\dagger(\mathbf{y},t)u_{r\beta}(\mathbf{q})\, , \tag{6.72}$$

$$d_r(\mathbf{q};t) = \left(\frac{m}{(2\pi)^3\omega_q}\right)^{1/2} \sum_\alpha \int d^3y\, e^{-i\mathbf{q}\cdot\mathbf{y}} \psi_\beta^\dagger(\mathbf{y},t)v_{r\beta}(\mathbf{q})\, , \tag{6.73}$$

$$d_r^\dagger(\mathbf{q};t) = \left(\frac{m}{(2\pi)^3\omega_q}\right)^{1/2} \sum_\alpha \int d^3x\, e^{i\mathbf{q}\cdot\mathbf{x}} v_{r\alpha}^\dagger(\mathbf{q})\psi_\alpha(\mathbf{x},t)\, . \tag{6.74}$$

Naturally, as we will now show, these are the usual creation and destruction operators for particles and antiparticles.

From (6.69), (6.70), and the orthogonality properties (6.66), we obtain

$$\langle 0| \, c_r(\mathbf{q}; t_x) \, \psi_\beta^\dagger(y) \, |0\rangle = \left(\frac{m}{(2\pi)^3 \omega_q} \right)^{1/2} e^{-i\omega_q t} e^{iq\,y} \, u_{r\beta}^\dagger(\mathbf{q}) \,, \tag{6.75}$$

$$\langle 0| \, \psi_\beta^\dagger(y) \, d_r^\dagger(\mathbf{q}; t_x) \, |0\rangle = \left(\frac{m}{(2\pi)^3 \omega_q} \right)^{1/2} e^{i\omega_q t} e^{-iq\,y} \, v_{r\beta}^\dagger(\mathbf{q}) \,, \tag{6.76}$$

as is easily confirmed by substituting into c and $d^\dagger$ the expressions (6.71), (6.74), while

$$\langle 0| \, \psi_\beta^\dagger(y) \, c_r(\mathbf{q}; t_x) \, |0\rangle = \langle 0| \, d_r^\dagger(\mathbf{q}; t_x) \, \psi_\beta^\dagger(y) \, |0\rangle = 0 \,. \tag{6.77}$$

With analogous steps, from (6.75) and (6.76) we also obtain

$$\begin{aligned}
\langle 0| \, c_r(\mathbf{q}; t_x) \, c_s^\dagger(\mathbf{p}; t_y) \, |0\rangle &= \delta^3(\mathbf{q} - \mathbf{p}) \, \delta_{rs} e^{-i\omega_q(t_x - t_y)} \,, \\
\langle 0| \, d_s(\mathbf{p}; t_y) \, d_r^\dagger(\mathbf{q}; t_x) \, |0\rangle &= \delta^3(\mathbf{q} - \mathbf{p}) \, \delta_{rs} e^{-i\omega_q(t_y - t_x)} \,.
\end{aligned} \tag{6.78}$$

Furthermore, from (6.77) it follows that

$$\langle 0| \, c_s^\dagger(\mathbf{p}; t_y) \, c_r(\mathbf{q}; t_x) \, |0\rangle = \langle 0| \, d_r^\dagger(\mathbf{q}; t_x) \, d_s(\mathbf{p}; t_y) \, |0\rangle = 0 \,,$$

from which, with $s = r$, $\mathbf{p} = \mathbf{q}$, $t_y = t_x = 0$, and by introducing a complete set of states,

$$\sum_X |\langle X|c_r(\mathbf{q}) \, |0\rangle|^2 = \sum_X |\langle X|d_r(\mathbf{q}) \, |0\rangle|^2 = 0 \,,$$

where $c_r(\mathbf{q}) = c_r(\mathbf{q}; t = 0)$, $d_r(\mathbf{q}) = d_r(\mathbf{q}; t = 0)$. Hence

$$c_r(\mathbf{q})|0\rangle = d_r(\mathbf{q})|0\rangle = 0 \,. \tag{6.79}$$

If we now define

$$|P; \mathbf{p}, r\rangle = c_r^\dagger(\mathbf{p})|0\rangle \,, \quad |A; \mathbf{p}, r\rangle = d_r^\dagger(\mathbf{p})|0\rangle \,, \tag{6.80}$$

by substituting into (6.75) (of which we take the complex conjugate) and into (6.76), in both cases with $t_x = 0$, we obtain

$$\langle 0| \, \psi_\beta(y) \, |P; \mathbf{p}, r\rangle = \left(\frac{m}{(2\pi)^3 \omega_q} \right)^{1/2} e^{-ip\,y} \, u_{r\beta}(\mathbf{p}) \,, \tag{6.81}$$

$$\langle 0| \, \psi_\beta^\dagger(y) \, |A; \mathbf{p}, r\rangle = \left(\frac{m}{(2\pi)^3 \omega_q} \right)^{1/2} e^{-ip\,y} \, v_{r\beta}^\dagger(\mathbf{p}) \,. \tag{6.82}$$

These equations tell us that $|P; \mathbf{p}, r\rangle$ and $|A; \mathbf{p}, r\rangle$ are states of momentum

$\mathbf{p}$ and energy $\omega_p = (\mathbf{p}^2 + m^2)^{1/2}$, and hence are single particle states. The two polarisation states associated with the variables $r = 1, 2$ can be chosen as states of definite helicity. Equation (6.78), with $t_x = t_y = 0$, fixes the normalisation of the states:

$$\langle P; \mathbf{p}, r \mid P; \mathbf{q}, s \rangle = \langle A; \mathbf{p}, r \mid A; \mathbf{q}, s \rangle = \delta_{rs} \delta(\mathbf{p} - \mathbf{q}) . \qquad (6.83)$$

The states $|P\rangle$ and $|A\rangle$ are necessarily different; indeed the two-point function $\langle 0|T(\psi\,\psi)|0\rangle$ vanishes [equation (6.61)], and from this we can obtain, with similar work to that carried out up to now, that

$$\langle 0|\, c_s(\mathbf{p}; t_y)\, d_r^\dagger(\mathbf{q}; t_x)\, |0\rangle = \langle P; \mathbf{p}, s|A; \mathbf{q}, r \rangle = 0 . \qquad (6.84)$$

The particle states $|P\rangle$ and antiparticle states $|A\rangle$ are therefore orthogonal and necessarily different.

In conclusion, we have seen that functional formalism allows the construction of the spectrum of one-particle states from the theory, as well as the identification of the already familiar role of the creation and destruction operators.

CHAPTER 7

SCATTERING PROCESSES AND THE S-MATRIX

CONTENTS

We can describe a process of collision between two particles in the following way.

At the initial instant, which we denote by $-T/2$, the particles are prepared in a state corresponding to two wave packets localised in regions of space that are remote from each other. For all practical purposes, we can neglect any interaction between the two particles at time $-T/2$.

Subsequently the system evolves without external influence for a time T, during which the particles interact with each other and give rise to the products of the collision. These are detected at time $+T/2$ by suitable experimental apparatus. At the moment of detection, the system consists of two particles (or possibly more, if the collisions are at relativistic energies) very far from each other and also therefore not interacting.

The interval T, during which the experiment takes place, is in general much longer than the typical times over which the interaction occurs. T is determined by the linear dimensions of the apparatus of particle production and detection which are macroscopic, of order, perhaps, of metres. Therefore $T \simeq 10^{-9} - 10^{-8}$ seconds. Instead, the interaction occurs when the particles are some fermis in distance from each other (1 fermi = 10^{-15} m) and hence over times of the order of 10^{-23} seconds, much less than T. For all practical purposes, the process occurs between the times $t = \pm\infty$.

In contrast to what happens in classical mechanics, even if we start from a perfectly defined state we cannot in general predict the result of a specific experiment. All that quantum mechanics can provide are the scattering amplitudes, complex numbers whose moduli squared give the probability of the different results of the experiment. The scattering amplitudes, for their part, are described by the matrix elements of an operator, the *S-matrix*, or scattering matrix, which is therefore the central element in scattering theory.

The study of collisions between particles represents almost the only method of inquiry at our disposal to investigate the structure of the fundamental interactions. The experimental study of scattering and the theoretical calculation of the *S*-matrix are therefore the point of contact between theory and experiment.

7.1 "IN" STATES AND "OUT" STATES

As we explained, at time $-T/2$ the initial state of the scattering process is constituted by particles separated from each other and effectively noninteracting. We can describe these states with a superposition of plane waves, characterised by the momentum of each particle

$$|p_1, \alpha; p_2, \beta > , \quad \text{(at time } t = -T/2), \tag{7.1}$$

where α and β represent other quantum numbers which are necessary, in addition to the momentum, to fully characterise the state of motion of the individual particles (for example, the components of the spins along the direction of motion).

In the Schrödinger representation, the state (7.1) evolves in time in a complicated way, describing a trajectory in the space of the states, a trajectory which is determined by the exact Hamiltonian, H, of the problem. The same trajectory, in the Heisenberg representation, is specified by a fixed vector, also determined by the momenta p_1 and p_2, which characterise the conditions of the system at time $-T/2 \to -\infty$. We will call this state "in" and denote it with

$$|p_1, \alpha; p_2, \beta; in > . \tag{7.2}$$

The overall set of all states of this type, i.e. the Heisenberg states with an arbitrary number of particles,[1] which can be expressed as plane waves for $t = -T/2 \to -\infty$, is the "in" basis (see, for example, [8, 9]). The set of "in" states obviously constitutes an orthonormal basis. We can ask if this basis is also complete.

In the final analysis, this question is equivalent to asking if we can reach all the momentum states of the system starting from particle states far away from them. The scattering processes are the only way in which we can study microscopic systems, hence the response cannot be other than in the affirmative, but with an important qualification which we will explain in a moment.

The hypothesis that the "in" states should be a complete set is known as

---

[1] The state which does not contain any particles, the so-called "vacuum state", is denoted with $|0\rangle$ and is the state of minimum energy, $E = 0$. As discussed diffusely in this book, the vacuum state plays a fundamental role in field theory.

the *asymptotic hypothesis* and is characterised with the relation

$$\sum_i |i; in >< i; in| = \mathbf{1} , \quad (\text{``in'' completeness}) . \tag{7.3}$$

Completeness of the "in" states. We first consider the case of a non-relativistic particle in a defined potential, with a discrete spectrum for $E < 0$ and continuous spectrum for $E > 0$ (for example, an electron in the potential of a proton, considered as a fixed source).

We can construct normalised wave packets by superposing eigenstates with $E > 0$. It is easy to show that, for these states, the motion takes place mainly at infinity, since the average value of the probability in time to find the particle in any finite region of space is zero (see the discussion in [10]). Actually

$$|\psi(x,t)|^2 = \int \int dE \, dE' c(E)^* c(E') \Psi_E(x)^* \Psi_{E'}(x) e^{+i(E-E')t} , \tag{7.4}$$

and furthermore, for $T \to \infty$

$$\frac{1}{T} \int_{-T/2}^{T/2} dt \, e^{+i(E-E')t} \to \frac{2\pi}{T} \delta(E' - E) , \tag{7.5}$$

from which it follows that

$$\frac{1}{T} \int_{-T/2}^{T/2} dt \int_V dx \, |\psi(x,t)|^2 = \frac{2\pi}{T} \int dE \, |c(E)|^2 \int_V dx \, |\Psi_E(x)|^2 \to 0 . \tag{7.6}$$

An arbitrary state of type (7.4), for times sufficiently long, will be represented by a superposition of free states, and hence reachable from the "in" states.

Conversely, if the wave packet is constructed as a superposition of the eigenstates of the discrete spectrum

$$|\psi(x,t)|^2 = \sum_{n,n'} c_n(E)^* c_{n'}(E') \Psi_n(x)^* \Psi_{n'}(x) e^{+i(E_n - E_{n'})t} , \tag{7.7}$$

equation (7.5) will be replaced by

$$\frac{1}{T} \int_{-T/2}^{T/2} dt e^{+i(E_n - E_{n'})t} = \delta_{n,n'} , \tag{7.8}$$

and the average probability (7.6) is different from zero

$$\frac{1}{T} \int_{-T/2}^{T/2} dt \int_V dx |\psi(x,t)|^2 = \sum_n |c_n(E)|^2 \int_V |\Psi_n(x)|^2 \neq 0 . \tag{7.9}$$

Consequently, the motion develops permanently in the region where the particle is bound by the potential.

In this case, the basis set of "in" states is evidently not complete, but rather (the index i represents the set of quantum numbers which characterise the various states):

$$\sum_i |i; in> < i; in| = \mathbf{1} - \sum_n |E_n> < E_n| , \tag{7.10}$$

where the projector of the bound states appears in the right-hand side.

In a theory invariant under translations, in which we also permit the proton to move, the total energy spectrum is always continuous and all the localised states end, sooner or later, at infinity. However, states with an electron and a proton far from each other do not provide a complete basis; the two particles can move towards infinity while remaining bound together.

If, however, we also include in the "in" basis also those states which, at time $-T/2 \to -\infty$, comprise the bound states (for example the hydrogen atom in the ground state), we can assume that the "in" states form a complete set[2]:

$$\sum_i |i; in> < i; in| = \mathbf{1} , \quad (\text{"in" completeness}) . \tag{7.11}$$

In concrete terms, this means that, to completely determine the physics of the electron–proton system, we must also study scattering experiments which include the hydrogen atom among the initial states, for example:

$$e + H \to e + e + p . \tag{7.12}$$

Naturally, in cases for which we can solve the equations of motion starting from the Hamiltonian, we can also decide the composition of the "in" states and ascertain whether relation (7.11) is valid or not. In some formulations of field theory, condition (7.11) is considered to be one of the fundamental axioms.

Alongside the "in" basis, we can now introduce the complete basis set of "out" states. These describe momentum states, in the Heisenberg representation, which reduce, at time $t = +T/2 \to +\infty$, to states with a certain number of free particles, each with a defined momentum.

Analogous with (7.2), we denote these states with

$$|p_1, \alpha; p_2, \beta; out> , \tag{7.13}$$

in the case of two particles.

The discussion on the completeness of the "out" basis set proceeds in the same manner as for the "in" states, and hence, with the same qualifications as before, we conclude that we must have

$$\sum_n |n; out> < n; out| = \mathbf{1} , \quad (\text{"out" completeness}) . \tag{7.14}$$

---

[2]In general, a unique time $-T/2$ does not exist for which all the possible states are reduced to states of non-interacting particles. The convergence of the states to the asymptotic conditions is not uniform, and the set of "in" states becomes complete only in the limit $-T/2 \to -\infty$. This point will be important in the discussion of the reduction formulae discussed later.

In the absence of interactions, the quantum numbers which characterise the "in" or "out" states, for example the momenta of the individual particles, are all conserved. In this case, the "in" and "out" states coincide.

However, even in the presence of interactions, the distinction between "in" and "out" states does not apply in the following cases:

(i) The state with no particles (the vacuum state) is stable, since it is the minimum energy state. Therefore

$$|0; in >= |0; out >= |0 > .\qquad(7.15)$$

(ii) For states which contain only one particle, the momentum and the spin component in the direction of the particle are conserved quantities, for which

$$|p, \alpha; in >= |p, \alpha; out >= |p, \alpha > .\qquad(7.16)$$

7.2 SCATTERING AMPLITUDES AND THE S-MATRIX

We can characterise the state $|p_1', \alpha'; p_2', \beta'; out >$ as that in which our detectors would definitely find two particles with momenta p_1' and p_2' and the other quantum numbers with values α' and β' respectively, at time $+T/2$. Similarly, the state $|p_1, \alpha; p_2, \beta; in >$ is that in which two particles with quantum numbers p_1, α and p_2, β are definitely present at time $-T/2$. Their scalar product therefore gives the probability amplitude of the reaction

$$(p_1, \alpha) + (p_2, \beta) \rightarrow (p_1', \alpha') + (p_2', \beta') .$$

More generally, the scattering amplitude is given by

$$S_{fi} =< f; out|i; in > ,\qquad(7.17)$$

where f and i are the values of the quantum numbers which characterise, respectively, the final and initial states. For normalised states, the modulus squared

$$| < f; out|i; in > |^2 = P(i \rightarrow f) ,\qquad(7.18)$$

gives the probability of the reaction. For the relation which links the probability (7.18) to the cross section, see [1].

The S-matrix can be written, in terms of "in" and "out" states, as

$$S = \sum_n |n; in\rangle\langle n, out| ,\qquad(7.19)$$

from which we find

$$S_{fi} = \langle f; out|i; in\rangle =< f; in|S|i; in >=< f; out|S|i; out > .\qquad(7.20)$$

From (7.19) it is immediately confirmed that the S operator transforms the "out" basis into the "in" basis

$$S|m; out \;>= |m; in>\;,\tag{7.21}$$

and hence, from the completeness of the two base sets, it follows that S is unitary

$$S^\dagger S = SS^\dagger = 1\;.\tag{7.22}$$

If we write the diagonal matrix element of (7.22) as

$$\begin{aligned}
1 &=<i; out|S^\dagger S|i; out>\\
&= \sum_f |<f; out|S|i; out>|^2 \tag{7.23}\\
&= \sum_f P(i \to f)\;,
\end{aligned}$$

we see that the unitarity of S is equivalent to requiring that the sum of the probabilities (7.18) over all the f states should be unity. The set of these states must coincide with all possible final states in a scattering experiment, as is precisely the case if the basis of these states forms a complete set.

The relations (7.20) show the connection between the definition of the S-matrix in the Heisenberg representation and the more elementary definition, in terms of the interaction representation, discussed in [1].

In (7.20) both the bra and the ket refer to the same time, which is either in the past or in the future, and thus does not refer to a specific representation. We can interpret $|f; out\rangle$ and $|i; out\rangle$ simply as the vectors of the Hilbert space, respectively $|f\rangle$ and $|i\rangle$, which identify the incoming and outgoing states of the scattering process. In this case, the vector $S|i\rangle$ represents the state in which $|i\rangle$ evolves in the interaction representation, and the projection $\langle f|S|i\rangle$ represents the probability amplitude to find this state in $|f\rangle$. In perturbation theory, S is given in the usual representation by means of the Dyson formula [1].

7.3 CONSERVED QUANTITIES

Consider a conserved quantity, Q. Since Q commutes with H, we can choose both the "in" and "out" states so that they are simultaneously eigenstates of Q and H. Because Q is a constant of the motion, the "in" states which correspond to a given value q of Q must transform, for $t \to +\infty$, into a state with the same eigenvalue, i.e.

$$<f, q'; out|i, q; in>= 0\;, \quad \text{if } q' \neq q\;.\tag{7.24}$$

Another way of saying the same thing consists of using the fact that S commutes with all the constants of the motion, for which

$$0 =<f, q'; in|[Q, S]|i, q; in>= (q' - q)<f, q'; in|S|i, q; in>\;.\tag{7.25}$$

Therefore the matrix element is zero if $q \neq q'$. For the same reason

$$U(R)SU(R)^\dagger = S ,\qquad (7.26)$$

if $U(R)$ is the unitary operator associated with an exact symmetry (which does not involve time reversal).

For systems invariant under translation, the S-matrix must be diagonal in the basis of states with defined momentum and energy; therefore its matrix elements have the form ($f \neq i$):

$$S_{fi} = (2\pi)^4 \delta^{(4)} \left(\sum P_{fin} - \sum P_{in} \right) \mathcal{M}_{fi} .\qquad (7.27)$$

7.4 THE LSZ REDUCTION FORMULAE

In this section we will derive the relations which link the q-point Green's functions to the S-matrix element for the reaction between p initial particles which give rise to q-p final particles (with $p \leq q - 2$). These relations were obtained by Lehman, Szymanzik and Zimmermann [11] and are known as the *LSZ reduction formulae*. For simplicity, we refer to the case of particles described by a neutral scalar field with $\lambda \phi^4$ interactions; the generalisation to more complex cases is straightforward. Subsequently, we will extend the Feynman rules derived for the Green's functions to those for the S-matrix elements.

Before proving the reduction formulae, however, we must introduce the so-called *asymptotic hypothesis* for fields in the Heisenberg representation and the corresponding asymptotic fields, $\phi_{in}(x)$ and $\phi_{out}(x)$.

The asymptotic fields, $\phi_{in}(x)$ and $\phi_{out}(x)$. To fix the quantum state in the Heisenberg representation corresponds, in classical mechanics, to determining the trajectory in the phase space of the system, by giving the initial conditions at a certain time t_0. It is therefore clear why the state vector, in this representation, does not change with time. To be more definite, let us assume the states are chosen in the "in" basis, i.e. $t_0 = -T/2$.

By contrast, what changes with time are the dynamic variables of the system, i.e. the fields, which are functions of position in space and time.

$$\phi = \phi(\mathbf{x}, t) = \phi(x) .\qquad (7.28)$$

In the free theory, the field applied to the vacuum creates a single particle state for any value of time. This is no longer true in the theory with interactions, in which the field also has non-zero matrix elements between the vacuum and states with three or more particles, as we will see in Section 8.4. However, we expect that the physical situation will tend to that of the free theory when the time tends to $\pm T/2 \simeq \pm\infty$. However, because the state is fixed, the previous requirement, which is denoted by the term *asymptotic condition*, must mean that the field, in these limits, should "in some sense" tend to the free field.

Let us consider the limit $t \to +\infty$. The asymptotic condition is formulated by requiring that the expectation values of the field, i.e. the matrix elements, converge to the corresponding matrix elements of a free field, which we denote as ϕ_{out}, if necessary multiplied by a proportionality constant.[3] The constant is defined so that ϕ_{out} is normalised like a canonical field. In formulae

$$\lim_{t \to +\infty} \langle \alpha | \phi(\mathbf{x}, t) | \beta \rangle = \sqrt{Z_+} \langle \alpha | \phi_{out}(\mathbf{x}, t) | \beta \rangle \ , \tag{7.29}$$

$$(-\Box - m^2)\phi_{out}(x) = 0 \ . \tag{7.30}$$

The limit $t \to -\infty$ is treated similarly and leads to the definition of the "in" field:

$$\lim_{t \to -\infty} \langle \alpha | \phi(\mathbf{x}, t) | \beta \rangle = \sqrt{Z_-} \langle \alpha | \phi_{in}(\mathbf{x}, t) | \beta \rangle \ , \tag{7.31}$$

$$(-\Box - m^2)\phi_{in}(x) = 0 \ . \tag{7.32}$$

The fields ϕ_{out} applied to the vacuum create many-particle non-interacting states at time $t = +\infty$, hence they generate the "out" states. Similarly, the application of ϕ_{in} to the vacuum generates the "in" basis. Restricting ourselves to the "out" case, we can expand the field in creation and destruction operators in the continuum [see equation (3.92)]:

$$\phi_{out}(x) = \int \frac{d^3k}{(2\pi)^{3/2}} \frac{1}{\sqrt{2\omega(\mathbf{k})}} \left[a_k \, e^{-ikx} + a_k^\dagger \, e^{ikx} \right] \ . \tag{7.33}$$

Equation (7.33) is easily inverted to obtain [compare to equations (3.94) and (3.95)]

$$a_{out}(\mathbf{p}) = \int d^3x \, f_{\mathbf{p}}(x)^* i \overset{\leftrightarrow}{\partial} \phi_{out}(x) \ , \quad f_{\mathbf{p}}(x) = \frac{1}{\sqrt{2\omega(\mathbf{k})(2\pi)^3}} e^{-ipx} \ . \tag{7.34}$$

The annihilation operators defined by (7.34) are independent of time, because of the Klein–Gordon equation.

Since the "out" field is normalised like a canonical field, a_{out} and $a_{out}^\dagger$ obey the canonical commutation rules and are just the creation and annihilation operators of the "out" states. Hence[4]

$$|p_1, p_2, \ldots ; out \rangle = a_{out}(p_1)^\dagger a_{out}(p_2)^\dagger \ldots |0\rangle \ , \tag{7.35}$$

and similar formulae hold for the "in" states.

---

[3]No more than a *weak convergence* can be required, as indicated in (7.29) and (7.32), otherwise the free theory would hold at all times.

[4]We recall that $|0; out \rangle = |0; in \rangle = |0\rangle$.

The one-particle states of the "in" and "out" bases coincide; for this reason we used the same mass in equations (7.30) and (7.32). It is easy to show that, for the same reason, the normalisation constants $Z_\pm$ are also equal to each other and to the constant introduced in the treatment of the two-point function in Appendix A of [2].

To obtain this result, we consider the matrix element of the "out" field between the vacuum and the state of one particle

$$\lim_{t\to+\infty} \langle 0|\phi(x)|\mathbf{p}\rangle = \sqrt{Z_+}\langle 0|\phi_{out}(x)|\mathbf{p}\rangle$$

$$= \sqrt{Z_+}\langle 0|\phi_{out}(x)|\mathbf{p};out\rangle = \sqrt{\frac{Z_+}{2\omega(\mathbf{p})(2\pi)^3}}\ e^{-ipx}\ ,\quad (7.36)$$

where we have used the expansion (7.33).

On the other hand, the matrix element of the field can be parameterised *at all times* as follows (cf. [2])

$$\langle 0|\phi(x)|\mathbf{p}\rangle = \langle 0|\phi(x)|\mathbf{p};out\rangle = \sqrt{\frac{Z}{2\omega(\mathbf{p})(2\pi)^3}}\ e^{-ipx}\ ,\quad (7.37)$$

and hence also in the limit $t\to+\infty$. Comparing with (7.36) we obtain $Z_+ = Z$. Repeating the argument in the limit $t\to-\infty$ we similarly obtain $Z_- = Z = Z_+$.

The asymptotic hypothesis and self-energy corrections. The asymptotic hypothesis at the foundation of S-matrix theory requires an important qualification. It is reasonable to expect, as time tends to $\pm\infty$, that the interactions between *different particles* should tend to zero. However, we cannot isolate a particle from the effect of its own field. This is the problem of the particle *self-energy*, already known from classical physics.

In the case of a classical electrically charged particle, the energy of the Coulomb field generated by the particle is easily calculated and depends inversely on its radius; it diverges for an exactly pointlike particle. Because of Einstein's equation stating the equivalence between mass and energy, the field generated by the particle makes an additional contribution to its inertial mass, which is hence *not the same* as it would be in the absence of its own field, i.e. in the limit in which the electric charge goes to zero.

The problem arises again in quantum mechanics and is solved through the procedure of renormalisation, of which we will see an example in the calculation of the two-point function in the next chapter.

The LSZ reduction formulae. To be specific, we consider the relevant four-point Green's function

$$G(x_1, x_2, x_3, x_4)\ ,\quad (7.38)$$

that we would like to connect to the S-matrix of the reaction

$$p_1 + p_2 \to p_3 + p_4 \ . \tag{7.39}$$

If we recall the Feynman rules by which G is constructed, we see that in every case a line originates from point x_1, which connects it to the rest of the diagram. In terms of amplitudes, this means that the dependence on x_1 arises from

$$G = \int d^4x \ \Delta_F(x_1 - x) \times \ldots \ , \tag{7.40}$$

where x is the coordinate of one of the vertices in the diagram. The Fourier transform of G with respect to x_1, with momentum p_1, must therefore contain a factor corresponding to the Fourier transform of Δ_F, i.e.

$$i\tilde{\Delta}_F(p_1) = \frac{i}{p_1^2 - m^2} \ . \tag{7.41}$$

The Fourier transform of G with respect to x_1 therefore has a pole when p_1 is on the "mass-shell", i.e. when $p_1^2 \to m^2$. We can isolate the residue at the pole by multiplying the Fourier transform of G by $p_1^2 - m^2$ and taking the limit $p_1^2 = m^2$.

In reality, as we saw in the study of the propagator, there are two poles corresponding to the possible signs of the frequency: $p_1^0 = \pm\omega(\mathbf{p}_1) = \pm\sqrt{\mathbf{p}_1^2 + m^2}$. We select the sign of p_1^0 by writing an exponential factor in the Fourier transform as $e^{-ip_1 x_1}$ for an incoming particle, as in equation (7.39), or $e^{+ip_1 x_1}$ for an outgoing particle, with $p_1^0 = +\omega(\mathbf{p})$ in both cases. As a formula

$$R(p_1, \ldots) = \lim_{p_1^2 \to m^2} (p_1^2 - m^2) \int d^4x_1 \ e^{-ip_1 x_1} G(x_1, \ldots) \ . \tag{7.42}$$

To connect the residue R to the S-matrix element, we write the Green's function explicitly as the vacuum expectation value of the T-product of the fields

$$G(x_1, x_2, x_3, x_4) = \langle 0|T\left[\phi(x_1)\phi(x_2)\phi(x_3)\phi(x_4)\right]|0\rangle \ , \tag{7.43}$$

and thus R becomes

$$R(p_1, \ldots) = \lim_{p_1^2 \to m^2} (p_1^2 - m^2) \int d^4x_1 \ e^{-ip_1 x_1} \langle 0|T\left[\phi(x_1)\ldots\right]|0\rangle \ . \tag{7.44}$$

The multiplication by p_1^2, in (7.44), is equivalent to the action of $-\Box$ on the exponential. Integrating by parts, we obtain (the taking of the limit, from now on, is understood)

$$R(p_1, \ldots) = \int d^4x_1 \ e^{-ip_1 x_1} (-\Box - m^2)\langle 0|T\left[\phi(x_1)\ldots\right]|0\rangle \ . \tag{7.45}$$

In the limit, the exponential itself satisfies the Klein–Gordon equation, for which reason we can also write

$$R(p_1,\ldots) = -\int d^4x_1 \ e^{-ip_1x_1} \overset{\leftrightarrow}{\square} \langle 0|T\left[\phi(x_1)\ldots\right]|0\rangle \tag{7.46}$$

where $\overset{\leftrightarrow}{\square}$ denotes $\overset{\rightarrow}{\square} - \overset{\leftarrow}{\square}$. Furthermore, the difference of the two $\square$ operators can be written as a total derivative

$$F(x)\overset{\leftrightarrow}{\square}G(x) = \partial_\mu\left[F(x)\overset{\leftrightarrow}{\partial^\mu}G(x)\right] . \tag{7.47}$$

Discarding the spatial derivatives, which in any case integrate to zero, we therefore obtain

$$R(p_1,\ldots) = -\int_{-T/2}^{+T/2} dt \ \frac{\partial}{\partial t} \int d^3x_1 \ e^{-ip_1x_1} \overset{\leftrightarrow}{\partial}_t \ \langle 0|T\left[\phi(x_1)\ldots\right]|0\rangle$$

$$= -\int d^3x_1 \ e^{-ip_1x_1} \overset{\leftrightarrow}{\partial}_t \ \langle 0|\phi(\mathbf{x}_1,+T/2)T\left[\phi(x_2)\ldots\right]|0\rangle$$

$$+ \int d^3x_1 \ e^{-ip_1x_1} \overset{\leftrightarrow}{\partial}_t \ \langle 0|T\left[\phi(x_2)\ldots\right]\phi(\mathbf{x}_1,-T/2)|0\rangle . \tag{7.48}$$

We have taken out of the T-product the fields $\phi(\mathbf{x}_1,\pm T/2)$ since the times in x_2, x_3, etc. are certainly larger than $-T/2$ and less than $+T/2$. The fields evaluated at $\pm T/2$ can be identified respectively with the "out" and "in" fields. Comparing with (7.34) we see that

$$R(p_1,\ldots) = i\sqrt{2\omega(2\pi)^3} \ \left[\langle 0|\sqrt{Z}a_{out}^\dagger(p_1)T\left[\phi(x_2)\ldots\right]|0\rangle\right.$$

$$\left. - \ \langle 0|T\left[\phi(x_2)\ldots\right]\sqrt{Z}a_{in}^\dagger(p_1)|0\rangle\right]$$

$$= -i\sqrt{Z} \ \sqrt{2\omega(2\pi)^3}\langle 0|T\left[\phi(x_2)\ldots\right]|p_1;in\rangle , \tag{7.49}$$

where we have used the relations

$$\langle 0|a_{out}(p_1)^\dagger = 0 \ , \ a_{in}(p_1)^\dagger|0\rangle = |p_1;in\rangle . \tag{7.50}$$

In conclusion, taking the residue of the pole of the Green's function of four fields for $p_1^2 = m^2$, we have obtained the matrix element of the T-product of three fields between the vacuum and the state of a particle of momentum p_1. Clearly, we can proceed with the process of *reduction* of the fields by taking the residues at the poles for $p_2^2 = m^2$, etc. (with the appropriate signs of the exponentials for initial and final particles) until the entire T-product has been eliminated. The final result is

$$\lim_{p_i^2 \to m^2} \prod i\frac{(p_i^2 - m^2)}{i\sqrt{Z}} \int \prod i\left(d^4x_i e^{\mp ip_ix_i}\right) \ G(x_1,x_2,\ldots)$$

$$= \prod i\left(\mp\sqrt{2\omega(p_i)(2\pi)^3}\right) \langle p_3,p_4;out|p_1,p_2;in\rangle \tag{7.51}$$

$$= \prod i\left(\mp\sqrt{2\omega(p_i)(2\pi)^3}\right) S_{fi} ,$$

where the minus (plus) sign in the exponential corresponds to incoming (outgoing) particles in reaction (7.39). The factor $\sqrt{Z}$ is obtained from the calculation of the two-point Green's function, as discussed in Section 8.4. Equation (7.51) can evidently be extended to a reaction with an arbitrary number of incoming and outgoing particles.

The reduction formulae (7.51) give us the desired relations between Green's functions and the S-matrix elements. A few observations follow.

Lorentz invariance of the cross section. The Green's function is Lorentz invariant and is, furthermore, invariant under translation. The latter statement means that G depends only on the differences x_1–x_2, etc. For the Fourier transform, this implies (see [1]) that

$$\tilde{G}(p_1, p_2, \ldots) = (2\pi)^4 \delta^{(4)}\left(\sum p_{in} - \sum p_{fin}\right)\mathcal{G}(p_1, p_2, \ldots) , \qquad (7.52)$$

where $\mathcal{G}$ is a regular, Lorentz-invariant function of the momenta. Clearly, the same applies to the residue of G at its poles. Hence, neglecting non-essential signs, we can write

$$S_{fi} = \langle p_3, p_4, \ldots; out | p_1, p_2; in \rangle$$

$$= \Pi \left[\frac{1}{\sqrt{2\omega(2\pi)^3}} \right] (2\pi)^4 \delta^{(4)}\left(\sum p_{in} - \sum p_{fin}\right)\mathcal{M}(p_1, p_2, \ldots) , \quad (7.53)$$

where $\mathcal{M}$ is the *Feynman amplitude*, a Lorentz-invariant function of its arguments. The cross section of the process $p_1 + p_2 \to p_3 + \ldots$ is calculated from (7.53) starting from the general formula (again, see [1])

$$(\rho v_{rel} N)\, d\sigma = \sum_{fin} \frac{|S_{fi}|^2}{T} . \qquad (7.54)$$

Our states are *not* normalised to a single particle in a volume V, hence the factor N on the left-hand side, which denotes the number of target particles in V. With the continuum normalisation, $N = V/(2\pi)^3$ [see equation (3.93)], while $\rho = 1/(2\pi)^3$ is the density of the incident particles, and $v_{rel} = |v_1 - v_2|$ is the relative speed of the initial particles.

We leave the proof of the following formulae to the reader.

- Cross section:

$$d\sigma = \frac{1}{4\omega(\mathbf{p}_1)\omega(\mathbf{p}_2)v_{rel}} (2\pi)^4 \delta^{(4)}\left(\sum p_{in} - \sum p_{fin}\right)$$

$$\times \Pi_{i=3,\ldots} \left[\frac{d^3 p_i}{(2\pi)^3 2\omega(\mathbf{p_i})} \right] |\mathcal{M}|^2 . \quad (7.55)$$

- Invariant phase space:

$$\frac{d^3 p}{2\omega(\mathbf{p}_i)} = d^4 p \; \theta(p^0)\delta(p^2 - m^2) . \qquad (7.56)$$

- invariant flux factor:

$$\omega(\mathbf{p_1})\omega(\mathbf{p_2})v_{rel} = \sqrt{(p_1 p_2)^2 - p_1^2 p_2^2} \ . \qquad (7.57)$$

Equations (7.56) and (7.57), substituted into (7.55), show that $d\sigma$ is Lorentz invariant.

CHAPTER 8

PERTURBATIVE GREEN'S FUNCTIONS IN $\lambda\phi^4$

CONTENTS

In the previous chapter we applied the method of the sum over paths to very simple field theories, in particular to real or complex scalar, *free* fields.

By *free* fields we mean fields describing particles which do not interact with each other. This translates into the requirement that the Lagrangian should not contain terms more than quadratic in the fields, and correspondingly equations of motion linear in the fields: the Klein–Gordon equation, for the scalar field, and the equations of Dirac and Maxwell, for fields of spin $\frac{1}{2}$ and (massless) spin 1, respectively.

In all these cases, the general solution to the equations of motion is given by a superposition of plane waves, which correspond to different possible definite-momentum states of the particles. In all interesting cases, the situation is more complicated; the Lagrangian contains terms of third or fourth (or even higher) order, the equations of motion are not linear, and theories are obtained for which we are unable to obtain exact solutions, in general. Approximate methods are needed, and perturbation theory is in first place.

In this course we shall focus on quantum electrodynamics and gauge theories. But for simplicity of illustration, we begin by considering the simple case of a real scalar field with a $\lambda\phi^4$ interaction and thus a Lagrangian

$$\mathcal{L} = \frac{1}{2}\partial_\mu\phi(x)\,\partial^\mu\phi(x) - \frac{1}{2}m^2\phi^2 - \frac{1}{4!}\lambda\phi^4 \,, \qquad (8.1)$$

which corresponds to a non-linear equation of motion,

$$(\Box + m^2)\phi(x) = -\frac{1}{3!}\lambda\phi^3 \,,$$

whose general solution is not known even at the classical level.

8.1 THE PERTURBATIVE GENERATING FUNCTIONAL

We can express, at least formally, physically interesting quantities as series of powers of the *coupling constant* λ, the so-called *perturbative series*. In particular, we can express the Green's functions as a series of powers of λ. If λ is small, the first few terms of this series can give a good approximation of the quantity of physical interest.

To proceed in this way, the Lagrangian is divided into two terms,

$$\mathcal{L}(\phi, \partial_\mu \phi) = \mathcal{L}^0(\phi, \partial_\mu \phi) + \mathcal{L}^1(\phi) , \qquad (8.2)$$

where $\mathcal{L}^0$ is the free Lagrangian studied in Section 3.3, and $\mathcal{L}^1$ an interaction term. For simplicity, we consider here the case of a single scalar field.[1]

The generating functional is [see equation (3.20)]

$$Z[J] = \int d[\phi(x)] \exp\left(i \int d^4x\, \mathcal{L}^1(\phi) \right)$$
$$\times \exp\left(i \int d^4x \left(\mathcal{L}^0(\phi, \partial_\mu\phi) - \phi(x)J(x) \right) \right) , \qquad (8.3)$$

which we can rewrite (see equation 3.23) as

$$Z[J] = \exp\left(i \int d^4x\, \mathcal{L}^1 \left(i\frac{\delta}{\delta J(x)} \right) \right) Z^0[J] , \qquad (8.4)$$

where $Z^0[J]$ is the functional associated with the action obtained from the unperturbed Lagrangian $\mathcal{L}^0$. In the case of the $\lambda\phi^4$ theory, equation (8.1), $Z^0[J]$ is given by (3.63), which we repeat for convenience

$$Z^0[J] = \exp\left(\frac{-i}{2} \iint d^4x\, d^4y\, J(x)\, \Delta_F(x-y)\, J(y) \right) , \qquad (8.5)$$

while a formal expansion of the generating functional in powers of λ is given by

$$Z[J] = \exp\left(\frac{-i\lambda}{4!} \int d^4x \left(i\frac{\delta}{\delta J(x)} \right)^4 \right) Z^0[J] \qquad (8.6)$$

$$= \sum_{n=0}^{\infty} \frac{(-i\lambda)^n}{(4!)^n n!} \left(\int d^4x \left(i\frac{\delta}{\delta J(x)} \right)^4 \right)^n Z^0[J] \qquad (8.7)$$

$$= Z^0[J] - i\frac{\lambda}{4!} \int d^4x \left(i\frac{\delta}{\delta J(x)} \right)^4 Z^0[J]$$

$$- \frac{\lambda^2}{2(4!)^2} \iint d^4x\, d^4y \left(i\frac{\delta}{\delta J(x)} \right)^4 \left(i\frac{\delta}{\delta J(y)} \right)^4 Z^0[J] + \dots . \qquad (8.8)$$

---

[1]We have assumed that $\mathcal{L}^1$ depends on the fields and not on their derivatives, a limitation which simplifies the following formal expansions but can be overcome without particular difficulty.

We recall that [see equation (8.5)]

$$i\frac{\delta}{\delta J(x)} Z^0[J] = \int d^4y\, \Delta_F(x - y)\, J(y)\, Z^0[J] \tag{8.9}$$

The first derivative that is taken "pulls down" a factor $\Delta_F J$ from the exponential, while subsequent ones can pull down further factors, or "capture" the J from a factor $\Delta_F J$ pulled down by a previous derivative. For example, for the second derivative we find

$$i\frac{\delta}{\delta J(x')} \left(i\frac{\delta}{\delta J(x)} Z^0[J] \right) \tag{8.10}$$

$$= \left(\int d^4y'\, \Delta_F(x' - y')\, J(y') \right) \left(\int d^4y\, \Delta_F(x - y)\, J(y) \right) Z^0[J]$$

$$+ i\Delta_F(x - x')\, Z^0[J] \, .$$

To better control the complexity of these calculations, which grows rapidly with an increasing number of derivatives, a graphical representation is helpful. A point, called a "vertex", from which four lines emerge, corresponds to each factor $(i\delta/\delta J(x))^4$. Each line can end in another vertex from which a J has been "captured" (or in the same vertex, if it captures a J produced by a previous derivative), or in a bubble, which represents a J not captured by other derivatives. In the first case the line (which we will call "an internal line") corresponds to a propagator $i\Delta_F$, in the second ("external line") to a factor $\Delta_F J$. For example, the two terms of the second derivative can be represented by the two diagrams of Figure 8.1. Naturally, in this case only a single line emerges from each of the two points x and x'. To unify the description of the two types of line it is helpful to rewrite the terms which correspond to a line ending in a bubble as $\Delta_F J = (i\Delta_F)(-iJ)$ so that each, internal or external, line corresponds to a factor $i\Delta_F$.

Figure 8.1   Diagram representing the two terms of equation (8.10). The hatched bubbles each correspond to a factor $-iJ$.

Diagrams with a single vertex correspond to terms of $Z[J]$ [equation (8.6)] of order λ, and it is easily shown that the only possibilities are the diagrams (a), (b) and (c) of Figure 8.2. Diagram (a), where all the Js have been captured, and the diagram is devoid of external lines, is called a *vacuum-vacuum* diagram.

Diagram (b) represents a modification to a single particle propagator,

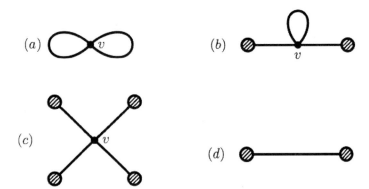

Figure 8.2 (a), (b), (c): diagrams to first perturbative order; diagram (d) represents the propagator to order λ^0 and is reported for comparison with the order λ correction embodied in (b).

which to order λ^0, is what we calculated previously (see Section 3.3), and which we can represent as a diagram without vertices, as in (d). We will return to this point later. Meanwhile, however, we take a look at diagrams of the same type which are encountered to second perturbative order, displayed in Figure 8.3. The suspicion arises that diagram (a) of this figure, together with (d) and (b) of the previous figure, is the beginning of an interesting series. A suspicion which, as we will see, is fully justified.

Finally, diagram (c) of Figure 8.2 represents the scattering of two particles. As we will see, this is directly connected to the S-matrix element which describes this process, in fact it represents the first order approximation to this process. Higher-order corrections also exist in this case, which to second order are those of Figure 8.4.

The correspondence between the diagrams of the $\lambda\phi^4$ theory and terms in the perturbative expansion of $Z[J]$ is obtained with a few simple rules

$$\text{bubble}: \qquad -i \int d^4x\, J(x) ,$$

$$\text{vertex}: \qquad \frac{-i\lambda}{4!} \int d^4v , \qquad (8.11)$$

$$\text{line from } x \text{ to } y: \qquad i\Delta_F(x-y) ,$$

to which must be added a rule to calculate the combinatorial factor to apply to the contribution of each diagram. In the case of $\lambda\phi^4$ theory, the latter rule is somewhat more complicated than in the case of quantum electrodynamics. In our context, in which $\lambda\phi^4$ serves as a simple model for QED, it is better to postpone this discussion. It is always possible, once the form of a particular

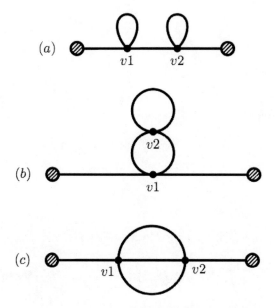

(a) $v1$ $v2$

(b) $v2$ $v1$

(c) $v1$ $v2$

Figure 8.3 Propagator corrections to second perturbative order.

contribution to $Z[J]$ has been identified, to return to (8.6) – with the rules we have given – to obtain the correct combinatorial factor.

A point v or x in space-time is associated with each vertex or bubble, at which an integration is carried out, and the argument of each Δ_F is the distance to each end-point, either vertex or bubble. Therefore diagram (c) of Figure 8.2 corresponds to a term in $Z[J]$ which we can call D_1 (and is the first diagram we will calculate!):

$$D_1[J] = (-i)^4 \int d^4x_1 \, J(x_1) \int d^4x_2 \, J(x_2) \int d^4x_3 \, J(x_3) \int d^4x_4 \, J(x_4)$$

$$\times \frac{-i\lambda}{4!} \int d^4v \, (i)^4 \Delta_F(x_1 - v)\Delta_F(x_2 - v)\Delta_F(x_3 - v)\Delta_F(x_4 - v) \,, \quad (8.12)$$

which represents, to first order in λ, the terms in $Z[J]$ that are proportional to J^4. The corrections of order λ^2 are given by the diagrams of Figure 8.4; (a)

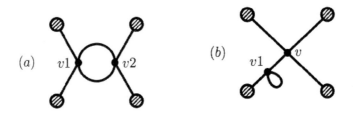

Figure 8.4   Vertex corrections to second perturbative order.

corresponds to

$$D_2[J] = (-i)^4 (6^2) \int d^4x_1\, J(x_1) \int d^4x_2\, J(x_2) \int d^4x_3\, J(x_3) \int d^4x_4\, J(x_4)$$

$$\times \left(\frac{-i\lambda}{4!}\right)^2 \iint d^4v_1 d^4v_2 (i)^6 \Delta_F(x_1 - v_1)\Delta_F(x_2 - v_1)\Delta_F(x_3 - v_2)$$

$$\times \Delta_F(x_4 - v_2)\Delta_F(v_1 - v_2)\Delta_F(v_1 - v_2)\,, \tag{8.13}$$

where (6^2) is a combinatorial factor. We could have obtained this result directly, including the factor (6^2), by isolating the J^4 terms in equation (8.6). We leave this task to the reader as an exercise. We will discuss diagram (b) of Figure 8.4 separately; it represents a correction to the external lines.

8.2   FEYNMAN RULES FOR GREEN'S FUNCTIONS

We consider the Green's function of a real scalar field at $2k$ points (we note that in the $\lambda\phi^4$ theory with interactions, a non-zero result is obtained only for an even number of points.[2] Employing the previous formulae, we have

$$G(x_1, \ldots, x_{2k}) = \int D[\phi(x)]\, \phi(x_1)\phi(x_2)\ldots\phi(x_{2k}) \exp[i \int d^4x \mathcal{L}(\phi, \partial_\mu \phi)]$$

$$= \left[(i\frac{\delta}{\delta J(x_1)})(i\frac{\delta}{\delta J(x_2)})\ldots(i\frac{\delta}{\delta J(x_{2k})})Z[J]\right]_{J=0}, \tag{8.14}$$

---

[2]The Lagrangian, equation (8.1), is symmetric under the transformation $\phi \to -\phi$. A unitary operator $\mathbf{P}$ on the space of states must therefore exist, such that $\mathbf{P}\phi\mathbf{P} = -\phi$. Since the vacuum $(n=0)$ is even, it follows that $< 0|T[\phi(x_1)\cdots\phi(x_n)]|0 >= 0$ unless n is even.

and furthermore, according to the perturbative expansion,

$$Z[J] = \int D[\phi(x)] \exp[i \int d^4x \, \mathcal{L}^1(\phi)] \exp[i \int d^4x \, \mathcal{L}^0 - i \int d^4x J(x)\phi(x)]$$

$$= \exp[i \int d^4x \, \mathcal{L}^1(i\frac{\delta}{\delta J(x)})] \, Z^0[J]$$

$$= \sum_n (\frac{i}{n!})^n \left[\int d^4x \, \mathcal{L}^1(i\frac{\delta}{\delta J(x)}) \right]^n Z^0[J] \, , \tag{8.15}$$

with

$$Z^0[J] = \exp \left[\frac{-i}{2} \int\int d^4x \, d^4y \, J(x)\Delta_F(x-y)J(y) \right]$$

$$= \exp \left[\frac{-i}{2} (J\Delta_F J) \right] . \tag{8.16}$$

The perturbative expansion of G hence takes the form

$$G(x_1, \ldots, x_{2k}) = \sum_n G^{(n)}(x_1, x_2, \ldots, x_{2k}) \, ,$$

with

$$G^{(n)}(x_1, \ldots, x_{2k})$$
$$= \left\{ (i\frac{\delta}{\delta J(x_1)}) \ldots (i\frac{\delta}{\delta J(x_{2k})}) \frac{1}{n!} \left[i \int d^4x \, \mathcal{L}^1(i\frac{\delta}{\delta J(x)}) \right]^n Z^0[J] \right\}_{J=0} . \tag{8.17}$$

Since we must finally set $J = 0$, in the expansion of Z^0 in powers of $(J\Delta_F J)$, only terms of order equal to the number of functional derivatives which appear in equation (8.17) contribute. Restricting ourselves to the case of the $\lambda\phi^4$ interaction, we can write, in conclusion

$$G^{(n)}(x_1, \ldots, x_{2k}) = (i\frac{\delta}{\delta J(x_1)}) \ldots (i\frac{\delta}{\delta J(x_{2k})}) \frac{1}{n!}$$

$$\times \left[i \int d^4x \, \frac{\lambda}{4!}(i\frac{\delta}{\delta J(x)})^4 \right]^n \frac{1}{M!} \left[\frac{-i}{2} (J\Delta_F J) \right]^M , \tag{8.18}$$

with $M = k + 2n$.

The functional derivatives can naturally be taken in many different ways. Each one can be represented by a *Feynman diagram*, a diagram in space-time in which M lines, the propagators $i\Delta_F$, connect the points $x_1, \ldots, x_{2k}$ to each other and/or with the n points where the interactions are localised, the *vertices*. Each of these diagrams is associated with an amplitude, which is the product of propagators, coupling constants λ, and numerical factors, determined by the structure of (8.18). Explicitly, we can see from (8.18) that:

- each time that two derivatives act on the same propagator, we obtain a factor 2 and hence the result is a factor $i\Delta_F$ calculated at the points associated with the functional derivatives in question;

- the roles of the propagators can be permuted without changing the result; this simplifies the factor $(1/M!)$;

- similarly, the roles of the vertices can be permuted obtaining $n!$ identical contributions, which simplifies the factor $(1/n!)$.

The considerations just explained are summarised in simple and elegant prescriptions to obtain the $2k$-point Green's function to the n-th order of perturbation theory, which are called of *Feynman rules*.

Feynman diagrams. We begin with the identification of the independent amplitudes which contribute to the Green's function, by constructing the corresponding Feynman diagrams.

- We fix the external points $(x_1, .., x_{2k})$ and the points $(x, y, z, ...)$ where the interactions are localised (vertices).

- We draw the diagrams in which M lines connect external points and vertices to each other in all topologically independent ways.

Feynman rules. An amplitude is associated with each diagram, according to the following rules:

- There is a factor $i\Delta_F(u - v)$ for every line which begins at point u and finishes at v;

- There is a factor $i\lambda$ for every vertex.

- There is a numerical factor, to be calculated case by case, which originates from the incomplete cancellation of the $1/4!$ factors at the vertices with the multiplicity of ways in which the derivatives corresponding to the diagram can be taken.

- We integrate the amplitude over the coordinates of all the vertices.

- The Green's function is the sum of the amplitudes corresponding to each diagram.

We give, as an example, the calculation of the two-point and four-point functions to first order in λ.

Calculation of $G^{(1)}(x_1, x_2)$. With one vertex and two external points we can construct two independent diagrams, shown in Figure 8.5. We now calculate the relevant numerical factors for the two diagrams by explicitly taking the functional derivatives (for brevity, we write 1 or 2 in place of x_1 or x_2)

$$G^{(1)}(1,2) = (i\frac{\delta}{\delta J(1)})(i\frac{\delta}{\delta J(2)})\frac{i\lambda}{4!}\int d^4x \ (i\frac{\delta}{\delta J(x)})^4 \frac{1}{3!}[\frac{-i}{2} \ (J\Delta_F J)]^3 \ . \quad (8.19)$$

We take the derivatives with respect to $J(1)$ and $J(2)$. The two derivatives can act (a) on the same propagator, or (b) on different propagators. We obtain

$$\widetilde{G}_{2a} = i\Delta_F(1-2)\frac{i\lambda}{4!}\int d^4x \ (i\frac{\delta}{\delta J(x)})^4 \frac{1}{2!}(\frac{-i}{2} \ J\Delta_F J)^2 \ , \quad (8.20)$$

$$\widetilde{G}_{2b} = \frac{i\lambda}{4!}\int d^4x \ (i\frac{\delta}{\delta J(x)})^4 \int\int d^4u d^4v \ [i\Delta_F(1-u)(-iJ(u))]$$

$$\times \ [i\Delta_F(2-v)(-iJ(v))] \ (\frac{-i}{2} \ J\Delta_F J) \ . \quad (8.21)$$

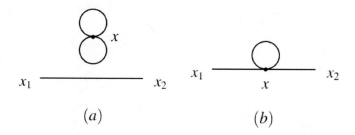

x_1 x_2 x_1 x x_2

(a)   (b)

Figure 8.5 First-order diagrams for the two-point Green's function.

In the case of $\widetilde{G}_{2a}$, we take the first derivative and find

$$\widetilde{G}_{2a} = i\Delta_F(1-2)\frac{i\lambda}{4!}\int d^4x \ (i\frac{\delta}{\delta J(x)})^3$$

$$\times \int d^4u \ \Delta_F(x-u)J(u)[\frac{-i}{2} \ (J\Delta_F J)] \ . \quad (8.22)$$

One of the other three derivatives must act on $J(u)$, which can happen in three different ways; the other two cancel the current in $(J\Delta_F J)$. Hence

$$\widetilde{G}_{2a} = C_a[i\Delta_F(1-2)] \ (i\lambda)\int d^4x \ [i\Delta_F(x-x)]^2 \ ,$$

$$C_a = \frac{3}{4!} = \frac{1}{8} \ . \quad (8.23)$$

In the case of $\widetilde{G}_{2b}$, two derivatives must operate on the currents $J(u)$ and

$J(v)$, which can happen in 4×3 different ways, while the other two act on $(J\Delta_F J)$. We therefore find, in total,

$$\tilde{G}_{2b} = C_b \, i\lambda \int d^4x \, [i\Delta_F(1-x)] \, [i\Delta_F(2-x)][i\Delta_F(x-x)] \,,$$

$$C_b = \frac{12}{4!} = \frac{1}{2} \,. \tag{8.24}$$

Calculation of $G^{(1)}(x_1, x_2, x_3, x_4)$. We show the Feynman diagrams for the four-point function to first order in Figure 8.6.

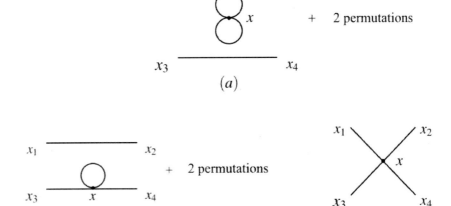

Figure 8.6 First-order diagrams for the four-point Green's function.

The explicit calculation is carried out starting from the expression

$$G^{(1)}(1,2,3,4) = (i\frac{\delta}{\delta J(1)})(i\frac{\delta}{\delta J(2)})(i\frac{\delta}{\delta J(3)})(i\frac{\delta}{\delta J(4)})\frac{i\lambda}{4!}$$

$$\times \int d^4x \, (i\frac{\delta}{\delta J(x)})^4 \, \frac{1}{4!}[\frac{-i}{2} \, (J\Delta_F J)]^4 \,. \tag{8.25}$$

Diagram (a) corresponds to making the derivatives with respect to the currents at the external points act 2×2 times on the two propagators. Thus we obtain

$$\widetilde{G}_{4a} = \frac{i\lambda}{4!} \, i\Delta_F(1-2) \, i\Delta_F(3-4)$$

$$\times \int d^4x \, (i\frac{\delta}{\delta J(x)})^4 \frac{1}{2!} [\frac{-i}{2} \, (J\Delta_F J)]^2 + (2 \leftrightarrow 3) + (2 \leftrightarrow 4) \, . \quad (8.26)$$

Repeating the argument which led to (19.31), we therefore find

$$\widetilde{G}_{4a} = \frac{3}{4!} \, i\lambda \, i\Delta_F(1-2) \, i\Delta_F(3-4)$$

$$\times \int d^4x \, [i\Delta_F(x-x)]^2 + (2 \leftrightarrow 3) + (2 \leftrightarrow 4) \, . \quad (8.27)$$

For diagram (b), only two of the derivatives with respect to the currents at the external points act on the same propagator. The result is

$$\widetilde{G}_{4b} = \frac{i\lambda}{4!} \, i\Delta_F(1-2) \int d^4x \, (i\frac{\delta}{\delta J(x)})^4$$

$$\times \int \int d^4u \, d^4v \, \Delta_F(3-u)\Delta_F(4-v)J(u)J(v) \, [\frac{-i}{2} \, (J\Delta_F J)]$$

$$+ (2 \leftrightarrow 3) + (2 \leftrightarrow 4) \, . \quad (8.28)$$

Proceeding as for (8.24), we therefore obtain

$$\widetilde{G}_{4b} = \frac{12}{4!} \, i\lambda \, i\Delta_F(1-2) \, i\Delta_F(3-x) \, i\Delta_F(4-x) \quad (8.29)$$

$$\times \int d^4x \, i\Delta_F(x-x) + (2 \leftrightarrow 3) + (2 \leftrightarrow 4) \, . \quad (8.30)$$

Finally, we calculate diagram (c). In this case the derivatives with respect to the currents at the external points must all act on different propagators. The resulting expression is

$$\widetilde{G}_{4c} = \frac{i\lambda}{4!} \int d^4x \, (i\frac{\delta}{\delta J(x)})^4 \int \int \int \int d^4u \, d^4v \, d^4w \, d^4z \quad (8.31)$$

$$\times \Delta_F(1-u)\Delta_F(2-v)\Delta_F(3-w)\Delta_F(4-z) \, J(u)J(v)J(w)J(z) \, .$$

In this case the cancellation of the 1/4! factor is complete, and we find:

$$\widetilde{G}_{4c} = i\lambda \int d^4x \, i\Delta_F(1-x)i\Delta_F(2-x)i\Delta_F(3-x)i\Delta_F(4-x) \, . \quad (8.32)$$

To close this section, we make a few comments of a general nature.

Feynman diagrams and the sum over paths. The analysis of the Green's functions in terms of Feynman diagrams corresponds fully to the idea of the sum over paths which we introduced at the start of the book. We can, for example, interpret the four-point Green's function as the quantum amplitude of a space-time process in which two particles are created at x_1 and x_2 and absorbed at x_3 and x_4. Each independent diagram, with fixed positions of the vertices, corresponds to a possible "path" and the corresponding amplitude is given by the product of the amplitudes of the different components of the path: propagation amplitudes, $i\Delta_F(x - y)$, and interactions, $i\lambda$. The perturbative expansion corresponds to having one, two, ..., n interactions along the paths of the different particles.

The coordinates of the external points do not determine how many interactions occur in the process nor do they fix the space-time points where the interactions actually take place. According to the general principles of quantum mechanics, we must integrate over the space-time coordinates of the interactions (vertices) for each diagram and sum the amplitudes of the independent diagrams.

Contractions in the same vertex. When two functional derivatives in an interaction Lagrangian capture the two Js from a $(J\Delta_F J)$ term, a line appears in the corresponding diagram which ends on the same vertex, as in Figure 8.2, (b). The closed loop corresponds to the insertion of an interaction term with two fewer powers of the fields, in this case of the form $constant \cdot \phi^2$. Because the Lagrangian *already contains* terms in all powers of the field up to four, the insertion corresponds to redefining the constants already included in the Lagrangian itself. The result is a new, simple, rule:

- Contractions in the same vertex can be ignored.

In Dyson's formulation of the S-matrix, this rule is automatically satisfied if the interaction Lagrangian is taken as a *normal product* of the fields; see [1].

8.3   CONNECTED PARTS AND VACUUM DIAGRAMS

The diagrams that we showed in Figures 8.2–8.4 are all topologically connected. In each of them it is possible to move from any vertex or bubble to another along the lines of the diagram. Disconnected diagrams also exist. For example, among the second-order terms in λ in equation (8.6) there is one in which each of the eight derivatives "pulls down" a factor $\Delta_F J$, which simply results in the square of the term from equation (8.12). This term is represented by Figure 8.7, a diagram comprised of two topologically distinct parts centred on two vertices v_1 and v_2. A disconnected diagram corresponds to a term (a functional of J) which factorises into the product of two or more functionals of J, and in the case of Figure 8.7 leads to[3] $(D_1[J])^2/2$.

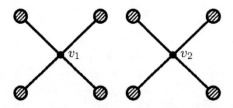

Figure 8.7   A disconnected diagram to second order in λ.

From the physical point of view, each of the two parts of the diagram in Figure 8.7 represents a process of scattering between two particles. The combination of the two parts represents two scattering processes independent of each other; we can imagine that the first takes place at CERN, the second at Frascati. The probability amplitude for the combination of further independent processes is simply the product of the amplitudes of each of them, and the probability that all occur is the product of the individual probabilities. There is no more to learn from the study of independent processes, hence it is better to concentrate on individual processes, which correspond to connected diagrams.

We have seen that by means of perturbation theory the functional $Z[J]$ can be expressed as a sum of diagrams, some of which are connected, and others disconnected. It is possible to define a functional $W[J]$ which generates only the connected diagrams

$$W[J] = \sum_{\text{(connected diagrams)}} D_i[J] \, . \tag{8.33}$$

The relation between $W[J]$ and $Z[J]$ is given simply by

$$Z[J] = \exp(W[J]) \, . \tag{8.34}$$

The proof of this result is given in Appendix B.

We note again that for $J = 0$ we obtain

$$Z[0] = \exp(W[0]) \, , \tag{8.35}$$

where $W[0]$ corresponds to the sum of connected vacuum-vacuum diagrams,

---

[3]See Appendix B, equation (B.14).

those which do not have "bubbles", the only ones which do not vanish for $J = 0$. We can therefore write

$$Z[J] = Z[0] \exp(W'[J]) , \qquad (8.36)$$

where $W'[J]$ is the sum of connected diagrams which are not of the vacuum-vacuum type, hence diagrams with external legs. The effect of the vacuum-vacuum diagrams on the generating functional $Z[J]$ consists therefore of a multiplicative constant $Z[0]$ that does not contribute to the calculation of the Green's functions, and we can simply omit it.

We can apply this concept directly to the Green's functions, by defining the *connected* Green's functions by means of the derivative of the functional $W[J]$, or in an equivalent way by the functional $W'[J]$ since the two differ by a constant $W[0]$, which does not contribute to the derivatives:

$$\langle 0|T\left(\phi(x_1)\cdots\phi(x_N)\right)|0\rangle\Big|_{\text{conn.}} = \frac{i\delta}{\delta J(x_1)}\cdots\frac{i\delta}{\delta J_N(x_N)}W[J]\Big|_{J=0} . \qquad (8.37)$$

The connected diagrams with exactly N external lines contribute to the N-point connected Green's function; hence N factors of J. If we look at the rules for the diagrams summarized by equations (8.11), we see that the functional derivatives $i\delta/\delta J(x_k)$ from (8.37) suppress the factors $-iJ(x)$ of the external lines, and fix the termination of each line at x_k. Hence, one goes from the diagrams for the functional generator to those for the Green's function simply by suppressing the bubbles and the relevant $-i\int d^4x\, J(x)$ factor, and fixing the end of each line at the coordinate of one of the fields present in the Green's function. We consider for example the four-point function; the lowest-order diagram that contributes is Figure 8.2(c), from which we obtain [see equation (8.12)]

$$\langle 0|T\left(\phi(x_1)\cdots\phi(x_4)\right)|0\rangle\Big|_{\text{conn.}} \qquad (8.38)$$

$$= -i\lambda\int d^4v\,(i)^4\Delta_F(x_1 - v)\Delta_F(x_2 - v)\Delta_F(x_3 - v)\Delta_F(x_4 - v) + \mathcal{O}(\lambda^2) .$$

We note that the fourth derivative on the left-hand side has also eliminated the factor $1/4!$ from (8.12). At graphical level we can represent the diagrams for the Green's function by substituting the "bubbles" with an indication of the coordinates of the points at which the external lines terminate, as shown for example in Figure 8.8.

8.4   PERTURBATIVE TWO-POINT GREEN'S FUNCTION

We already derived in [2], the exact representation of the two-point function in field theory, known as the Källén–Lehmann (KL) representation.

For the theory of a scalar field corresponding to a particle of mass m, the

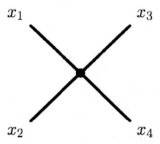

Figure 8.8 The diagram of order λ for the four-point Green's function.

KL representation takes the form

$$G(x,y) = iZ\Delta_F(x-y,m^2) + \int_{(3m)^2}^{+\infty} dM^2\, \sigma(M^2)\, i\Delta_F(x-y,M^2)\,, \quad (8.39)$$

where $i\Delta_F(x-y,m^2)$ is the propagator of a free scalar particle of spin zero and mass m. The function $\sigma(M^2)$, the so-called *spectral function*, is positive definite. In this section we discuss the two-point function of the scalar $\lambda\phi^4$ theory at second order in λ. We repeat here the Lagrangian given in equation (8.1):

$$\mathcal{L} = \frac{1}{2}\partial_\mu\phi\partial^\mu\phi - \frac{1}{2}m_0^2\phi^2 - \frac{1}{4!}\lambda\phi^4\,, \quad (8.40)$$

denoting, however, the coefficient of the quadratic term in ϕ by m_0, to distinguish it from the physical mass m which appears in the formula (8.39).

We can define m_0 as the mass which the particle would have *in the limit* $\lambda \to 0$, a clearly unobservable quantity, while m is the inertial mass of an isolated particle, in principle measurable and determined by the position of the pole in the Fourier transform of the two-point function (8.39). The calculation which follows shows how, starting from Feynman diagrams in terms of the bare mass, the two-point function can be reconstructed in terms of the physical mass, or renormalised. This is the simplest case of *renormalisation* in field theory. The relevant connected Feynman diagrams are shown in Figure 8.9.

We consider the Fourier transform

$$\widetilde{G}(p_1,p_2) = \int d^4x_1 d^4x_2\, e^{-i(p_1 x_1)} e^{+i(p_2 x_2)}\, G(x_1,x_2)\,, \quad (8.41)$$

limiting ourselves initially to first-order terms in λ, Figure 8.9, (a) and (b). Following equations (19.31) and (8.24), we can immediately write:

$$\widetilde{G}(p_1,p_2) = (2\pi)^4 \delta^{(4)}(p_1-p_2)\widetilde{G}_{(a+b)}(p_1)\,, \quad (8.42)$$

with

$$\widetilde{G}_{(a+b)}(p) = \frac{i}{p^2-m_0^2} + \frac{i}{p^2-m_0^2}(i\lambda C_1)\frac{i}{p^2-m_0^2} + \mathcal{O}(\lambda^2)\,, \quad (8.43)$$

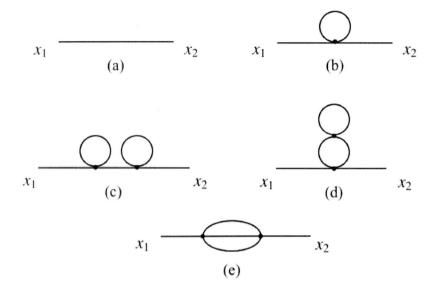

Figure 8.9 Connected diagrams for the two-point function, to second perturbative order.

where we have set

$$C_1 = i\frac{1}{2}\Delta_F(0) . \tag{8.44}$$

In (8.43) the *bare mass*, m_0, appears. We note that the propagator has a double pole at m_0. We now introduce the *renormalised mass*, m, by writing

$$m_0^2 = m^2 + \delta m^2 . \tag{8.45}$$

The quantity δm^2 must depend on λ, in particular $\delta m^2 = \mathcal{O}(\lambda)$, because $m^2 \to m_0^2$ if we let $\lambda \to 0$. We can expand the first term of (8.43) by using the relation

$$\frac{1}{A+\epsilon} = \frac{1}{A} - \frac{1}{A^2}\epsilon + \frac{1}{A^3}\epsilon^2 + \dots , \tag{8.46}$$

and stopping at the first order in δm^2. In this way we obtain

$$\frac{i}{p^2 - m_0^2} = \frac{i}{p^2 - m^2} + \frac{i}{(p^2 - m^2)^2}\,\delta m^2 + \mathcal{O}(\lambda^2) . \tag{8.47}$$

If we limit ourselves to the first order in λ, neglecting terms of order λ^2, we can identify m_0^2 with m^2 in the second term of (8.43). In total we therefore have

$$\widetilde{G}_{(a+b)}(p) = \frac{i}{p^2 - m^2} + \frac{i}{(p^2 - m^2)^2}[\delta m^2 - \lambda C_1] + \mathcal{O}(\lambda^2) . \tag{8.48}$$

According to the KL representation, the Green's function must have a simple pole for $p^2 = m^2$, where m is the *physical mass*. This fixes $\delta m^2 = \lambda C_1 + \mathcal{O}(\lambda^2)$. To this order, we then obtain equation (8.39) with

$$Z = 1 \ , \quad \sigma(M^2) = 0 \ . \tag{8.49}$$

Extension to order λ^2. We can determine the form of the propagator to second order easily enough, as follows.

The diagrams in Figure 8.9 $(c), (d), (e)$ correspond respectively to adding to (8.43) the terms

$$\widetilde{G}_c = \frac{i}{p^2 - m_0^2} \, (i\lambda C_1) \, \frac{i}{p^2 - m_0^2} \, (i\lambda C_1) \, \frac{i}{p^2 - m_0^2} \ ,$$

$$\widetilde{G}_d = \frac{i}{p^2 - m_0^2} \, (i\lambda^2 C_2) \, \frac{i}{p^2 - m_0^2} \ ,$$

$$\widetilde{G}_e = \frac{i}{p^2 - m_0^2} \, [i\lambda^2 F(p^2)] \frac{i}{p^2 - m_0^2} \ . \tag{8.50}$$

In these terms, which are already of second order, we can identify m_0 with m. We denote, as before, the constant corresponding to the single bubble of Figures 8.9(b) and (c) as $i\lambda C_1$, and the contribution, again a constant, of the double bubble in Figure 8.9(d) as $i\lambda^2 C_2$. The explicit calculation of $\widetilde{G}_e$ is rather more difficult and we are not yet ready to do it. However, it is enough to know that the amplitude corresponding to the loop in Figure 8.9(e) can be written as $i\lambda^2 F(p^2)$, with F a non-trivial function of p^2, regular at $p^2 = m^2$.

We can represent the function $F(p^2)$ in the following way

$$F(p^2) = F(m^2) + (p^2 - m^2)F'(m^2) + (p^2 - m^2)R(p^2) \ , \tag{8.51}$$

where $R(p^2)$ is a regular function that *vanishes*[4] for $p^2 = m^2$.

For (8.43), we must now expand the first term to second order in δm^2 and the second term to first order. We obtain

$$\widetilde{G}_{a+b}(p^2) = \frac{i}{p^2 - m^2} + \frac{i}{(p^2 - m^2)^2} \, [\delta m^2 - \lambda C_1]$$

$$+ \frac{i}{(p^2 - m^2)^3} \, [(\delta m^2)^2 - 2\delta m^2(\lambda C_1)] \ . \tag{8.52}$$

For $\widetilde{G}_e$, by using the representation (8.51), we find

$$\widetilde{G}_e = \frac{i}{(p^2 - m^2)^2} \, [-\lambda^2 F(m^2)]$$

$$+ \frac{i}{(p^2 - m^2)} \, [-\lambda^2 F'(m^2)] + \frac{i}{p^2 - m^2} \, [-\lambda^2 R(p^2)] \ . \tag{8.53}$$

---

[4]$(p^2 - m^2)R(p^2)$ is the remainder of the Taylor expansion of $F(p^2)$ in $p^2 - m^2$ to first order, and is an infinitesimal of higher order with respect to $p^2 - m^2$.

Putting everything together, we arrive at the result

$$G(p^2) = \frac{i[1 - \lambda^2\, F'(m^2)]}{p^2 - m^2} + \frac{i}{(p^2 - m^2)^2}\, [\delta m^2 - \lambda C_1 - \lambda^2 C_2 - \lambda^2 F(m^2)]$$

$$+ \frac{i}{(p^2 - m^2)^3}\, [(\delta m^2)^2 - 2\delta m^2\, (\lambda C_1) + \lambda^2 C_1^2] \qquad (8.54)$$

$$+ i\frac{-\lambda^2 R(p^2)}{p^2 - m^2} + \mathcal{O}(\lambda^3)\;.$$

To cancel the residue at the double pole we must now require that

$$\delta m^2 = \lambda C_1 - \lambda^2 C_2 - \lambda^2 F(m^2) + \mathcal{O}(\lambda^3)\;, \qquad (8.55)$$

and with this position the residue of the triple pole is also cancelled. The final result is therefore

$$G(p^2) = \frac{i[1 - \lambda^2\, F'(m^2)]}{p^2 - m^2} + i\frac{-\lambda^2 R(p^2)}{p^2 - m^2} + \mathcal{O}(\lambda^3)\;. \qquad (8.56)$$

We have now obtained a representation of the two-point function comprised of a pole term (the first) and a term regular in $p^2 - m^2$, since $R(m^2) = 0$. The result has the form given in (8.39) if we identify

$$1 - \lambda^2\, F'(m^2) = Z\;, \qquad \frac{-\lambda^2 R(p^2)}{p^2 - m^2} = \mathcal{R}(p^2) = \int_{(3m)^2}^{+\infty} \frac{\sigma(M^2)}{p^2 - M^2}\, dM^2\;. \qquad (8.57)$$

A deeper study shows that the function on the left-hand side of (8.57) is an analytic function of the complex variable p^2 with a branch-cut on the positive real p^2 axis which starts from $p^2 = (3m)^2$ and extends to $+\infty$. The right-hand side of (8.57) is therefore the familiar Cauchy representation of an analytic function, as the integral of the discontinuity over the cut. The discontinuity is actually the spectral function $\sigma(M^2)$ which, as expected, is of $\mathcal{O}(\lambda^2)$. Explicitly

$$\sigma(M^2) = \frac{i}{2\pi} \lim_{\epsilon \to 0} [\mathcal{R}(M^2 + i\epsilon) - \mathcal{R}(M^2 - i\epsilon)]\;. \qquad (8.58)$$

CHAPTER 9

S-MATRIX FEYNMAN DIAGRAMS IN $\lambda\phi^4$

CONTENTS

Using the reduction formulae, we can translate the rules for the calculation of Feynman diagrams for the Green's functions, discussed in Section 8.2, into rules for the calculation of S-matrix elements. The starting point is the Fourier transform which appears in the reduction formula, equation (7.51). To simplify the notation we consider all particles as if they belonged to the initial state. For a particle in the final state, it is sufficient to change the sign of the momentum and the energy. We must therefore calculate the following expression

$$\prod_{k=1}^{n}\left(\frac{(p_k^2 - m^2)}{i\sqrt{Z}}\right)\int \prod_{k=1}^{n}d^4x_k \prod_{k=1}^{n}e^{-ip_k x_k}\langle 0|T\left(\phi(x_1)\ldots\phi(x_n)\right)|0\rangle . \quad (9.1)$$

To obtain (7.51) it suffices just to put $p = q$ for the particles in the initial state, and $p = -q$ for those in the final state.

9.1　ONE-PARTICLE IRREDUCIBLE DIAGRAMS

For the Green's functions, we must consider only the connected Feynman diagrams with $n \geq 3$ external lines. We can further divide these diagrams into two categories, according to whether they are *one-particle irreducible*, 1PI, or not.

A diagram is one-particle reducible on the external lines if, by cutting only a single line, we can separate a diagram of the two-point function with a non-trivial interaction. In Figure 9.1 we show the connected diagrams for the four-point function, to second order in λ^2. Diagrams (a) and (c) are 1PI, while diagram (b) and its permutations are one-particle reducible on the external lines. Naturally, diagrams of higher order can also be one-particle reducible relative to more external lines, like those shown in Figure 9.2.

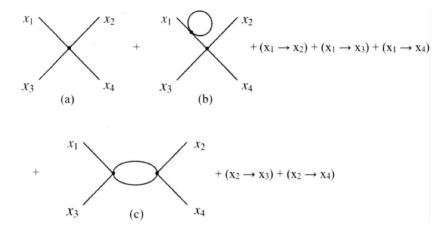

Figure 9.1 Connected diagrams for the four-point Green's function, to second perturbative order.

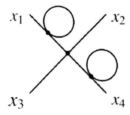

Figure 9.2 Example of a connected diagram one-particle reducible on two external lines.

Keeping in mind that the amplitude of a diagram is the product of its different components, we can represent the combination of all the connected diagrams with a block diagram in which every external line is replaced by the exact two-point function while the remaining Green's function is 1PI on all the external lines; see Figure 9.3 in the case of the four-point function.

The expansion of the 1PI function on the external lines is shown in Figure 9.4, again for the four-point function.

Taking the result from Section 8.4, equation (8.39), we can write

$$G(x,y)|_{conn} = iZ\Delta_F(x - y, m) + i\mathcal{R}(x - y) , \qquad (9.2)$$

where m is the *renormalised* mass and $\mathcal{R}(x - y)$ is a function whose Fourier transform is regular when $p^2 \to m^2$. In the block diagram of Figure 9.3, the coordinates of the fields appear as terminations of the external lines. Therefore

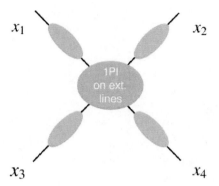

Figure 9.3 Block representation of the connected four-point Green's function. The external lines represent the exact two-point functions, and the central block represents the four-point Green's function which is 1PI on the external lines.

Figure 9.4 Expansion of the connected four-point Green's function, which is 1PI on the external lines.

each of the external lines is represented by [see also equations (3.61) and (3.62)]

$$i\Delta_F(v - x_k) = \frac{iZ}{(2\pi)^4} \int d^4p \, \frac{e^{-ipv} \, e^{ip_k x_k}}{p^2 - m^2 + i\epsilon} + i\mathcal{R}(v - x_k) \,, \tag{9.3}$$

where v is the coordinate of the vertex to which the line is attached, and x_k is the coordinate of one of the points of the Green's function. Applying the Fourier transform and multiplying by a factor $-i \, (p_k^2 - m^2)/\sqrt{Z}$ selects the residue at the pole, and leads to

$$\lim_{p_k^2 \to m^2} \frac{(p_k^2 - m^2)}{i\sqrt{Z}} \int d^4x_k \, e^{-ip_k x_k} \left(iZ\Delta_F(v - x_k)\right) = \sqrt{Z}e^{-ip_k v} \,. \tag{9.4}$$

To take the limit required by the reduction formula requires multiplication by $\sqrt{Z}$ and to assign the momentum of the relevant physical particle to the external line, with $p_k = q_k$ for particles in the initial state, and $p_k = -q_k$ for particles in the final state. A factor $1/(\sqrt{2\omega(q)}(2\pi)^3)$ also remains associated with the external line in the S-matrix element.

Diagrams which are 1PI on the external lines contain internal lines which are associated with the *bare propagator*, $i\Delta_F(x-y, m_0)$. An internal line which connects two vertices with coordinates v_1, v_2 corresponds to a factor

$$i\Delta_F(v_2 - v_1) = \frac{i}{(2\pi)^4} \int d^4p \, \frac{e^{-ipv_2} \, e^{ipv_1}}{p^2 - m^2 + i\epsilon} . \tag{9.5}$$

As we have seen for the external lines, a factor e^{-ipv_2} (e^{+ipv_1}) corresponds to a momentum p which enters the vertex v_2 (exits from the vertex v_1). At each vertex, there is a factor $e^{\pm ipv}$ for each, internal or external, line associated with the vertex itself. Integration over the vertex coordinate leads to a δ-function that represents conservation of energy and momentum carried by the lines which join at each vertex.

$$\int d^4x \, e^{ix \sum(\pm p_i)} = (2\pi)^4 \delta^4 \left(\sum(\pm p_i) \right) . \tag{9.6}$$

For a diagram with n vertices, the n δ-functions allow elimination of the integration over $(n-1)$ internal momenta, since we have to extract a δ-function which corresponds to conservation of 4-momentum of the overall reaction and therefore does not depend on the internal momenta.

9.2 FEYNMAN RULES FOR THE S-MATRIX ELEMENTS

We summarise the rules for diagrams which represent the S-matrix elements.

- The connected Feynman diagrams which are 1PI on the external lines should be considered;

- amplitudes are associated with the different elements of each diagram (vertices, internal lines, external lines) as follows:

$$\text{Vertex}: \quad \frac{-i\lambda}{4!}(2\pi)^4 \, \delta^4(\sum p_{\text{in}} - \sum p_{\text{out}}) ,$$

$$\text{Internal line}: \quad \frac{i}{(2\pi)^4} \int \frac{d^4p}{p^2 - m_0^2 + i\epsilon} , \tag{9.7}$$

$$\text{External line}: \quad \frac{\sqrt{Z}}{\sqrt{2\omega_p (2\pi)^3}} .$$

$$\tag{9.8}$$

It remains to define the combinatorial coefficients to assign to each diagram, of which we have seen specific examples in the preceding section, but we will not discuss this problem in general for the case of $\lambda\phi^4$ theory.

As an example, we consider the amplitude for the scattering of two particles. We must start with the diagrams of Figure 9.4. We will consider p_1, p_2 as initial (incoming) particles, and p_3, p_4 as final (outgoing) particles. The value of the first diagram is simply

$$D_1 = (2\pi)^4 \delta^4(p_1 + p_2 - p_3 - p_4) \, Z^2(-i\lambda) \prod \frac{1}{\sqrt{2\omega_{q_i}}} \; . \tag{9.9}$$

For the second diagram we find the result

$$D_2 = C(-i\lambda)^2 \, Z^2 \prod \frac{1}{\sqrt{2\omega_{q_i}}} \int \frac{d^4 q_1}{(2\pi)^4} \frac{d^4 q_2}{(2\pi)^4} \frac{i}{q_1^2 - m_0^2 + i\epsilon} \frac{i}{q_2^2 - m_0^2 + i\epsilon}$$

$$\times (2\pi)^4 \, \delta^4(p_1 + p_2 - q_1 - q_2)(2\pi)^4 \, \delta^4(q_1 + q_2 - p_3 - p_4) \, , \tag{9.10}$$

where q_1 and q_2 are the momenta associated with the internal lines and C represents a combinatorial factor in which we have also incorporated the $1/4!$ factors. The two δ-functions can be combined into one which guarantees conservation of momentum and energy between the incoming and outgoing particles, and a second which eliminates one of the integrations, by fixing $q_2 = p_1 + p_2 - q_1$. This leads to

$$D_2 = (2\pi)^4 \, \delta^{(4)}(p_1 + p_2 - p_3 - p_4) \, Z^2 \, C(-i\lambda)^2 \tag{9.11}$$

$$\times \prod \frac{1}{\sqrt{2\omega_{q_i}}} \int \frac{d^4 q_1}{(2\pi)^4} \frac{i}{q_1^2 - m_0^2 + i\epsilon} \frac{i}{(p_1 + p_2 - q_1)^2 - m_0^2 + i\epsilon} \; .$$

We note that the residual integral is divergent; for large values of q_1 it behaves like $\int d^4 q/q^4$, a logarithmic divergence. Shortly, we will talk about divergences in perturbation theory, and how to cure them, in the case of electrodynamics. At the same time, we will also discuss calculation of the renormalisation constant Z and how to eliminate the *bare* mass, m_0, and coupling constant, e_0, in a systematic way in favour of physically measurable constants, the *renormalised* quantities.

QUANTUM ELECTRODYNAMICS

CONTENTS

In this chapter we will apply the methods developed in previous chapters to the construction of a perturbation theory applicable to quantum electrodynamics, the theory of interactions between electrons, described by the Dirac field, and the electromagnetic field. This system is described by the Lagrangian

$$\mathcal{L} = \bar{\psi} \left[i\gamma^{\mu}(\partial_{\mu} - ie_0 A_{\mu}) - m_0 \right] \psi - \frac{1}{4} F^{\mu\nu} F_{\mu\nu} \ . \tag{10.1}$$

As we know [1, 2], a relevant characteristic of (10.1) is its invariance under gauge transformations, which are simultaneous transformations of the electromagnetic field and the field of the electron,

$$A_{\mu}(x) \rightarrow A_{\mu}(x) + \partial_{\mu} f(x) \ , \tag{10.2}$$

$$\psi(x) \rightarrow e^{ie_0 f(x)} \psi(x) \ . \tag{10.3}$$

The gauge invariance establishes a tight link between the field of the electron and the field of the photon. An invariance under transformations of the electron field by a phase factor which depends arbitrarily on location—equation (10.3)—would not be possible without the electromagnetic field. Gauge invariance is therefore not a secondary characteristic of the electromagnetic field, but its reason for existence. A consequence of gauge invariance is the existence of a symmetry under a transformation of the field of the electron with a constant phase factor,

$$\psi(x) \rightarrow e^{i\alpha} \psi(x) \ .$$

This is a "global" symmetry (in contrast to "local" symmetry—the full gauge symmetry) from which follows, through Noether's theorem, the conservation of current,

$$\partial_\mu(\bar{\psi}\gamma^\mu\psi) = 0 , \tag{10.4}$$

which is therefore itself a consequence of the gauge invariance.

In this chapter we adopt the Feynman gauge, in which the Lagrangian is written

$$\mathcal{L} = \bar{\psi}[i\gamma^\mu(\partial_\mu - ie_0 A_\mu) - m_0]\psi - \frac{1}{2}\partial_\nu A_\mu \partial^\nu A^\mu . \tag{10.5}$$

To construct a perturbation theory, we can divide the Lagrangian (10.5) into two parts according to

$$\mathcal{L} = \mathcal{L}_0 + \mathcal{L}_1 , \tag{10.6}$$

where $\mathcal{L}_0$ describes free fields, and is therefore completely solvable, while $\mathcal{L}_1$ is the interaction term

$$\mathcal{L}_0 = \bar{\psi}(i\gamma^\mu\partial_\mu - m_0)\psi - \frac{1}{2}\partial_\nu A_\mu \partial^\nu A^\mu ,$$
$$\mathcal{L}_1 = e_0(\bar{\psi}\gamma^\mu\psi)A_\mu .$$

As we will see later, m_0 is not the physical mass of the electron, but the value that the mass would take in the absence of interactions, i.e. for $e_0 = 0$. We indicate with δm the difference

$$m = m_0 + \delta m .$$

The term δm can be considered as the energy of the electric (and magnetic) field produced by the charge (and magnetic moment) of the electron. Since we will use perturbation theory to calculate the transition amplitudes associated with collision processes—the S-matrix elements—it is convenient to use a different definition of $\mathcal{L}_0$ and $\mathcal{L}_1$, such that $\mathcal{L}_0$ describes electrons of mass m, the mass of the particles described by the complete Lagrangian $\mathcal{L}$. Therefore we set

$$\mathcal{L}_0 = \bar{\psi}(i\gamma^\mu\partial_\mu - m)\psi - \frac{1}{2}\partial_\nu A_\mu \partial^\nu A^\mu , \tag{10.7}$$
$$\mathcal{L}_1 = e_0(\bar{\psi}\gamma^\mu\psi)A_\mu + \delta m(\bar{\psi}\psi) . \tag{10.8}$$

The term $\delta m(\bar{\psi}\psi)$ in $\mathcal{L}_1$ is called the mass *counterterm*. While the mass of the electron is modified by interactions, the mass of the photon, as we will see later, is *protected* by the gauge invariance and remains zero. Similarly, the parameter e_0 is not, as we will see, the measured electric charge of the electron, e. This is obtained from e_0 by applying the corrections which arise from the presence of interactions. We note that neither e_0 nor m_0 are observable quantities, but parameters of the Lagrangian. The renormalisation process, which will be discussed in Chapter 11, consists essentially of expressing all the results of the theory in terms of the measurable quantities e and m.

The construction of the perturbative series for the S-matrix proceeds following the same steps illustrated in the case of the scalar field:

1. Construct the perturbative series for the generating functional of the Green's functions in terms of Feynman diagrams.

2. Determine the reduction formulae, which relate the S-matrix elements to the Green's functions.

3. Construct the perturbative series for the S-matrix.

10.1 FEYNMAN DIAGRAMS FOR THE GENERATING FUNCTIONAL

The generating functional for electrodynamics depends on three auxiliary functions: $J(x)$ and $\bar{J}(x)$ for the electron field, and J_μ for the photon. In the limit $e \to 0$, i.e. in the absence of interactions,[1] the generating functional Z is simply the product of the generating functionals for the electrons and the photons,

$$Z^0[J, \bar{J}, J_\mu] = Z^0[J, \bar{J}] \, Z^0[J_\mu] \, ,$$

$$Z^0[J, \bar{J}] = \exp\left(-i \int d^4x \, d^4y \, \bar{J}(x) \, S_F(x-y) \, J(y)\right) \, , \qquad (10.9)$$

$$Z^0[J_\mu] = \exp\left(\frac{i}{2} \int d^4x \, d^4y \, J^\mu(x) \Delta_F(x-y) \, J_\mu(y)\right) \, ,$$

where $\Delta_F(x-y)$ is the Feynman propagator associated with a real scalar field with mass $m = 0$, $\Delta_F(x-y; 0)$.

In the presence of the interaction Lagrangian (10.8), we can express the generating functional as (see Section 8.1)

$$Z[J, \bar{J}, J_\mu] = e^V Z^0[J, \bar{J}, J_\mu] = \sum_{n=0}^{\infty} \frac{V^n}{n!} Z^0[J, \bar{J}, J_\mu] \, . \qquad (10.10)$$

The vertex operator, V, is simply $i \int d^4x \, \mathcal{L}_1$, which translated with the correspondence rules (6.59) and (5.13) becomes

$$i\mathcal{L}_1(x) = ie \left(-i\frac{\delta}{\delta J(x)}\right) \gamma^\mu \left(i\frac{\delta}{\delta \bar{J}(x)}\right) \left(i\frac{\delta}{\delta J^\mu(x)}\right) \qquad (10.11)$$

$$+ i\delta m \left(-i\frac{\delta}{\delta J(x)}\right) \left(i\frac{\delta}{\delta \bar{J}(x)}\right) \, .$$

---

[1] Recall that in this case $\delta m = 0$.

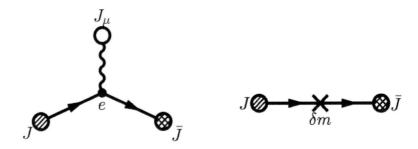

Figure 10.1   The two diagrams with a single vertex.

This expression may seem more complicated than what was obtained in the scalar case but is actually simpler, at least in one respect. Each of the two terms in $\mathcal{L}_1$ has only one derivative of each type, and this considerably simplifies the combinatorial analysis.

The term V^n in the expansion of Z can be calculated directly, and results in a series of contributions representable by Feynman diagrams. As in the scalar case, each derivative can act on Z^0, and in this case brings down a "line with a bubble", i.e. a term $\Delta_F J_\mu$, $S_F J$ or $\bar{J} S_F$, or it is taken on a "line" produced by a preceding derivative, in which case the "line with a bubble" becomes an internal line. If the bubble is not captured by a subsequent derivative, what remains is an external line.

All this can be expressed in terms of diagrams. For example, to first order in V we have the two diagrams of Figure 10.1. Here we state some of the rules for their interpretation:

- Photons are represented by wavy lines and an empty bubble represents a J_μ.

- A dashed bubble represents a J and a hatched bubble represents a $\bar{J}$.

- The fermion lines have a direction indicated on the diagram, and move[2] from one J to a $\bar{J}$. An external line of type $\bar{J}$ is called "outgoing" from the diagram, one of type J "incoming".

- There are two types of vertex, one associated with the electron–photon interaction, the other with the mass counterterm, represented with an x.

We examine more carefully the correspondence between the various elements of the diagram and the elements of the result.

---

[2]This definition of direction is obviously arbitrary—we could have chosen the direction of $\bar{J}$ to J—but this is in common use.

External fermion lines. A derivative produces an external line,

$$\left(-i\frac{\delta}{\delta J(x)}\right) Z^0 = +\int d^4y\, \bar{J}(y)S(y-x)\, Z^0\ ,$$

$$\left(i\frac{\delta}{\delta \bar{J}(y)}\right) Z^0 = +\int d^4x\, S(y-x)J(x)\, Z^0\ . \tag{10.12}$$

Internal fermion lines. The result of applying a second derivative depends on the order (as we expect since the derivatives *anticommute*),

$$\left(i\frac{\delta}{\delta \bar{J}(y)}\right)\left(-i\frac{\delta}{\delta J(x)}\right) Z^0 = iS(y-x)\, Z^0\ , \tag{10.13}$$

but, changing the order,

$$\left(-i\frac{\delta}{\delta J(x)}\right)\left(i\frac{\delta}{\delta \bar{J}(y)}\right) Z^0 = -iS(y-x)\, Z_0\ . \tag{10.14}$$

External photon lines.

$$\left(i\frac{\delta}{\delta J^\mu(x)}\right) Z^0 = -\int d^4y\, \Delta_F(x-y)\, J_\mu(y)\, Z^0\ . \tag{10.15}$$

Internal photon lines. Keeping in mind that

$$\frac{\delta}{\delta J^\nu(x)}J_\mu(x) = g^{\mu\nu}\delta^4(x-y)\ ,$$

we find

$$\left(i\frac{\delta}{\delta J^\nu(x)}\right)\left(i\frac{\delta}{\delta J^\mu(y)}\right) Z^0 = -ig^{\mu\nu}\Delta_F(x-y)\, Z^0\ . \tag{10.16}$$

We can directly formulate the rules for the interpretation of diagrams in momentum space.

We recall the expression for Δ_F, (9.5), and S_F, (6.56), from which it follows that for an internal photon and fermion line we have respectively

$$-ig_{\mu\nu}\Delta_F(x-y) = \frac{-ig_{\mu\nu}}{(2\pi)^4}\int d^4p\, \frac{1}{p^2+i\epsilon}\, e^{-ipx}e^{ipy}, \tag{10.17}$$

$$iS_F(x-y) = \frac{i}{(2\pi)^4}\int d^4p\, e^{-ipx}e^{ipy}\frac{\not{p}+m}{p^2-m^2+i\epsilon}\ . \tag{10.18}$$

For an external line that finishes in a J [see the second of (10.12)] we have

$$-\int d^4x\, S(v-x)J(x)$$

$$= \frac{i}{(2\pi)^4}\int d^4p\, e^{-ipv}\frac{\not{p}+m}{E^2-\vec{p}^2-m^2+i\epsilon}\int d^4x\, e^{ipx}\, i\, J(x)\ , \tag{10.19}$$

where v is the coordinate of one of the vertices in the diagram. Similarly for a line of type $\bar{J}$

$$- \int d^4x \, \bar{J}(x) S(x - v)$$

$$= \frac{i}{(2\pi)^4} \int d^4p \, e^{ipv} \int d^4x \, e^{-ipx} \, i \, \bar{J}(x) \frac{\not{p} + m}{p^2 - m^2 + i\epsilon} \; . \quad (10.20)$$

Finally, for an external photon line

$$\int d^4x \, \Delta_F(v - x) \, J_\mu(x)$$

$$= \frac{-i}{(2\pi)^4} \int d^4p \, \frac{1}{p^2 + i\epsilon} \, e^{-ipv} \int d^4x \, e^{ipx} \, i \, J_\mu(x) \; . \quad (10.21)$$

The various factors $e^{\pm ipv}$ merge in the vertices in which the lines terminate, and are integrated, producing a factor $(2\pi)^4 \delta^4(\sum p_i)$ which guarantees conservation of 4-momentum at every individual vertex.

10.2 TWO-POINT FUNCTIONS

We can easily establish a spectral representation for the two-point functions of the fermion fields and of the vector potential, similar to that found for the scalar field.

We explicitly consider the fermion field and the two-point function

$$i[G_F(x)]_{\alpha\beta}(x) = \langle 0|T\left[\psi_\alpha(x)\bar{\psi}_\beta(0)\right]|0\rangle \; , \quad (10.22)$$

and analyse the case $x^0 > 0$. As we did previously, we insert a complete system of states into the product of the fields, separating the contribution of the one-particle state, which in this case is an electron created from the vacuum by $\bar{\psi}$:

$$i[G_F(x)]_{\alpha\beta} \quad (10.23)$$

$$= \sum_s \int d^3p \, e^{-ipx} \, \langle 0|\psi_\alpha(0)|\mathbf{p}, s\rangle\langle\mathbf{p}, s|\bar{\psi}_\beta(y)|0\rangle + \langle 0|\psi_\alpha(x)\bar{\psi}_\beta(0)|0\rangle_{(2+)} \; .$$

The single particle contribution is easily calculated (see [?, ?] Appendix A for the case of the scalar field).

$$\langle 0|\psi_\alpha(x)\bar{\psi}_\beta(0)|0\rangle_1 = \int \frac{d^3p}{(2\pi)^3} \frac{m}{E(p)} \, e^{-ipx} \, Z_2 \sum_s (u_s(p))_\alpha (\bar{u}_s(p))_\beta$$

$$= \frac{Z_2}{(2\pi)^3} \int \frac{d^3p}{2E(p)} e^{-ipx} (\not{p} + m)_{\alpha\beta} \quad (10.24)$$

$$= (i\not{\partial} + m)_{\alpha\beta} \left[\frac{Z_2}{(2\pi)^3} \int \frac{d^3p}{2E(p)} e^{i\mathbf{p}\cdot\mathbf{x}} e^{-iE(p)t} \right] \; .$$

Calculating the T-product restricted to the single particle for $x^0 < 0$ in a similar way, we find

$$-\langle 0|\bar{\psi}_\beta(0)\psi_\alpha(x)|0\rangle_1 = -\int \frac{d^3p}{(2\pi)^3}\frac{m}{E(p)}\,e^{+ipx}\,Z_2(p^2)\sum_s (v_s(p))_\alpha(\bar{v}_s(p))_\beta$$

(10.25)

$$= -\frac{Z_2}{(2\pi)^3}\int \frac{d^3p}{2E(p)}e^{+ipx}(\not{p}-m)_{\alpha\beta}$$

$$= (i\not{\partial}+m)_{\alpha\beta}\frac{Z_2}{(2\pi)^3}\int \frac{d^3p}{2E(p)}\,e^{i\mathbf{p}\cdot\mathbf{x}}e^{+iE(p)t}\,.$$

(10.26)

Putting everything together, we therefore arrive at

$$\langle 0|T\left[\psi_\alpha(x)\bar{\psi}_\beta(0)\right]|0\rangle_1$$

$$= (i\not{\partial}+m)_{\alpha\beta}\left\{\frac{Z_2}{(2\pi)^3}\int \frac{d^3p}{2E(p)}e^{i\mathbf{p}\cdot\mathbf{x}}\left[\theta(x^0)e^{-iE(p)t}+\theta(-x^0)e^{+iE(p)t}\right]\right\}$$

$$= Z_2 iS_F(x,m)_{\alpha\beta}$$

, (10.27)

where S_F is the propagator of the free fermion with mass equal to the physical mass of the electron, m.

For the terms with intermediate states with two or more particles, we proceed as in Appendix A of [2] and find

$$\langle 0|T\left[\psi(x)\bar{\psi}(0)\right]|0\rangle_{(2+)} = \int dM^2\left[i\not{\partial}\rho_1(M^2)+M\rho_2(M^2)\right]_{\alpha\beta}i[\Delta_F(x,M)]\,.$$

(10.28)

We note that in this case it is necessary to introduce two spectral functions, defined by the relation

$$\sum_{n,2+}(2\pi)^4\delta^{(4)}(P_n-p)\langle 0|\psi_\alpha(0)|n\rangle\langle n|\bar{\psi}_\beta(0)|0\rangle$$

$$= (2\pi)\left[\not{p}\rho_1(M^2)+M\rho_2(M^2)\right]_{\alpha\beta}\,.$$

(10.29)

The spectral functions, being Lorentz invariant, are functions of p^2, which is fixed at the value $p^2 = M^2$.

Putting everything together, we find the spectral representation of the electron propagator in the form

$$i[G_F(x)]_{\alpha\beta}(x) = \langle 0|T\left[\psi_\alpha(x)\bar{\psi}_\beta(0)\right]|0\rangle = Z_2 i[S_F(x)]_{\alpha\beta}$$

(10.30)

$$+\int dM^2\left[i\not{\partial}\rho_1(M^2)+M\rho_2(M^2)\right]_{\alpha\beta}i[\Delta_F(x,M)]\,,$$

from which it follows that the Fourier transform of the two-point function has a pole at $\not{p} = m$ with residue iZ_2.

We leave the derivation of the spectral representation of the photon field to the reader. The result is

$$i[G_F(x)]^{\mu\nu}(x) = \langle 0|T\left[A^\mu(x)A^\nu(0)\right]|0\rangle \tag{10.31}$$

$$= -g^{\mu\nu}\left[Z_3 i\Delta_F(x, M=0) + \int dM^2\sigma_3(M^2)i\Delta_F(x, M^2)\right]$$

$$+ \cdots ,$$

where the dots denote terms proportional to partial derivatives of functions of x, with respect to x^μ and/or x^ν. These terms can be omitted from the propagator because of conservation of the currents, by which the propagator is multiplied in the calculation of the physical amplitudes.

10.3   REDUCTION FORMULAE

The arguments which led to the reduction formula (7.51) can be applied directly to electrodynamics, with the only complication originating from the spin of the electron and the photon.

As before we can define the "in" and "out" states, which are now states with a specified number of electrons, positrons and photons, with assigned values of momenta, spin and polarisation. These states are created from the vacuum, or destroyed, by the "in" and "out" fields, which are obtained in the limit of $t \to \pm\infty$ and involve the renormalisation constants of the electron, Z_2, and of the photon, Z_3[3]. For example

$$\lim_{t\to+\infty} \psi(x) = \sqrt{Z_2}\psi_{out}(x) ,$$

$$\lim_{t\to+\infty} A^\mu(x) = \sqrt{Z_3}A^\mu_{out}(x) . \tag{10.32}$$

The "in" and "out" fields are expanded in plane waves, with coefficients which are the creation and annihilation operators, which in their turn can be obtained by projecting the fields onto the system of plane waves and of spinors or appropriate polarisation vectors.

For example, for electrons and positrons, we write (omitting the "in" and "out" suffixes for brevity):

$$\psi_\alpha(x) = \int d^3p \sqrt{\frac{m}{E(\mathbf{p})(2\pi)^3}}\left[a_r(\mathbf{p})[u_r(\mathbf{p})]_\alpha\, e^{-ipx} + [c_r(\mathbf{p})]^\dagger[v_r(\mathbf{p})]_\alpha\, e^{+ipx}\right] ,$$

$$\bar\psi_\alpha(x) = \int d^3p \sqrt{\frac{m}{E(\mathbf{p})(2\pi)^3}}\left[b_r(\mathbf{p})[\bar v_r(\mathbf{p})]_\alpha\, e^{-ipx} + [a_r(\mathbf{p})]^\dagger[\bar u_r(\mathbf{p})]_\alpha\, e^{+ipx}\right] ,$$

$$\tag{10.33}$$

---

[3]This is standard notation, according to which Z_1 is the vertex renormalisation constant.

and we find

$$\sqrt{\frac{E(\mathbf{p})(2\pi)^3}{m}}\, a_r(\mathbf{p}) = \int d^3x\, e^{+ipx}\, [\bar{u}_r(\mathbf{p})]_\alpha [\gamma^0]_{\alpha\beta} \psi_\beta(x) \qquad (e^- \text{ annih.}),$$

$$\sqrt{\frac{E(\mathbf{p})(2\pi)^3}{m}}\, [c_r(\mathbf{p})]^\dagger = \int d^3x\, e^{-ipx}\, [\bar{v}_r(\mathbf{p})]_\alpha [\gamma^0]_{\alpha\beta} \psi_\beta(x) \qquad (e^+ \text{ crea.}),$$

$$\sqrt{\frac{E(\mathbf{p})(2\pi)^3}{m}}\, [a_r(\mathbf{p})]^\dagger = \int d^3x\, e^{-ipx}\, [\bar{\psi}(\mathbf{p})]_\alpha [\gamma^0]_{\alpha\beta} [u_r(\mathbf{p})]_\beta \qquad (e^- \text{ crea.}),$$

$$\sqrt{\frac{E(\mathbf{p})(2\pi)^3}{m}}\, c_r(\mathbf{p}) = \int d^3x\, e^{+ipx}\, [\bar{\psi}(\mathbf{p})]_\alpha [\gamma^0]_{\alpha\beta} [v_r(\mathbf{p})]_\beta \qquad (e^+ \text{ annih.}).$$

$$(10.34)$$

We leave the reader to prove, following the procedure described in Section 7.4 and using (10.34) and the Dirac equation, the basic reduction formula of the electron in the initial state:

$$S_{if} = \langle k'r', p's'; out | kr, ps; in \rangle \qquad (10.35)$$

$$= \langle k'r', p's'; out | a^\dagger_{in}(p, s) | kr; in \rangle$$

$$= i\, \sqrt{\frac{m}{(2\pi)^3 E(\mathbf{p}) Z_2}}$$

$$\times \int d^4x_1 \langle k'r', p's'; out | \overline{\psi}(x_1) | k, r; in \rangle (i\overleftarrow{\partial}_1 + m) u_s(\mathbf{p})\, e^{-ipx_1}\, .$$

Equation (10.35) and the analogous expression for the electron in the final state allow the S-matrix element to be obtained from the corresponding Green's function.

Similar relations hold for the one-photon states, for example

$$S_{if} = \langle k'r', p's'; out | kr, ps; in \rangle \qquad (10.36)$$

$$= \langle k'r', p's'; out | a^\dagger_{in}(k, r) | ps; in \rangle$$

$$= \frac{i}{\sqrt{(2\pi)^3 E(\mathbf{p}) Z_3}} \int d^4x_1 \langle k'r', p's'; out | A^\mu(x_1) | p, s; in \rangle \overleftarrow{(\Box_1)} \epsilon^{(r)}_\mu(\mathbf{k}) e^{-ikx_1}.$$

Comparing these relations with those that hold for a scalar field, we see that the differences in the case of electrodynamics are minimal. Beyond what is due to the presence of spinor or polarisation vectors, we must only take into account the different multiplicative factors in the case of fermions, equation (10.35), and that of bosons, (10.36).

In conclusion, we can write the reduction formula for a process with any number of particles in the initial and final states in a (very) schematic form as [to be compared with (7.51)]

$$\langle f|S|i\rangle = \langle f; out|i; in\rangle = \left(\prod_{\text{bos}} \sqrt{\frac{1}{Z_3\, 2\omega_q(2\pi)^3}}\right) \left(\prod_{\text{ferm}} \sqrt{\frac{m}{Z_2\, E(\mathbf{q})(2\pi)^3}}\right)$$

$$\prod_{e^-\text{ out},\, e^+\text{ in}} [\bar{u}(\slashed{q}_k - m)]_\alpha \cdots [\bar{v}(\slashed{q}_s - m)]_\delta \cdots \prod_{\text{bos}} \cdots \epsilon^{(s)}_\mu(q_\ell)q_\ell^2 \cdots$$

$$\int \prod_{\text{all}} [d^4 x] \prod_{\text{final}} e^{iq_k x_k} \prod_{\text{initial}} e^{-iq_k x_k} \langle 0|T\left(\psi_\alpha \cdots \bar\psi_\beta \cdots \bar\psi_\gamma \cdots \psi_\delta \cdots A^\mu\right)|0\rangle$$

$$\prod_{e^-\text{ in},\, e^+\text{ out}} [(\slashed{q}_l + m)u]_\beta \cdots [(\slashed{q}_t + m)v]_\gamma \cdots , \tag{10.37}$$

where the various factors or manipulations apply to the type of particles denoted. The Green's function must contain a field projected onto the appropriate spinor or polarisation vector corresponding to each initial or final particle:

$u_r(\mathbf{q})$	initial fermion,
$v_r(\mathbf{q})$	final anti-fermion,
$\bar{v}_r(\mathbf{q})$	initial anti-fermion,
$\bar{u}_r(\mathbf{q})$	final fermion,
$\epsilon^\mu_r(\mathbf{q})$	initial or final photon.

10.4   FEYNMAN DIAGRAMS FOR THE S-MATRIX

As we did in the case of the scalar field, we can pass directly from diagrams which describe the generating functional to those which describe S-matrix elements.

We must restrict ourselves to Green's functions which correspond to *connected* diagrams, with external lines associated with the initial and final particles.

By inserting the Green's functions into the reduction formulae and taking residues at the poles corresponding to the external momenta, we completely eliminate the propagators of the external lines and we obtain a factor $\sqrt{Z_2}$ for each electron or positron and a factor $\sqrt{Z_3}$ for every photon.

In this way we arrive at the general expression for the S-matrix element,

$$\langle f|S|i\rangle \tag{10.38}$$

$$= \left(\prod_{\text{ferm.}} \sqrt{2m}\right) \left(\prod_{\text{all}} \sqrt{\frac{Z}{2\omega_q(2\pi)^3}}\right) (2\pi)^4\, i\, \delta^4\Big(\sum_{\text{in}} q_i - \sum_{\text{fin}} q_i\Big) \mathcal{M}_{fi}\,.$$

As we will see, the $\sqrt{Z}$ factors associated with the external lines disappear when the renormalisation is carried out. We have shown the kinematic factors and also the factor

$$(2\pi)^4 \ i \ \delta^4(\sum_{in} q_i \ - \ \sum_{fin} q_i) \ ,$$

which guarantees conservation of energy and momentum. The factor "i" follows the convention according to which the S-matrix is expressed as $\mathbf{1} + iT$, where T is the transition matrix, or T- matrix.

In terms of diagrams we must limit ourselves to connected diagrams which are, furthermore, *one-particle irreducible on the external lines*. We will therefore have

$$(2\pi)^4 i \ \delta^4(\sum_{in} q_i \ - \ \sum_{fin} q_i)\mathcal{M}_{fi} = \sum_i D_i \ . \tag{10.39}$$

Feynman rules. To better define the rules for the calculation of diagrams we must say something about their structure. First we observe that [see equation (10.11)] at every vertex, two fermion lines meet, one of type "$\bar{J}$" produced by the derivative $\delta/\delta J$, which we can consider *outgoing* from the vertex, the other of type "J", produced by $\delta/\delta\bar{J}$, which we can consider *incoming*. As a consequence of (10.13), an outgoing line from a first vertex becomes incoming at a second. Therefore if, starting from one vertex, we follow a fermion line along its direction from vertex to vertex, two situations can arise: either we arrive at an outgoing line of the diagram (or following it in reverse direction, an incoming line of the diagram), or we return to the starting point. In the first case we have an open line, in the second we have a closed line, or *loop*. Since we are dealing with fermions, we must pay attention to the signs. In the two terms of the vertex (10.11), the derivative $\delta/\delta\bar{J}$ (incoming line) is to the right of $\delta/\delta J$ (outgoing line). For an open line we can order the V operators which contribute so they are adjacent and that the contractions always arise between the factor $\delta/\delta\bar{J}$ at one of the two vertices and the $\delta/\delta J$ in the one to the right, a situation which we will call *normal*. For example, for an open line with three vertices in x, y, z,

$$\left[\left(-i\frac{\delta}{\delta J(x)}\right)\gamma^\mu\left(i\frac{\delta}{\delta\bar{J}(x)}\right)\right]\left[\left(-i\frac{\delta}{\delta J(y)}\right)\gamma^\nu\left(i\frac{\delta}{\delta\bar{J}(y)}\right)\right]$$
$$\left[\left(-i\frac{\delta}{\delta J(z)}\right)\gamma^\sigma\left(i\frac{\delta}{\delta\bar{J}(z)}\right)\right] \ ,$$

so that we always have the case of equation (10.13). In the case of a closed fermion line we necessarily find an odd number of cases in which two derivatives are contracted in the reverse order, i.e. $\delta/\delta\bar{J}$ to the right of $\delta/\delta J$. In these cases equation (10.14) applies, which has the opposite sign with respect to (10.13). For example, for a closed line with two vertices

$$\left[\left(-i\frac{\delta}{\delta J(x)}\right)\gamma^\mu\left(-i\frac{\delta}{\delta\bar{J}(x)}\right)\right]\left[\left(i\frac{\delta}{\delta J(y)}\right)\gamma^\nu\left(-i\frac{\delta}{\delta\bar{J}(y)}\right)\right] \ ,$$

the internal contraction is in the normal order, the external contraction in the opposite order. Therefore we can always use equation (10.13) adding a further (-1) factor for each closed line. We have omitted notation for the spinor indices, since the adjacent indices are summed. In the closed line we also have a sum between the first and last index, a trace; the example which we have given corresponds, including the (-1) factor, to

$$(-1)Tr\left[\gamma^{\mu}iS_F(x-y)\gamma^{\nu}iS_F(y-x)\right] \ .$$

For the open lines, the corresponding expression is written from left to right *beginning with the outgoing line*, which can represent a fermion in the final state, or an antifermion in the initial state, and finishing with the incoming line—an initial fermion or an antifermion in the final state. Therefore for an open fermion line the rule is to *go back along the fermion line, starting from the end*. In the case of a closed loop, the rule of *going back along the fermion line* still holds, but the starting point can be any vertex, since the trace is invariant for circular permutations. In view of this, the elements of a diagram are:

electron–photon vertex	$ie\,(2\pi)^4\,\delta^4(\sum q_i)\,\gamma^{\mu}$,
mass counterterm vertex	$i\delta m\,(2\pi)^4\,\delta^4(\sum q_i)$,
internal fermion line	$\dfrac{1}{(2\pi)^4}\displaystyle\int d^4p\,\dfrac{i(\not{p}+m)}{p^2-m^2+i\epsilon}$,
internal photon line	$\dfrac{1}{(2\pi)^4}\displaystyle\int d^4p\,\dfrac{-ig_{\mu\nu}}{p^2+i\epsilon}$,
external photon line	ϵ_{μ} ,
incoming fermion line	$\begin{cases}\text{initial fermion} & u_r(\mathbf{q}) \\ \text{final antifermion} & v_r(\mathbf{q})\end{cases}$,
outgoing fermion line	$\begin{cases}\text{final fermion} & \bar{u}_r(\mathbf{q}) \\ \text{initial antifermion} & \bar{v}_r(\mathbf{q})\end{cases}$,
closed fermion line	add a factor(-1) .

(10.40)

It is important to note that the *relative* signs between different diagrams which contribute to the same process are physically relevant. This is the case of the (-1) factor associated with closed fermion lines, since different diagrams for the same process can have a different number of closed lines.

By way of an exercise, in Figure 10.2 we show the second-order Feynman diagrams of the perturbative expansion related to the four-point function

$$G(x_1,x_2;y_1,y_2) = <0|T\left[\psi(x_1)\bar{\psi}(x_2)A^{\mu}(y_1)A^{\nu}(y_2)\right]|0> \ . \qquad (10.41)$$

All the diagrams must have an open fermion line and can be ordered according to the number of vertices $V = 1, 2, 3$ which fall on that line.

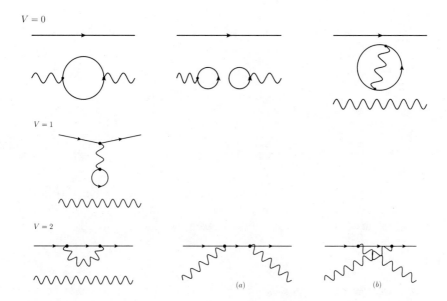

$V = 0$

$V = 1$

$V = 2$

(a) (b)

Figure 10.2 Second-order perturbative diagrams with number of vertices $V = 0, 1, 2$ along the open fermion lines. The diagrams which contribute to a scattering process are the connected diagrams (a) and (b).

We note that second-order corrections to the propagation of the photon (first diagram with $V= 0$) and the electron (first diagram with $V= 2$) appear as connected sub-diagrams. In the third diagram with $V= 0$, the first perturbative correction to the *vacuum–vacuum* transition amplitude appears as a connected sub-diagram. The second figure with $V= 0$ and the figure with $V= 1$ both contain an amplitude related to the transition of a photon in the vacuum. This amplitude vanishes due to a particular symmetry of QED, charge conjugation[4] [1].

The diagrams which contribute to a scattering process, for example Compton scattering,

$$\gamma(k, r) + e(p, s) \to \gamma(k', r') + e(p', s') , \qquad (10.42)$$

where (k, r), (p, s), (k', r') and (p', s') denote 4-momentum, spin and polarisation of the particles in, respectively, the initial and final states, are only the connected diagrams, denoted by (a) and (b).

For Compton scattering, the invariant amplitude of equation (10.39), obtained by using the Feynman rules derived in this chapter, takes the form

---

[4]The vanishing of all amplitudes with an odd number of external photons (one in our case) is called Furry's theorem.

$$\mathcal{M}_{fi} = \mathcal{M}_{fi}^{(a)} + \mathcal{M}_{fi}^{(b)} , \qquad (10.43)$$

where

$$\mathcal{M}_{fi}^{(a)} = e^2 \ \bar{u}_{s'}(p')\gamma_\mu\epsilon_{r'}^\mu(k')\frac{i}{p+k-m}\gamma_\nu\epsilon_r^\nu(k)u_s(p) ,$$

$$\mathcal{M}_{fi}^{(b)} = e^2 \ \bar{u}_{s'}(p')\gamma_\nu\epsilon_{r'}^\nu(k)\frac{i}{p-k'-m}\gamma_\mu\epsilon_r^\mu(k')u_s(p) . \qquad (10.44)$$

Note that, for simplicity, in the above equations we have assumed real polarisation vectors and we have left the spin indices implied. The details of the calculations necessary to obtain (10.44) are given in Appendix D.

10.5   COMBINATORIALS

In short, *there are no combinatorial problems*. For the diagrams of interest, i.e. all connected diagrams other than those of the vacuum–vacuum type, the factor $1/n!$ which appears in the expansion of Z in powers of the differential operator V [equation (10.10)] is balanced exactly. The reason is that in a connected diagram every vertex has a unique role compared other vertices. Let us suppose that we have proved this result, and that we have assigned to the n vertices "roles" $\{r_1, r_2, \cdots r_n\}$. The diagram with n vertices will be produced by the term $V^n/n!$, and there exist $n!$ ways of assigning the roles to the n copies of V. This exactly cancels the factor $1/n!$.

It remains to show that in every diagram the roles of the vertices are all different. To do this it is sufficient to show that an algorithm exists to assign to each vertex a number of ascending order, for example:

Step 1. We suppose that there are $a > 0$ open fermion lines and $l \geq 0$ closed fermion lines. First we order the open lines, each of which has an identity defined by the incoming and outgoing momenta. If $A_1, A_2, \cdots A_a$ are the open lines, we can open the list of vertices by first putting those of A_1, beginning (so as not to lose the training) from the end of the line, then that of A_2, and so on. In this way we have a list which contains all the vertices of the open lines, each with its numerical position.

Step 2. If $l > 0$, the list must still be completed with the vertices with l closed lines. At least some of these will be connected by a photon to one of the vertices already in the list, otherwise the diagram would be disconnected, and from these we choose as next in the list the one connected to the vertex with the lowest numerical position. Starting from the vertex just added to the list, we add to the list the other vertices which are encountered on the same closed line, going back along the line (again, to stay in training).

If $l - 1 > 0$, the list must still be completed with the vertices with $l - 1$ closed lines. At least some of these will be connected by a photon to one of the vertices already in the list, otherwise the diagram would be disconnected, and from these we choose as next in the list the one connected to the vertex with the lowest numerical position. Starting from the vertex just added to the list, we add the other vertices which are encountered on the same closed line, going back along the line (once more, to maintain the training).

Step 3 Continue until no vertices remain outside the list.

If there are no open lines (as for example in the process $\gamma + \gamma \rightarrow \gamma + \gamma$), we order the incoming particles (necessarily photons), and open the list with the vertex at which the first of these arrives. We add the vertices which are found on the same closed line first (must we say in what order?). If $l > 0$ closed lines with non-catalogued vertices remain, return to Step 2.

At this point every vertex has its role, for example "vertex 7 is the third starting from the end of the second open line", and the proof is complete.

RENORMALISATION OF QED

CONTENTS

Renormalisation provides a solution to the problem of divergences occurring in the calculation of Feynman diagrams that contain closed *loops*, but is a more general concept that will be encountered even in the absence of divergences. The necessity for renormalisation actually arises from the existence of interactions, which means that the masses which appear in the Lagrangian are not those of the particles which the various fields describe, and that the fields themselves are not "well normalised", as it clearly appears from the presence of Z factors in the contribution of the single particle states to the relevant propagators.

In field theory the presence of divergences represents the norm, but two radically different situations can arise. The most interesting is that of renormalisable theories, which include quantum electrodynamics and more generally the Standard Theory of the fundamental interactions. To the second category belong non-renormalisable theories.

In a renormalisable theory, divergences are encountered only when trying to establish relationships between quantities which appear in the Lagrangian, in the case of QED the mass m_0 and the charge e_0, and the corresponding physical quantities, m and e. In theories of this type, the divergences can be swept under the carpet, by expressing the results in terms of physically observable quantities. If we consider the example of the electron mass, the mass correction δm, introduced in the previous chapter, becomes divergent in perturbation theory. As we saw, however, it is possible to reorganise the perturbation theory so that the δm term is cancelled exactly by an appropriate counterterm, so that it does not appear, for example, in the calculation of the S-matrix elements.

An example of a non-renormalisable theory is the Fermi theory of weak interactions. In a non-renormalisable theory, divergences are present in the calculation of any physical quantity, for example in the calculation of any S-matrix element.

In any case, the divergences appear in integrals over virtual particle momenta extending to infinity. Concerning the meaning of the divergences, two hypotheses can be advanced: the first is that these divergences are a characteristic of the perturbative method, which would not be present in a hypothetical non-perturbative approach. The second is that the theory is only a first approximation of physical reality, and not valid for extremely high momenta which, because of the uncertainty principle, correspond to extremely short distances. Conversely, for convergent integrals, the results depend on the behaviour of the integrands (and therefore the behaviour of the theory to which the integrals refer) for finite momentum values. Therefore in a renormalisable theory, once the divergences have been eliminated by a redefinition of the parameters of the theory, the results of the perturbative calculation should depend on the behaviour of the theory for finite momenta (distances not infinitesimally small) and can give a good approximation even if the theory loses its validity in the infinite momentum limit.

This reasoning, necessarily qualitative at this stage, can be made quantitative once a deeper understanding of the theory has been obtained. To give an example using research underway today, there is actually a slight discrepancy between the experimentally measured value of the anomalous magnetic moment[1] of the muon and the theoretical prediction obtained using the Standard Model. The presence of this discrepancy could be a hint of the existence of new physical phenomena in the TeV energy scale, explorable by the LHC experiments.

To hide the divergences by the redefinition of some parameters of the theory (mass, electric charge, etc.) constitutes what is called *renormalisation* and it is a procedure which requires some caution. The manipulation of quantities which are mathematically divergent is suspect and should clearly be avoided. The method of avoiding suspect manipulations consists of what is called *regularisation* of the theory. The idea is very simple: if T is the theory which is of interest (QED in our case), then construct a family of theories $T(\eta)$ which depend on a parameter η, such that:

1. In the limit $\eta \to 0$, $T(\eta) \to T$.

2. $T(\eta)$ has all the "important" properties of T. In the case of QED (or of

---

[1] The anomalous magnetic moment of a spin $\frac{1}{2}$ particle represents the deviation from the magnetic moment value (equal to one Bohr magneton) predicted by the Dirac equation. Later we will give a more precise definition of this quantity and calculate it to first order using perturbation theory.

the Standard Model) the principal concern is maintenance of the gauge invariance.

3. For $\eta \neq 0$, $T(\eta)$ does not have divergences.

In this case $T(\eta)$ is called a *regularised* version of T. In $T(\eta)$ the manipulations necessary for the renormalisation concern finite quantities, and are allowed. Only after having carried out the renormalisation is the limit $\eta \to 0$ taken. Since the renormalisation has hidden all the potential divergences, the limit is finite.

The method actually used is that of "dimensional regularisation". In simple terms this means that we can consider QED as defined by the set of Feynman diagrams which describe to each order in α the various processes, and by the rules which allow the calculation of each diagram. In the calculation of diagrams with closed loops, integrals which are logarithmically divergent appear, i.e. integrals of the type

$$\int d^4k \, I(k) \, , \quad I(k) \sim \frac{1}{k^4} \quad \text{for } |k| \to \infty \, . \tag{11.1}$$

These integrals would *not* be divergent in a space of less than four dimensions, for example in three dimensions, where we would have d^3k instead of d^4k. Dimensional regularisation consists of considering a theory described by the same diagrams as QED, with the only difference that all the integrals of the type (11.1) are not carried out in four dimensions but in $4 - \eta$ dimensions. This can be considered in some sense as an analytic continuation of the number of spatial dimensions. For $\eta > 0$ the theory is free of divergences. After having carried out the renormalisation, the limit $\eta \to 0$ can be taken. For a still elementary but more detailed illustration of the method, we refer the reader to the book by Mandl and Shaw [3].

The interest of dimensional regularisation is that it does not disturb the validity of some fundamental relations such as the Ward identity. Even if in the case of QED alternative regularisation methods exist, dimensional regularisation is the only method which has allowed perturbative treatment of theories based on non-Abelian gauge invariance, as for example in the unified description of the electromagnetic and weak interactions of the Standard Model.

The divergences which appear in the integrals for momenta $p \to \infty$ are called *ultraviolet* divergences. In QED, a second type of divergence, known as *infrared*, appears when the photon momentum, either real (emitted photon) or virtual (propagator), tends to zero. Infrared divergences have a very precise physical meaning: emission of electromagnetic waves, with an energy spectrum $dW/d\nu$ which tends to a constant for $\nu \to 0$, is already associated at the classical level, with each process, e.g. scattering, in which a charged particle changes direction in an abrupt manner. But since this radiation is composed of photons of energy $h\nu$, the spectral density of the number of photons varies like $dN/d\nu = (1/h\nu)dW/d\nu$, and tends to infinity for $\nu \to 0$.

The infrared divergence is not present in actual measured quantities since any experimental apparatus has a finite energy resolution; for example, a measurement of a scattering process is not able to distinguish the ideal scattering process from one which is accompanied by the (inevitable) emission of one or more low-energy photons. The probability (or cross section) of the scattering process, summed with those for the same process accompanied by one or more low-energy photons, is finite. The presence of infrared divergences introduces noteworthy technical complications in the comparison between theory and experiment, but no problem at the conceptual level.

In this chapter we briefly discuss the three ultraviolet divergences present in QED. They appear in the calculation of the propagators of the photon and the electron, and at the electron–electron–photon vertex. We will also discuss the Ward identity which links these corrections to the electron propagator and to the vertex. In each case the treatment in this chapter concerns corrections of order α. There exist valid proofs of renormalisability of the theory and Ward identity to all perturbative orders, but these arguments go beyond the introductory level of this course. We will conclude with an explicit calculation of the corrections of order α to the magnetic moment of the electron.

11.1 THE PHOTON PROPAGATOR

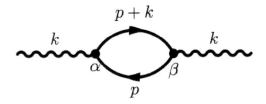

Figure 11.1   The photon propagator to order α.

The order α corrections to the photon propagator are illustrated by Figure 11.1. These corrections are added to the zeroth order propagator; hence the propagator correct to order α can be written as

$$iD_{\mu\nu}(k) = \frac{-ig_{\mu\nu}}{k^2 + i\epsilon} \rightarrow \frac{-ig_{\mu\nu}}{k^2 + i\epsilon} + \frac{-ig_{\mu\alpha}}{k^2 + i\epsilon} \, ie_0^2 \, \Pi^{\alpha\beta}(k) \, \frac{-ig_{\beta\nu}}{k^2 + i\epsilon} \, . \qquad (11.2)$$

where $\Pi^{\alpha\beta}(k)$ represents the closed electron loop, and is defined by

$$ie_0^2 \, \Pi^{\alpha\beta}(k) = \frac{(-1)(ie_0)^2}{(2\pi)^4} \int d^4p \, \mathrm{Tr}\left[\gamma^\alpha \frac{i(\not{p} + m)}{(p)^2 - m^2 + i\epsilon} \gamma^\beta \frac{i(\not{p} + \not{k} + m)}{(p+k)^2 - m^2 + i\epsilon}\right] .$$

We note the presence of the factor (-1) and of the trace, typical of closed fermion loops. We have written the terms inside the trace going backward along the fermion line. Simplifying the expression (factors i, -1, e_0^2), we obtain

$$\Pi^{\alpha\beta}(k) = \frac{i}{(2\pi)^4} \int d^4p \, \frac{\text{Tr}\left[\gamma^\alpha(\not{p}+m)\gamma^\beta(\not{p}+\not{k}+m)\right]}{((p)^2 - m^2 + i\epsilon)\,((p+k)^2 - m^2 + i\epsilon)} \ . \tag{11.3}$$

The integral is divergent, and we must resort to regularisation. We apply a dimensional regularisation, which consists of changing from the four-dimensional space to one of D dimensions,

$$\Pi^{\alpha\beta}(k) = \frac{i}{(2\pi)^D} \int d^D p \, \frac{\text{Tr}\left[\gamma^\alpha(\not{p}+m)\gamma^\beta(\not{p}+\not{k}+m)\right]}{((p)^2 - m^2 + i\epsilon)\,((p+k)^2 - m^2 + i\epsilon)} \ , \tag{11.4}$$

where an analytic continuation to non-integer D values is implied, carried out for example with the methods outlined in Appendix E. The integral is then finite, except for integer values of D, in particular $D = 4$, where the integral diverges like $1/(D-4)$. For any non-integer value of D the usual manipulations are allowed, for example the change of variables which we will use to prove an important property of $\Pi^{\alpha\beta}(k)$,

$$k_\beta \Pi^{\alpha\beta}(k) = 0 \ . \tag{11.5}$$

This identity is another consequence of gauge invariance, or more simply of conservation of the current $j^\beta = \bar{\psi}\gamma^\beta\psi$. In fact, the vertex γ^β in (11.3), or in Figure 11.1, represents the action of a current j^β and multiplying by k_β amounts to taking the four-divergence of the current, which vanishes. The proof of (11.5) originates from the identity

$$\text{Tr}\left[\gamma^\alpha(\not{p}+m)\not{k}(\not{p}+\not{k}+m)\right]$$
$$= \text{Tr}\left[\gamma^\alpha(\not{p}+m)\left((\not{p}+\not{k}-m)-(\not{p}-m)\right)(\not{p}+\not{k}+m)\right]$$
$$= \text{Tr}\left[\gamma^\alpha(\not{p}+m)\right]\left((p+k)^2 - m^2\right) - \text{Tr}\left[\gamma^\alpha(\not{p}+\not{k}+m)\right]\left((p)^2 - m^2\right)$$
$$= 4p^\alpha\left((p+k)^2 - m^2\right) - 4(p+k)^\alpha\left((p)^2 - m^2\right) \ ,$$

from which

$$k_\beta \Pi^{\alpha\beta}(k) = \frac{4i}{(2\pi)^4}\left[\int d^D p \, \frac{p^\alpha}{((p)^2 - m^2 + i\epsilon)} - \int d^D p \, \frac{(p+k)^\alpha}{((p+k)^2 - m^2 i\epsilon)}\right] = 0 \ ,$$

as shown by a change of variables, $(p+k) \to p$, in the second integral.[2] This result is confirmed by an explicit calculation of $\Pi^{\alpha\beta}(k)$, for which we refer the reader to Chapter 10 of Mandl and Shaw [3].

$\Pi^{\alpha\beta}(k)$ is a symmetric tensor with two indices, a function of the vector k, hence its most general form is necessarily

$$\Pi^{\alpha\beta}(k) = g^{\alpha\beta}A(k^2) + k^\alpha k^\beta B(k^2) \ , \tag{11.6}$$

---

[2] We note that this manipulation is only allowed in the regularised theory, in which the integral is convergent.

where $A(k^2)$ and $B(k^2)$ are functions of the scalar k^2, but the condition (11.5) fixes a relation between the two functions,

$$A(k^2) = -k^2 B(k^2) \ .$$

Therefore[3] $A(0) = 0$, and expanding in powers of k^2,

$$A(k^2) = k^2 A'(0) + k^2 \Pi_c(k^2) \ , \tag{11.7}$$

where $\Pi_c(k^2)$ contains the higher orders in the expansion in powers of k^2, and therefore

$$\Pi_c(0) = 0 \ . \tag{11.8}$$

The divergence of $\Pi^{\alpha\beta}(k)$ is quadratic; for high values of p the integral (11.3) varies as $\int d^4p/p^2$. In the expansion in powers of k, however, the rate of divergence decreases by one unit for each power of k, or by two units for each power of k^2. For example, to obtain the value of A' we must take the second derivative of Π with respect to k, and thus obtain an integrand which varies as p^{-4}, which leads to a logarithmic divergence. The remaining terms in the expansion of $\Pi(k^2)$ are convergent. These considerations apply to any Feynman diagram; even if a diagram diverges, in an expansion in powers of the external momenta, the divergences are found only in the coefficients of the first terms of the expansion.

A further simplification of the expression for $\Pi^{\alpha\beta}(k)$ is obtained from the consideration that, since in any diagram the propagator (11.2) is connected to the currents which run along the fermion lines,

$$j^\mu D_{\mu\nu}(k) \, j^\nu \ ,$$

and that these currents are conserved, the term of $\Pi^{\alpha\beta}(k)$ proportional to $k^\alpha k^\beta$ gives zero contribution and can be omitted. In conclusion we can write

$$\Pi^{\alpha\beta}(k) = g^{\alpha\beta} \left[k^2 A'(0) + k^2 \Pi_c(k^2) \right] \ , \tag{11.9}$$

and therefore the photon propagator, correct to order α is

$$iD_{\mu\nu}(k) = \frac{-ig_{\mu\nu}}{k^2 + i\epsilon} \left[1 + e_0^2 A'(0) \right] + \frac{-ig_{\mu\nu} e_0^2 \Pi_c(k^2)}{k^2 + i\epsilon} \ . \tag{11.10}$$

In this expression the first term corresponds to the photon propagator, but multiplied by a factor $\left(1 + e_0^2 A'(0)\right)$, while the second, thanks to (11.8), has no singularity for $k^2 = 0$. If we compare this situation with the spectral representation discussed in [2], also cf. Section 8.4, we see that the first term corresponds to propagation of the free photon, while the second corresponds to

---

[3] As shown in [3], it follows from $A(0) = 0$ that the photon also remains massless following radiative corrections, another consequence of gauge invariance.

intermediate states with more particles, in this case an electron–positron pair. The factor $\left(1 + e_0^2 A'(0)\right)$ can therefore be interpreted as a renormalisation constant.

$$Z_3 = 1 + e_0^2 A'(0) \tag{11.11}$$

As already noted, a consequence of gauge invariance, via equation (11.7), is that $A(0) = 0$, which means in turn that there is no photon mass renormalisation. Indeed, the corrected propagator, equation (11.10), has a pole at $q^2 = 0$, like the uncorrected one.

11.2 RENORMALISATION OF THE CHARGE

Neglecting for a moment the effect of renormalisation owing to the vertex corrections or to the electron propagator, which as we will see cancel because of the Ward identity, we briefly discuss renormalisation of the charge due to the corrections to the photon propagator.

The physical significance of the renormalisation constant can be understood by recalling that the photon propagator, with its singularity at $k^2 = 0$ not only describes the exchange of photons between two electrons, but also their Coulomb interaction. The correction to the propagator which we have found is equivalent to modifying the Coulomb interaction between two electrons.

$$\frac{e_0^2}{r} \rightarrow \frac{e_0^2 Z_3}{r} \ .$$

The electric charge of the electron is operationally defined by means of the Coulomb interaction, hence the effective electron charge is not e_0 but $e = e_0 \sqrt{Z_3}$.

It is helpful to re-express the perturbative series in powers of the physical charge e, and to do this we note that the photon propagator always appears in the combination $e_0^2 D_{\mu\nu}(k)$ which we can rewrite as

$$e_0^2 D_{\mu\nu}(k) = e^2 D_{R\mu\nu}(k) \ , \tag{11.12}$$

where $D_{R\mu\nu}(k)$ is the "renormalised" propagator,

$$D_{R\mu\nu}(k) = \frac{1}{Z_3} D_{\mu\nu}(k) = \frac{-ig_{\mu\nu}}{k^2 + i\epsilon} + \frac{-ig_{\mu\nu} e^2 \Pi_c(k^2)}{k^2 + i\epsilon} + \mathcal{O}(e^4) \ . \tag{11.13}$$

It remains to confirm that this rule also applies to external photon lines, corresponding to photons present in the initial or final state of the process.

We recall that in the expression for the S-matrix in terms of diagrams, (10.38), we must consider only one-particle irreducible diagrams on the external lines and associate a factor $\sqrt{Z_3}$ with each external photon. This factor originates from the partial simplification of the factor $1/\sqrt{Z_3}$ in the reduction formula, with the numerator of the two-point function of the photon, the pole term, that as we have just seen is equal to Z_3, equations (11.10) and (11.11). Naturally the line corresponding to the external photon ends on a vertex of the diagram from which two fermion lines (of which at least one is internal) leave. The external photon line–vertex amplitude can be written as

$$(\cdots e_0\gamma^\mu\cdots)\,\epsilon_\mu\,\sqrt{Z_3} = (\cdots e\gamma^\mu\cdots)\,\epsilon_\mu\,, \tag{11.14}$$

and also in this case the factor $\sqrt{Z_3}$ transforms the parameter e_0 into the physical charge e.

We note that what we have done is equivalent to adding the following rules for calculating diagrams:

- In the calculation of diagrams, use the physical charge e instead of the parameter e_0 which appears in the Lagrangian.

- After having calculated the value of $\Pi^{\alpha\beta}(k)$ (in a regularised theory), subtract the contribution of $A'(0)$ or, in other words, set the value of $A'(0)$ to zero.

- Eliminate the $\sqrt{Z_3}$ factors from (10.38).

11.3 THE ELECTRON PROPAGATOR

In this section we study the corrections of order α to the electron propagator, described by the two diagrams of Figure 11.2. We note that δm represents the mass adjustment due to interactions, which in perturbation theory can be expressed as a series of powers in $\alpha = e^2/4\pi$,

$$\delta m = \delta_2 e_0^2 + \delta_4 e_0^4 + \cdots . \tag{11.15}$$

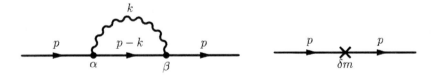

Figure 11.2 Corrections of order α to the electron propagator.

To calculate the order α corrections, we must include the $\mathcal{O}(e_0^2)$ term of δm in the propagator, neglecting higher-order terms.

As a consequence of these corrections, the electron propagator is modified as

$$iS_F(p) \to \frac{i}{\not{p} - m + i\epsilon} + \frac{i}{\not{p} - m + i\epsilon}[ie_0^2\Sigma(p) + ie_0^2\delta_2]\frac{i}{\not{p} - m + i\epsilon}\,, \tag{11.16}$$

with $\Sigma(p)$ given by the expression

$$ie_0^2\Sigma(p) = \frac{(ie_0)^2}{(2\pi)^4} \int d^4k \frac{-ig_{\alpha\beta}}{k^2 + i\epsilon} \left[\gamma^\alpha \frac{i}{\not p - \not k - m + i\epsilon} \gamma^\beta \right] ,$$

which with some simplifications becomes

$$\Sigma(p) = \frac{i}{(2\pi)^4} \int \frac{d^4k}{k^2 + i\epsilon} \left[\gamma^\alpha \frac{1}{\not p - \not k - m + i\epsilon} \gamma_\alpha \right] . \tag{11.17}$$

We have used a compact notation for the electron propagator; in a slightly more explicit form, the factor in square brackets in the integrand can be rewritten, by using well-known properties of the γ matrices, as

$$[\cdots] = \frac{\gamma^\alpha(\not p - \not k + m)\gamma_\alpha}{(p-k)^2 - m^2 + i\epsilon} = \frac{-2(\not p - \not k) + 4m}{(p-k)^2 - m^2 + i\epsilon} .$$

Therefore we can write $\Sigma(p) = s(p) + v_\mu(p)\gamma^\mu$ where $s(p)$ is a scalar, hence necessarily a function of p^2, and $v_\mu(p)$ is a vector, necessarily of the form $v_\mu(p) = p_\mu s'(p)$, where $s'(p)$ is also a scalar function of p^2. However, since $p^2 = \not p^2$ we can combine the two terms in a single function of $\not p$, which it is useful to expand in powers of $(\not p - m)$,

$$\Sigma(p) = A + B(\not p - m) + \Sigma_c(p)(\not p - m) , \tag{11.18}$$

and since the expression for $\Sigma(p)$ is linearly divergent, the first two coefficients A and B of this expansion are divergent, while the subsequent terms, collected in $\Sigma_c(p)$, are convergent. Since $\Sigma_c(p)(\not p - m)$ includes all the powers ≥ 2 of $(\not p - m)$, it must be true that

$$\Sigma_c(p)\Big|_{\not p = m} = 0 . \tag{11.19}$$

Substitution into (11.16) leads to

$$S_F(p) \to \frac{1}{\not p - m + i\epsilon} - \left(\frac{1}{\not p - m + i\epsilon} \right)^2 e_0^2(A + \delta_2) - \frac{e_0^2 B}{\not p - m + i\epsilon} - \frac{e_0^2 \Sigma_c(p)}{\not p - m + i\epsilon} .$$

The correction $\propto (A + \delta_2)$ represents a change of mass, as is seen for example by considering it as the first term of an expansion[4]

$$\frac{1}{\not p - m + e_0^2(A + \delta_2) + i\epsilon}$$

$$= \frac{1}{\not p - m + i\epsilon} - \left(\frac{1}{\not p - m + i\epsilon} \right)^2 e_0^2(A + \delta_2) + \dots , \tag{11.20}$$

---

[4]We will return to the justification of this step in the following section.

but since m is the "true" mass of the electron, we must choose δ_2 so that it exactly cancels the term A, i.e.

$$A + \delta_2 = 0 \ . \tag{11.21}$$

In conclusion, the propagator to order e^2 becomes

$$S_F(p) \rightarrow \frac{1 - e_0^2 B}{\not{p} - m + i\epsilon} - \frac{e_0^2 \Sigma_c(p)}{\not{p} - m + i\epsilon} \ . \tag{11.22}$$

We note that the first term is singular for $\not{p} = m$, and therefore corresponds to the propagation of a single particle of mass m, while thanks to (11.19) the second term is regular in $\not{p} = m$, and must correspond to the propagation of states with more than one particle—in our case an electron plus a photon, as seen from the diagram of Figure 11.2. If we compare this expression with the general formula for the two-point Green's function, (10.31), we see that the factor $(1 - e_0^2 B)$ can be interpreted as a renormalisation constant of the electron, Z_2 or, better, as its approximation to order α

$$Z_2 = 1 - e_0^2 B + \mathcal{O}(e_0^4) \ . \tag{11.23}$$

The corrections of Figure 11.2 apply also to the case of an external line. The discussion of the corrections follows the line of the similar discussion of the corrections to an external photon line in the previous section.

1. In the expression of the S-matrix in terms of diagrams, (10.38), we must consider only diagrams which are one-particle irreducible on the external lines.

2. A factor $\sqrt{Z_2}$ is associated with each external electron line, which originates from the partial simplification of the $1/\sqrt{Z_2}$ factor in the reduction formula, with the numerator of the two-point function of the electron, the pole term, which is equal to Z_2, equation (10.31).

3. The line corresponding to the external electron ends on a vertex of the diagram from which a fermion line and a photon line (of which at least one is internal) leave.

We can interpret the corrections to the fermion lines, both internal and external, as a further renormalisation of the charge. For an external line which joins a vertex, the correction is by a factor $\sqrt{Z_2}$. For an internal line the factor Z_2 is subdivided between the two vertices which the line joins, hence a factor $\sqrt{Z_2}$ (neglecting terms $\propto e^4$) for each vertex. In conclusion, since at each vertex two, external or internal, fermion lines merge, as a consequence of the corrections to the fermion lines, the vertex $e_0 \gamma^\mu$ will be multiplied by $1 - e_0^2 B = Z_2$. This is equivalent to a further renormalisation of the electric charge

$$e_0 \rightarrow e_0 Z_2 = e_0(1 - e_0^2 B) \ . \tag{11.24}$$

Thanks to the Ward identity, as we will see, this correction will be exactly balanced by a similar correction to the vertex function, so that the electric charge is only renormalised by the correction to the photon propagator studied in the previous section.

11.3.1 The propagator to all orders

In this section we briefly discuss the general structure of the electron propagator as obtained by perturbation theory. These results are important for a discussion of renormalisation to all orders, but in the present context they will be used only to justify the identification of the term A in the expansion of Σ [equation (11.18)] as the mass correction.

It is possible to reorder the perturbative series for the propagator by summing the contributions of diagrams which, like those of Figure 11.3, represent an iteration of corrections already encountered at lower orders. We say that the diagrams of Figure 11.3 have two *insertions*, while those of Figure 11.2 have a single insertion, and it is obvious that at higher orders we will find diagrams with three or more insertions.

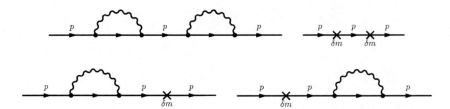

Figure 11.3   Corrections to the electron propagator with two insertions.

Going to fourth order we find three diagrams with a single insertion, shown in Figure 11.4, and analogous with what was done for the second-order insertion in (11.16) we denote their contribution with $e_0^4 \Sigma_4(p)$. It is possible to sum the contributions of diagrams with multiple insertions; if we define $\Delta(p)$ as the sum of all the single insertions

$$\Delta(p) = e_0^4 \Sigma(p) + e_0^4 \Sigma_4(p) + \ldots + \delta m , \qquad (11.25)$$

as a consequence of diagrams with single or multiple insertions, the propagator

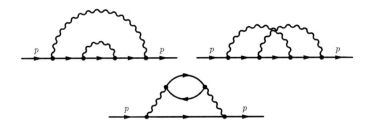

Figure 11.4 Insertions of order e_0^4.

will become[5]

$$
iS_F(p) \rightarrow \frac{i}{\not{p} - m + i\epsilon} + \frac{i}{\not{p} - m + i\epsilon} i\Delta(p) \frac{i}{\not{p} - m + i\epsilon}
$$

$$
+ \frac{i}{\not{p} - m + i\epsilon} i\Delta(p) \frac{i}{\not{p} - m + i\epsilon} i\Delta(p) \frac{i}{\not{p} - m + i\epsilon} + \dots \quad (11.26)
$$

$$
= \frac{i}{\not{p} - m + \Delta(p) + i\epsilon} .
$$

If, analogous with (11.18), we expand $\Delta(p)$ in powers[6] of $(\not{p} - m)$,

$$
\Delta(p) = \tilde{A} + \tilde{B}(\not{p} - m) + \Delta_c(p)(\not{p} - m) , \quad (11.27)
$$

then (11.26) leads to

$$
S_F(p) = \frac{i}{(\not{p} - m)\left(1 + \tilde{B} + \Delta_c(p)\right) + \tilde{A} + i\epsilon} , \quad (11.28)
$$

which is singular in $\not{p} = m$ only if $\tilde{A} = 0$. The condition that m should be the electron mass is hence translated into the condition $\tilde{A} = 0$, which to order e_0^2 reduces to (11.21).

---

[5] A generalisation of the geometric series, valid for any two operators X and Y, is obtained by writing

$$
S = \frac{1}{X} + \frac{1}{X} Y \frac{1}{X} + \frac{1}{X} Y \frac{1}{X} Y \frac{1}{X} + \dots = \frac{1}{X} + \frac{1}{X} Y S ,
$$

from which, multiplying from the left by X,

$$
XS = 1 + YS ; \quad \text{and hence,} \quad S = \frac{1}{X - Y} .
$$

[6] It can be proven that terms collected in $\Delta_c(p)$ are free of divergences. This is easily confirmed to order e_0^2.

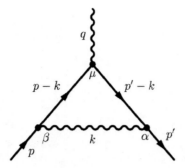

Figure 11.5 Vertex corrections of order e_0^2.

11.4   THE VERTEX

The correction to the vertex of order e_0^2 is given by the diagram of Figure 11.5. With this correction the vertex takes the form

$$ie_0\gamma^\mu \to ie_0\left(\gamma^\mu + e_0^2\Lambda^\mu(p',p)\right) , \qquad (11.29)$$

where

$$ie_0^3\Lambda^\mu(p',p) = \frac{(ie_0)^3}{(2\pi)^4}\int d^4k\,\frac{-ig_{\alpha\beta}}{k^2+i\epsilon}\gamma^\alpha\frac{i}{p\!\!\!/'-k\!\!\!/-m+i\epsilon}\gamma^\mu\frac{i}{p\!\!\!/-k\!\!\!/-m+i\epsilon}\gamma^\beta ,$$

and with some simplifications,

$$\Lambda^\mu(p',p) = \frac{-i}{(2\pi)^4}\int\frac{d^4k}{k^2+i\epsilon}\gamma^\alpha\frac{1}{p\!\!\!/'-k\!\!\!/-m+i\epsilon}\gamma^\mu\frac{1}{p\!\!\!/-k\!\!\!/-m+i\epsilon}\gamma_\alpha .$$

For large values of k the integral behaves as $\int d^4k/k^4$, and is therefore logarithmically divergent. Here is the source of a further complication because the integral is not only divergent for $k \to \infty$, but also for $k \to 0$ and therefore has an infrared divergence.

The infrared divergence will be discussed separately. For now let us say that it requires a regularisation process which consists of attributing a small mass λ to the photon which will be set to zero only at the end of the relevant calculations of physical quantities. Therefore we write

$$\Lambda^\mu(p',p) =$$
$$\frac{-i}{(2\pi)^4}\int\frac{d^4k}{k^2-\lambda^2+i\epsilon}\gamma^\alpha\frac{1}{p\!\!\!/'-k\!\!\!/-m+i\epsilon}\gamma^\mu\frac{1}{p\!\!\!/-k\!\!\!/-m+i\epsilon}\gamma_\alpha . \qquad (11.30)$$

It is easily shown that the divergent term is proportional to γ^μ. If we then write

$$\Lambda^\mu = \frac{-i}{(2\pi)^4}\int\frac{d^4k}{k^2-\lambda^2+i\epsilon}\gamma^\alpha\frac{p\!\!\!/'-k\!\!\!/+m}{(p'-k)^2-m^2+i\epsilon}\gamma^\mu\frac{p\!\!\!/-k\!\!\!/+m}{(p-k)^2-m^2+i\epsilon}\gamma_\alpha ,$$

the divergent term is the one which contains $\not{k} \ldots \not{k}$ in the numerator, therefore defining $\not{k} = k_\delta \gamma^\delta = k_\theta \gamma^\theta$ and omitting p, p' and m from the denominator, being negligible in the limit $k \to \infty$

$$\Lambda^\mu \Big|_{\text{divergent}} = \gamma^\alpha \gamma^\delta \gamma^\mu \gamma^\theta \gamma_\alpha \frac{-i}{(2\pi)^4} \int \frac{d^4 k}{k^2 - \lambda^2 + i\epsilon} \frac{k_\delta \, k_\theta}{(k^2 + i\epsilon)^2} .$$

The divergent part of this integral is a constant tensor (no vectors appear in the integral); it must therefore have the form $K g_{\delta\theta}$, where K is a constant (divergent, but as usual we must imagine carrying out the calculation in a *regularised* way). Therefore

$$\Lambda^\mu \Big|_{\text{divergent}} = K g_{\delta\theta} \gamma^\alpha \gamma^\delta \gamma^\mu \gamma^\theta \gamma_\alpha = K \gamma^\alpha \gamma^\delta \gamma^\mu \gamma_\delta \gamma_\alpha = 4 K \gamma^\mu ,$$

where we have twice used the well-known identity $\gamma^\delta \gamma^\mu \gamma_\delta = -2\gamma^\mu$.

Since the divergent part of Λ^μ is proportional to γ^μ we can write

$$\Lambda^\mu(p', p) = L\gamma^\mu + \Lambda_c^\mu(p', p) , \tag{11.31}$$

where $\Lambda_c^\mu(p', p)$ is free of divergences. However, $\Lambda_c^\mu(p', p)$ can itself contain terms $\propto \gamma^\mu$. Therefore, a second condition arises that fixes the separation between divergent and non-divergent parts. A possible choice of this condition starts from the consideration that $e_0 \bar{u}(\mathbf{p}) \gamma^\mu u(\mathbf{p})$ represents the electric current of an electron with momentum $\mathbf{p}$. As a consequence of the vertex corrections, this becomes[7]

$$e_0 \bar{u}(\mathbf{p}) \gamma^\mu u(\mathbf{p}) \to e_0 \bar{u}(\mathbf{p}) \left(\gamma^\mu + e_0^2 \Lambda^\mu(p, p) \right) u(\mathbf{p}) . \tag{11.32}$$

It is easily shown[8] that $\bar{u}(\mathbf{p}) \Lambda^\mu(p, p) u(\mathbf{p})$ is proportional to $\bar{u}(\mathbf{p}) \gamma^\mu u(\mathbf{p})$. We can therefore define the separation into divergent and non-divergent parts by $\Lambda^\mu(p', p)$, equation (11.31), through the condition

$$\bar{u}(\mathbf{p}) \Lambda^\mu(p, p) u(\mathbf{p}) = L \bar{u}(\mathbf{p}) \gamma^\mu u(\mathbf{p}) , \quad \text{or} \quad \bar{u}(\mathbf{p}) \Lambda_c^\mu(p, p) u(\mathbf{p}) = 0 . \tag{11.33}$$

Since equation (11.32) represents the current of an electron as modified by

---

[7] For the moment we neglect the corrections to the external lines and the correction to the photon propagator, both of which produce a renormalisation of the electric charge.

[8] We assume $\mathbf{p} = 0$, from which the general case follows with a Lorentz transformation. If we set

$$u(0) = \begin{pmatrix} \phi \\ 0 \end{pmatrix},$$

the only scalar quantity, candidate for the role of $\bar{u}(0) \Lambda^0 u(0)$, is $\phi^\star \phi = u^\star u = \bar{u}\gamma^0 u$, and therefore $\bar{u}(0)[\Lambda^0(p, p)]_{\mathbf{p}=0} u(0) \propto \bar{u}(0)\gamma^0 u(0)$. The only vector quantity is $\phi^\star \boldsymbol{\sigma} \phi$, but this is an axial vector while $\boldsymbol{\Lambda}$ must be a polar vector. Therefore $\bar{u}(0)[\boldsymbol{\Lambda}(p, p)]_{\mathbf{p}=0} u(0) = 0$. Finally from

$$\gamma = \begin{pmatrix} 0 & \boldsymbol{\sigma} \\ -\boldsymbol{\sigma} & 0 \end{pmatrix} \quad \text{it follows that} \quad \bar{u}(0)\boldsymbol{\gamma} u(0) = 0.$$

In conclusion $\bar{u}(0)[\Lambda^\mu(p, p)]_{\mathbf{p}=0} u(0) \propto \bar{u}(0)\gamma^\mu u(0)$, as we wished to prove.

the vertex correction, it follows that this correction implies a further renormalisation of the charge

$$e_0 \to \frac{1}{Z_1} e_0 = (1 + e_0^2 L) e_0 \; . \tag{11.34}$$

As we will see in the next section, the Ward identity establishes a relation between the correction to the electron propagator and the vertex correction, so that the two renormalisations of the charge, (11.24) and (11.34), exactly cancel.

11.5   WARD'S IDENTITY

The Ward identity,

$$\frac{d\Sigma(p)}{dp_\mu} = \Lambda^\mu(p,p) \; , \tag{11.35}$$

was derived in Chapter 4, equation (4.56), from the formulation of QED in terms of the sum over paths. Here we apply it to derive an identity for the renormalisation constants Z_1 and Z_2.

From (11.18) we obtain:

$$\frac{d\Sigma(p)}{dp_\mu} = B\gamma^\mu + \frac{d\Sigma_c(p)}{dp_\mu}(\not{p} - m) + \Sigma_c(p)\gamma^\mu \; .$$

If we take the matrix element of this between spinors $u(p)$ keeping in mind that $\Sigma_c(p)$ is itself proportional to $\not{p} - m$ and that $(\not{p} - m)u(p) = \bar{u}(p)(\not{p} - m) = 0$, we obtain

$$\bar{u}(p)\frac{d\Sigma(p)}{dp_\mu}u(p) = B\bar{u}(p)\gamma^\mu u(p) \; ,$$

while from (11.33) we obtain

$$\bar{u}(\mathbf{p})\Lambda^\mu(p,p)u(\mathbf{p}) = L\bar{u}(\mathbf{p})\gamma^\mu u(\mathbf{p}) \; .$$

The last two equations, with the identity (11.35), give

$$B = L \quad \text{i.e.} \quad Z_1 = Z_2 \; . \tag{11.36}$$

The Ward identity can be proved to order e_0^2 by directly calculating the derivative of $\Sigma(p)$ resulting from equation (11.17). Logically this operation does not make sense since the integral in equation (11.17) is divergent, as is, for that matter, that in (11.30) with which we must compare. Therefore we must suppose the theory to have been regularised, for example with a dimensional regularisation, by writing

$$\Sigma(p) = \frac{i}{(2\pi)^{4-\eta}} \int \frac{d^{4-\eta}k}{k^2 + i\epsilon} \left[\gamma^\alpha \frac{1}{\not{p} - \not{k} - m + i\epsilon} \gamma_\alpha \right] \; .$$

Since this integral is now regular, we can take the derivative of the integrand. We use the identity[9]:

$$\frac{d}{dp_\mu}\frac{1}{\not p - \not k - m + i\epsilon} = -\frac{1}{\not p - \not k - m + i\epsilon}\gamma^\mu\frac{1}{\not p - \not k - m + i\epsilon},$$

from which it follows that

$$\frac{d\Sigma(p)}{dp_\mu} = \frac{-i}{(2\pi)^4}\int\frac{d^{4-\eta}k}{k^2 + i\epsilon}\left[\gamma^\alpha\frac{1}{\not p - \not k - m + i\epsilon}\gamma^\mu\frac{1}{\not p - \not k - m + i\epsilon}\gamma_\alpha\right],$$

which agrees with the regularised expression for $\Lambda^\mu(p,p)$, equation (11.30), thus completing the proof.

Comparing with (11.34) and (11.24), we see that (11.36) implies the exact compensation of the charge renormalisations arising from the corrections to the vertex and to the electron propagator. This is an extremely important result. The corrections to the propagator and to the vertex evidently depend on the mass of the particle, and still more on the interactions to which the particle is subject. It would be reasonable to expect that the corrections to the propagators of the electron, μ and τ, not to mention corrections to the propagators of quarks or W bosons, should all be different from one another. In this situation, without the Ward identity, it would be extremely difficult to understand the universality of the constant e which describes both the charge of the electron and the proton, which are experimentally equal to extremely high precision. Conversely, we note that the charge renormalisation due to corrections to the photon propagator is the same for all the charged particles with which the photon couples.

---

[9]This is derived by multiplying the following identity by $1/(\not p - \not k - m + i\epsilon)$ from the left

$$0 = \frac{d}{dp_\mu}\left[(\not p - \not k - m + i\epsilon)\frac{1}{\not p - \not k - m + i\epsilon}\right]$$

$$= \gamma^\mu\frac{1}{\not p - \not k - m + i\epsilon} + (\not p - \not k - m + i\epsilon)\frac{d}{dp_\mu}\frac{1}{\not p - \not k - m + i\epsilon}.$$

APPLICATIONS OF QED

CONTENTS

In this chapter we will discuss in detail some applications of the concepts developed so far. We shall consider first the scattering in an external field, and the associated issues arising from the emission of low-energy photons—bremsstrahlung—and the occurrence of the infrared divergence. We will then analyse the calculation of order α corrections to the energy levels of atomic electrons obtained from solving the Dirac equation [1], to the photon propagator and to the magnetic moment of the electron. The agreement between the results of these calculations and the experimental data has confirmed unequivocally the validity of QED and, more generally, the concepts which are at the foundations of quantum field theory.

12.1   SCATTERING IN AN EXTERNAL FIELD

Scattering of an electron in an external field (see also [1]) provides an opportunity to review, in a simple context, the renormalisation process and to discuss with a concrete example the problem of infrared divergences which we only touched on in the previous chapter.

"Scattering by an external field" is in reality just an idealisation of the scattering of an electron in a collision with a massive target, typically a heavy nucleus, which we can, to a good approximation, consider as an electric charge localised at a precise position in space. A stationary nucleus with charge Ze at $\mathbf{x} = 0$ can, for example, be described by a "classical" current $j^\mu(x)$ independent of time such that

$$j^\mu(x) = \{\rho(\mathbf{x}), \mathbf{j} = 0\} , \qquad \int d^3x \, \rho(x) = Ze . \qquad (12.1)$$

With this idealisation we avoid providing a quantum description of this current, but naturally we also lose the possibility of describing phenomena involving its structure, such as the existence of excited nuclear states. In this chapter we will use this method to describe an electric charge, but we can also use it to describe emission of electromagnetic waves by a radio antenna, or synchrotron radiation emitted by electrons circulating in a storage ring.

The effect of such a "classical" current can be included by introducing a new term into the interaction Lagrangian (10.8), which describes the interaction of the electromagnetic field A^μ with the classical current j_μ

$$\mathcal{L}_1 = e_0(\bar\psi \gamma^\mu \psi)A_\mu - j^\mu A_\mu + \delta m(\bar\psi\psi) . \qquad (12.2)$$

With this addition, the photon Lagrangian is modified in the form

$$-\frac{1}{2}(\partial^\nu A^\mu \, \partial_\nu A_\mu) \to -\frac{1}{2}(\partial^\nu A^\mu \, \partial_\nu A_\mu) - j^\mu A_\mu . \qquad (12.3)$$

To describe the effect of the current j^μ in terms of diagrams, a simple modification to the procedure which leads from the generating functional to Feynman diagrams for the S-matrix is sufficient. Comparing (12.3) with (5.12) we see that the change of action produced by the external current j^μ is identical to that produced by an auxiliary function J^μ if $J^\mu = j^\mu$. We assume that the S-matrix element of interest is connected to a Green's function $\langle 0|T(\cdots \psi \cdots \bar\psi \cdots A^\mu \cdots)|0\rangle$ by the reduction formula; the calculation then requires four steps:

(1) Calculate $Z[J, \bar J, J_\mu]$ as a sum of diagrams.

(2) From it, calculate the partial derivative which corresponds to the Green's function

$$(\cdots \frac{i\delta}{\delta\bar J}, \cdots \frac{-i\delta}{\delta J} \cdots \frac{i\delta}{\delta J_\mu} \cdots)Z[J, \bar J, \mathbf{J}_\mu] .$$

After step (2) only those diagrams which have a number of "bubbles" survive, i.e. with factors J, $\bar J$ or J_μ equal to or larger than the number of derivatives of each type.

(3) If no external field: set $J = \bar J = J_\mu = 0$.

To calculate the same Green's function in the presence of the classical current j_μ it is enough to modify the third step:

(3') Set $J = \bar J = 0$ and set $J_\mu = j_\mu$.

(4) Apply the reduction formula to go from the Green's function to the S-matrix element.

Let us look at the difference between (3) and (3′); step (3) sets to zero all the diagrams in which some "bubbles" (factors J, $\bar{J}$ or J_μ) survive after taking the partial derivatives. In case (3′), in the presence of a classical current j_μ, the diagrams in which factors of type J or $\bar{J}$ exist (after the derivatives) are set to zero, while all the diagrams with external lines which terminate on a bubble of type J_μ remain alive, but with J_μ substituted by the classical current j_μ.

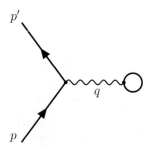

Figure 12.1 Feynman diagram for electron scattering in an external field.

Figure 12.1 represents the lowest-order diagram for scattering of an electron in an external field generated by the current j_μ, and in this case p and p' are the incoming and outgoing electron momenta. The same figure can also be considered as part of a more complex diagram in which p or p' (or both) are momenta of internal lines of the diagram. To the line with a bubble in this figure corresponds a factor

$$\int d^4y\, \Delta_F^{\mu\nu}(x-y)\, j_\nu(y) \;=\; \frac{1}{(2\pi)^4} \int d^4q\, \frac{-i}{q^2+i\epsilon}\, e^{-iqx}\, j^\nu(q)$$

$$j^\mu(q) \;=\; \int d^4y\, e^{iqy}\, j^\mu(y) \;,$$
(12.4)

the same as equation (10.21), but with the classical current j_μ in place of the auxiliary function J_μ . The factor e^{-iqx} combines with similar factors from the two fermion lines to produce the factor $(2\pi)^4\delta^4(p'-p-q)$ associated with the vertex. In conclusion, we must add to the Feynman rules (10.40):

$$\text{Line of external field} \qquad \frac{-i}{(2\pi)^4} \int d^4q\, \frac{j_\mu(q)}{q^2+i\epsilon} \;,$$
(12.5)

noting, however, that the factor $(2\pi)^4\delta^4(p'-p-q)$ associated with the vertex eliminates the integration over q by fixing the value of $q = p' - p$ and that the factors $(2\pi)^4$ also simplify.

We consider the example of a 4-current associated with a spherical charge centred on $\mathbf{x} = 0$,

$$j_0 = \rho(|\mathbf{x}|); \qquad \int d^3x \, \rho(|\mathbf{x}|) = Ze_0; \qquad \mathbf{j} = 0 \ . \qquad (12.6)$$

We then have

$$j_0(q) = \int d^4x \, e^{iqx} j_0(x) = 2\pi \, \delta(q_0) \, Ze_0 \, F(\mathbf{q}^2) \ , \qquad (12.7)$$

where the form factor $F(\mathbf{q}^2)$, defined by the Fourier transform of the spatial charge distribution,

$$Ze \, F(\mathbf{q}^2) = \int d^3x \, e^{-i\mathbf{q}\cdot\mathbf{x}} \rho(|\mathbf{x}|) \ , \qquad (12.8)$$

is normalised so that $F(0) = 1$. The amplitude for scattering of an electron by an external field is therefore, to lowest order, with $q = p' - p$,

$$
\begin{aligned}
\langle \mathbf{p}', s' | S | \mathbf{p}, s \rangle &= \frac{m}{(2\pi)^3 \sqrt{E \, E'}} \frac{ie_0}{q^2} j_\mu(q) \, \left(\bar{u}(p')\gamma^\mu u(p) \right) \\
&= \frac{m}{(2\pi)^3 E} \, 2\pi \, \delta(E' - E) \frac{iZe_0^2 \, F(\mathbf{q}^2)}{q^2} \left(\bar{u}(p') \, \gamma^0 \, u(p) \right) \ .
\end{aligned}
\qquad (12.9)
$$

Since the value of $q^2 = (p' - p)^2$ is fixed, we can omit the term $i\epsilon$ in the denominator. The factor $\delta(E' - E)$ guarantees conservation of energy by the electrons, while momentum is not conserved, since the presence of the external field causes a violation of translational invariance.[1]

It can also be interesting to express the result in terms of the classical field $\tilde{A}_\mu$ produced by the current j_μ. The equation $\Box \tilde{A}_\mu = j_\mu$ leads to

$$\tilde{A}_\mu(q) = \frac{-1}{q^2} \, j_\mu(q) \ ,$$

and we can rewrite the S-matrix element as

$$\langle \mathbf{p}', s' | S | \mathbf{p}, s \rangle = \frac{-ie_0 \, m}{(2\pi)^3 \sqrt{E \, E'}} \tilde{A}_\mu(q) \left(\bar{u}(p')\gamma^\mu u(p) \right) \ . \qquad (12.10)$$

We will use this form for discussion of the radiative corrections to the magnetic moment of the electron.

12.2 BREMSSTRAHLUNG AND INFRARED DIVERGENCE

The scattering of a charged particle is associated, in general, with the emission of photons, as illustrated in the Feynman diagrams of Figure 12.2.

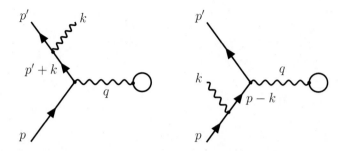

Figure 12.2 Feynman diagrams for bremsstrahlung processes.

The process is inelastic, because part of the energy of the incident particle is converted into energy of the photon present in the final state.

From the experimental point of view, the processes of electromagnetic radiation, called *bremsstrahlung*, are indistinguishable from elastic processes when the energy of the emitted photon is below the detection threshold of the measuring apparatus. In principle, it is therefore necessary to take into account the fact that the observed cross section is the sum of two contributions, associated with elastic scattering and with bremsstrahlung of photons below the detection threshold[2]

$$\left(\frac{d\sigma}{d\Omega_{p'}}\right)_{\text{expt}} = \left(\frac{d\sigma}{d\Omega_{p'}}\right)_{\text{el}} + \left(\frac{d\sigma}{d\Omega_{p'}}\right)_{\text{brem}}. \qquad (12.11)$$

In (12.11), $\Omega_{p'}$ is the solid angle which identifies the direction of the scattered electron, and the elastic cross section is that obtained from the amplitude (12.10).

In the case of the processes described by the diagrams of Figure 12.2, using the Feynman rules discussed in Section 12.1, we find

$$\langle \mathbf{p}'s'; \mathbf{k}r | S | \mathbf{p}s \rangle =$$

$$\frac{-ie_0^2\, m}{(2\pi)^3 \sqrt{E\, E'}\sqrt{2\omega}} \bar{u}(p') \left[\not{\epsilon}\frac{\not{p}+\not{k}+m}{2(pk)}\tilde{A} + \tilde{A}\frac{\not{p}-\not{k}+m}{-2(pk)}\not{\epsilon} \right] u(p) , \qquad (12.12)$$

where ω and $\mathbf{k}$ are, respectively, the energy and momentum of the photon, whose polarisation state is described by the vector ϵ.

---

[1] In a realistic treatment of the collision between an electron and a nucleus of mass M, if the nucleus is stationary in the initial state, after the collision it will have absorbed from the electron a momentum $\mathbf{p}' - \mathbf{p}$, and therefore a recoil energy $(\mathbf{p}' - \mathbf{p})^2/2M$. Obviously the picture of the nucleus as a fixed charge is only valid if the electron energy, and therefore the momentum transfer, is much smaller than the rest energy of the nuclear target.

[2] We note that cross sections should be summed, not amplitudes, because the two processes under consideration correspond to different final states.

In general, the cross-section calculation starting from the amplitude (12.12) is rather complicated. It is notably simplified, however, in the limit $\omega \approx 0$, which also implies $\mathbf{p} \approx \mathbf{p}'$ and $\mathbf{q} \approx \mathbf{p}' - \mathbf{p}$, in which we are interested. In this case the amplitude takes the form

$$\langle \mathbf{p}'s'; \mathbf{k}r|S|\mathbf{p}s\rangle = \langle \mathbf{p}', s'|S|\mathbf{p}, s\rangle \frac{e_0}{\sqrt{2\omega}} \left[\frac{p'\epsilon}{p'k} - \frac{p\epsilon}{pk} \right] , \qquad (12.13)$$

from which the cross section for the process—averaged over the electron spin in the initial state and summed over the spins of the final state particles—is easily obtained. The result can be cast in the form[3]

$$\left(\frac{d\sigma}{d\Omega_{p'}} \right)_{\text{brem}} = \left(\frac{d\sigma}{d\Omega_{p'}} \right)_{\text{el}} \frac{\alpha}{(2\pi)^2} \int_{|\mathbf{k}|<\Delta} \frac{d^3k}{\omega} \left\{ - \left[\frac{p'}{p'k} - \frac{p}{pk} \right]^2 \right\} , \qquad (12.14)$$

where the elastic contribution corresponds to the amplitude (12.10) and Δ is the energy threshold of the photon detecting apparatus.

Besides the simple factorised form, the cross section (12.14) exhibits a logarithmic divergence in the *infrared* limit $\omega \to 0$. The obtained expression would therefore seem to imply that the observed cross section, arising from the sum of the elastic contribution discussed in the previous section with the inelastic part given by (12.14), should diverge for $\omega \to 0$, an obviously unacceptable conclusion. However, we notice that the cross section for the bremsstrahlung process is of order α^3, while the elastic contribution is of order α^2. To calculate the observed cross section (12.11) in a consistent way in perturbation theory, we must therefore also take into account contributions of order α^3 to the elastic cross section.

As we saw in Chapter 11, the vertex correction of order α, in addition to the ultraviolet divergence which contributes to renormalisation of the charge, has a logarithmic divergence, similar to that of the cross section (12.14), in the infrared limit. The contribution of this correction to the process of elastic scattering is illustrated by the diagram of Figure 12.3.

Using the expression (11.30) for the vertex correction we can rewrite $\bar{u}(p')\Lambda^\mu u(p)$ keeping in mind that for $\omega \approx 0$ the terms linear in k in the numerators of the fermion propagators can be neglected. Hence we obtain

$$e_0^2 \bar{u}(p')\Lambda^\mu(p,p')u(p) = -i e_0^2 \bar{u}(p')\gamma^\mu u(p) \int \frac{d^4k}{(2\pi)^4} \frac{1}{k^2 - \lambda^2 + i\epsilon} \qquad (12.15)$$

$$\times \frac{4(pp')}{[(p'-k)^2 - m^2 + i\epsilon][(p-k)^2 - m^2 + i\epsilon]} .$$

The integration over k_0, the time component of k, is carried out with

---

[3] Being defined as the difference of two isochronous timelike four-vectors, the four-vector appearing in square brackets is spacelike and the cross section (12.14) turns out to be positive.

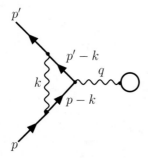

Figure 12.3 Feynman diagram for the vertex correction to electron scattering in an external field.

Cauchy's theorem, noting that for $\omega \approx 0$ the dominant contribution is that associated with the pole at $k_0 = \omega_\lambda = \sqrt{k^2 + \lambda^2}$.

Thus we find

$$e_0^2 \bar{u}(p')\Lambda^\mu(p,p')u(p) \qquad (12.16)$$

$$= e_0^2 \bar{u}(p')\gamma^\mu u(p) \frac{-1}{2(2\pi)^3} \int \frac{d^3k}{\omega_\lambda} \frac{4(pp')}{[-2(p'k)+\lambda^2][-2(pk)+\lambda^2]} + \dots$$

$$= e_0^2 \bar{u}(p')\gamma^\mu u(p) F(p,p') + \dots ,$$

where the three-dimensional integration is restricted to the region $k \approx 0$ and the ellipses indicate terms which remain finite in the $\lambda \to 0$ limit. The observable contribution to (12.16) can be easily isolated using (see Chapter 11)

$$\Lambda^\mu(p,p') = L\gamma^\mu + \Lambda_c^\mu(p,p') , \quad \bar{u}(p)\Lambda_c^\mu(p,p)u(p) = 0 , \qquad (12.17)$$

implying

$$L = e_0^2 F(p,p) + \dots = e_0^2 F(p',p') + \dots , \qquad (12.18)$$

and

$$e_0^2 \bar{u}(p')\Lambda_c^\mu(p,p')u(p) \qquad (12.19)$$

$$= e_0^2 \bar{u}(p')\gamma^\mu u(p) \left\{ F(p,p') - \frac{1}{2}\left[F(p,p) + F(p',p')\right] \right\} + \dots .$$

It follows that

$$e_0^2 \bar{u}(p')\Lambda_c^\mu(p,p')u(p) = e_0^2 \bar{u}(p')\gamma^\mu u(p)$$

$$\times \frac{1}{4(2\pi)^3} \int \frac{d^3k}{\omega_\lambda} \left[\frac{2p'}{[-2(p'k)+\lambda^2]} - \frac{2p}{[-2(pk)+\lambda^2]} \right]^2 + \dots . \qquad (12.20)$$

Using (12.20) we can write the amplitude for the process of elastic scattering to order e_0^3 in the form

$$\langle \mathbf{p}', s' | S | \mathbf{p}, s \rangle \tag{12.21}$$

$$\times \left\{ 1 + \frac{1}{2} \frac{\alpha}{(2\pi)^3} \int_{|\mathbf{k}| < \Delta} \frac{d^3 k}{\omega_\lambda} \left[\frac{2p'}{[-2(p'k) + \lambda^2]} - \frac{2p}{[-2(pk) + \lambda^2]} \right]^2 + \cdots \right\},$$

where, consistent with the simplifications introduced to obtain (12.20), the integration is restricted to the region $|\mathbf{k}| \approx 0$.

From (12.14) and (12.21) it immediately follows that the contribution to the elastic cross section arising from interference between the two amplitudes appearing in (12.21) exactly cancel the bremsstrahlung cross section (12.14), properly regularised giving the photon a mass λ. The remaining contributions are finite and independent of λ.

Bloch and Nordsieck have proved that the infrared divergences cancel to all orders of perturbation theory [12].

12.3 THE LAMB SHIFT

The energy eigenvalues obtained from the solution to the Dirac equation for the hydrogen atom (see [1], Section 6.2) depend on the principal quantum number n and the total angular momentum j. The states[4] $2S_{1/2}$ and $2P_{1/2}$ are therefore degenerate.

In 1947 Willis Lamb and Robert Retherford observed experimentally a difference in energy between these states, known as the *Lamb shift*, whose interpretation in terms of the effects of interactions between atomic electrons and the radiation field can be considered to be the first application of quantum electrodynamics.

A few weeks after the publication of these experimental results, Hans Bethe made a perturbative calculation of the self-energy of an atomic electron using a simple non-relativistic formalism [13]. As we saw in Chapter 11, the dominant effect of the self-interaction of the electron is renormalisation of the bare mass, m_0, which is modified from its value in the absence of interactions to the *observed* value $m = m_0 + \delta m$. The term $\Sigma_c(p)$, which in general gives rise to observable corrections, does not contribute to Coulomb scattering processes (see [1], Chapter 14) in which the electrons in the initial and final states are described by solutions to the free Dirac equation.[5]

According to Bethe, the mass renormalisation of the electron in the atom is different from the one in the vacuum by an amount, which depends upon the atomic state itself. We can write the renormalisation of the atomic electron

---

[4] Here we use standard spectroscopic notation, in which the states $2S_{1/2}$ and $2P_{1/2}$ correspond, respectively, to $n = 1$, $j = 1/2$ and orbital angular momentum, $\ell = 0$ and 1.

[5] If $u(p)$ is a solution to the free Dirac equation, the relation $\Sigma_c(p)u(p) = 0$ holds [see equation (11.19)].

mass in an explicit way as

$$m_{at} = m_0 + \delta m_{at} = m_0 + \delta m + (\delta m_{at} - \delta m) = m + (\delta m_{at} - \delta m) \; . \quad (12.22)$$

The effect of renormalisation that is obtained in this way is finite and it reproduces the Lamb shift in excellent agreement with the experimental data.

Bethe's calculation The starting point is the Hamiltonian which describes the system of an atomic electron interacting with an electromagnetic field described by the vector potential $\mathbf{A}(\mathbf{x})$:

$$H = H_e + H_{rad} + H_{int} \; , \quad (12.23)$$

with

$$H_e = \frac{\mathbf{p}^2}{2m_0} + V(\mathbf{x}) \; ,$$

$$H_{rad} = \frac{1}{2} \int d^3x \, |\mathbf{E}(\mathbf{x}) + \mathbf{B}(\mathbf{x})|^2 \; , \quad (12.24)$$

$$H_{int} = -\frac{e}{m_0} \, \mathbf{p} \cdot \mathbf{A}(\mathbf{x}) \; ,$$

where $\mathbf{E}(\mathbf{x}) = \partial_0 \mathbf{A}(\mathbf{x})$, $\mathbf{B}(\mathbf{x}) = \boldsymbol{\nabla} \times \mathbf{A}(\mathbf{x})$, $V(\mathbf{x})$ is the electrostatic potential of the atomic nucleus and the field $\mathbf{A}(\mathbf{x})$ satisfies the Coulomb gauge condition,[6] $\boldsymbol{\nabla} \cdot \mathbf{A}(\mathbf{x}) = 0$.

We consider the perturbative corrections to the energy of the state $|n, 0\rangle$, E_n, which describes an electron in the state n in the absence of radiation. The first-order term in H_{int} is zero, since the field $\mathbf{A}$ is linear in the photon creation and annihilation operators, $a_{\mathbf{k}\lambda}$ and $a_{\mathbf{k}\lambda}^\dagger$. The second-order term assumes a particularly simple form in the so-called dipole approximation, which amounts to neglecting the x dependence of the radiation field. The expression obtained in this approximation, which is allowed because the photons providing a significant contribution to the sum have wavelengths which are large compared to the Bohr radius, is

$$\Delta E_n = -\frac{e^2}{m_0^2} \times$$

$$\times \sum_{m, \lambda} \int \frac{d^3k}{2k(2\pi)^3} \frac{\langle n, 0|(\mathbf{p} \cdot \boldsymbol{\epsilon}_\lambda) a_{\mathbf{k}\lambda}|m, \mathbf{k}\lambda\rangle \langle m, \mathbf{k}\lambda|(\mathbf{p} \cdot \boldsymbol{\epsilon}_\lambda) a_{\mathbf{k}\lambda}^\dagger|n, 0\rangle}{E_m - E_n + k} ,$$

$$(12.25)$$

where the sums include all the states with one electron in the state m and

---

[6]In the Hamiltonian we have omitted the term in $\mathbf{A}^2$ which is obtained from the minimal substitution $\mathbf{p} \to \mathbf{p} - e\mathbf{A}$. The contribution of this term is independent of the electron momentum, and hence irrelevant to the calculation we are discussing.

a photon with wave vector $\mathbf{k}$ and polarisation λ, whose energy is equal to $E_m + k$ ($k = |\mathbf{k}|$). Going to the continuum limit and using the orthogonality between the polarisation vectors of the radiation field $\boldsymbol{\epsilon}_\lambda$ ($\lambda = 1, 2$) and the wave vector $\mathbf{k}$, implying

$$\sum_{\lambda=1}^{2} p^i p^j \epsilon_\lambda^i \epsilon_\lambda^j = p^i p^j \left(\delta^{ij} - \frac{k^i k^j}{\mathbf{k}^2} \right) = \mathbf{p}^2 - \frac{(\mathbf{p} \cdot \mathbf{k})^2}{\mathbf{k}^2}, \tag{12.26}$$

we obtain

$$\Delta E_n = -\frac{e^2}{m_0^2} \frac{1}{6\pi^2} \int k \, dk \sum_m \frac{|\mathbf{p}_{mn}|^2}{E_m - E_n + k}, \tag{12.27}$$

with $\mathbf{p}_{mn} = \langle m | \mathbf{p} | n \rangle$.

In the case of free electrons the expression we have obtained simplifies drastically, since only the diagonal matrix elements of the operator $\mathbf{p}$ are different from zero. Consequently, (12.27) becomes

$$\Delta E_n = \Delta E_p = -\frac{e^2}{m_0} \frac{1}{3\pi^2} \frac{\mathbf{p}^2}{2m_0} \int dk, \tag{12.28}$$

from which it follows that, to order e^2, we can write

$$E_p = \frac{\mathbf{p}^2}{2m_0} \left(1 - \frac{\delta m}{m_0} \right) = \frac{\mathbf{p}^2}{2m}, \tag{12.29}$$

with

$$\delta m = \frac{4\alpha}{3\pi} \int dk, \tag{12.30}$$

where $\alpha = e^2/4\pi$ is the fine structure constant.

Nature of the mass divergence of the electron.   The relation (12.30) shows that the self-energy of an electron in vacuum is linearly divergent with the value of the ultraviolet cutoff. This result is consistent with the classical consideration that the self-energy of an electric charge concentrated into a sphere of radius R diverges like R^{-1} and, apparently, also with the linear divergence obtained for $\Sigma(p)$ in equation (11.17). However, in QED the correction to the self-energy, i.e. to the rest mass, diverges only logarithmically. This is due to a special symmetry, called *chiral* symmetry, which is realised by QED in the $m_0 \to 0$ limit, and reflects invariance under the transformation

$$\psi(x) \to e^{i\alpha\gamma_5} \psi(x). \tag{12.31}$$

Higher-order corrections must also cancel for $m_0 \to 0$, so that the renormalised mass also vanishes in the limit of exact chiral symmetry. Consequently, δm must be proportional to m_0 and hence

$$\delta m = C \frac{\alpha}{\pi} m \log \frac{\Lambda}{m}, \tag{12.32}$$

where Λ is the cutoff and C is a constant which is calculated starting from (11.17). The conclusion of this argument is that in relativistic QED a single subtraction, exactly like that in (12.22), is sufficient to ensure that the electron self-energy is finite.

In Bethe's non-relativistic treatment, the integral in (12.27) is regularised by setting the upper integration limit to the value $K = m$, or the limit of the relativistic approximation, in line with the need for a single subtraction in relativistic QED.

Equation (12.28) can be generalised to the case of an electron in the state n not interacting with the field $\mathbf{A}$ by substituting

$$\mathbf{p}^2 \to \langle n|\mathbf{p}^2|n\rangle = \sum_m |\langle m|\mathbf{p}|n\rangle|^2 = \sum_m |\mathbf{p}_{mn}|^2 .$$

We now subtract from (12.27) the self-energy of the electron in vacuo, equation (12.30), which we denote with ΔE_n^0. Thus we obtain, to order e^2

$$\Delta E_n - \Delta E_n^0 = \frac{e^2}{m^2} \frac{1}{6\pi^2} \sum_m |\mathbf{p}_{mn}|^2 \int dk \, \frac{E_m - E_n}{E_m - E_n + k} . \tag{12.33}$$

First we carry out the integral over k, keeping in mind that the upper limit K is much larger than the difference $E_m - E_n$. Thus we find[7]

$$\int_0^K dk \frac{E_m - E_n}{E_m - E_n + k} = (E_m - E_n) \ln \frac{K}{|E_m - E_n|} . \tag{12.34}$$

We note that, thanks to the subtraction we have introduced, the rate of divergence of the result obtained, in the limit $K \to \infty$, is lowered from linear to logarithmic.

It remains to calculate the sum over the intermediate states, which is significantly simplified if the dependence on the intermediate state m is eliminated from the argument of the logarithm, by substituting for the difference $|E_m - E_n|$ its average value $\langle|E_m - E_n|\rangle$. We can then use the relation

$$|\mathbf{p}_{mn}|^2 (E_m - E_n) = \langle n|\mathbf{p}|m\rangle\langle m|\mathbf{p}|n\rangle (E_m - E_n) \tag{12.35}$$

$$= \frac{1}{2}\langle n|[\mathbf{p}, H_e]|m\rangle\langle m|\mathbf{p}|m\rangle - \frac{1}{2}\langle n|\mathbf{p}|m\rangle\langle m|[\mathbf{p}, H_e]|m\rangle ,$$

and

$$[\mathbf{p}, H_e] = [\mathbf{p}, V] = -i(\boldsymbol{\nabla} V) , \tag{12.36}$$

---

[7]The result is obtained immediately in the case in which $E_m - E_n \geq 0$. In the opposite case it is necessary to calculate the principal value of the integral.

to obtain

$$\sum_m |\mathbf{p}_{mn}|^2 (E_m - E_n) = -\frac{i}{2}\langle n|[\boldsymbol{\nabla}V, \mathbf{p}]|n\rangle = \frac{1}{2}\langle n|\boldsymbol{\nabla}^2 V|n\rangle$$

$$= \frac{1}{2}\int d^3x |\psi_n(x)|^2 \boldsymbol{\nabla}^2 V , \qquad (12.37)$$

where $\psi_n(x)$ is the wave function of the state n. From Poisson's equation for a pointlike charge distribution

$$\boldsymbol{\nabla}^2 V = -Ze^2 \delta(x) , \qquad (12.38)$$

where Z is the atomic number, it therefore follows that

$$\sum_m |\mathbf{p}_{mn}|^2 (E_m - E_n) = Ze^2 |\psi_n(0)|^2 . \qquad (12.39)$$

Putting everything together we arrive at the expression for the Lamb shift

$$\Delta E_{\text{Lamb}} = \Delta E_n - \Delta E_n^0 = Z\alpha^2 \frac{4}{3m^2} |\psi_n(0)|^2 \ln \frac{K}{\langle |E_m - E_n|\rangle} , \qquad (12.40)$$

which shows that the result should be zero for all states with orbital angular momentum different from zero, for which $\psi_n(0) = 0$. For the $2S_{1/2}$ state, using the values $\langle |E_m - E_n|\rangle = 226.3$ eV and $K = m$, Bethe obtained a frequency $\nu_{\text{Lamb}} = 1040$ MHz, in excellent agreement with the measured value of 1000 MHz.

The relativistic calculations subsequently carried out confirmed the validity of the method used by Bethe, showing that the relativistic effects, the vacuum polarisation and the vertex correction give rise to negligible contributions. The results of the theoretical calculations of the Lamb shift which include these effects reproduce the experimental results with a precision of 0.0001%.

12.4   VACUUM POLARISATION

In this section we derive the expression for the vacuum polarisation tensor to order α. The relevant integrals for the calculation are discussed in Appendix E.

12.4.1   Calculation of the tensor $\Pi^{\mu\nu}(k)$ to one loop

We start from (11.4), which provides the expression for the vacuum polarisation tensor in the dimensional regularisation

$$\Pi^{\mu\nu}(k) = \frac{i}{(2\pi)^D} \int d^D p \frac{\text{Tr}\left[\gamma^\mu(\not{p} + m)\gamma^\nu(\not{p} + \not{k} + m)\right]}{(p^2 - m^2 + i\epsilon)\left((p+k)^2 - m^2 + i\epsilon\right)} . \qquad (12.41)$$

Using the Feynman parameterisation (see Section E.2), we can rewrite (12.41) in the form

$$\Pi^{\mu\nu}(k) = \int_0^1 d\alpha \left[\frac{i}{(2\pi)^D} \int d^D p \frac{\text{Tr}\left[\gamma^\mu(\slashed{p}+m)\gamma^\nu(\slashed{p}+\slashed{k}+m)\right]}{[(p+\alpha k)^2 - m^2 + \alpha(1-\alpha)k^2 + i\epsilon]^2} \right]$$
$$= \int_0^1 d\alpha \left[\frac{i}{(2\pi)^D} \int d^D p \frac{\text{Tr}\left[\gamma^\mu(\slashed{p}-\alpha\slashed{k}+m)\gamma^\nu(\slashed{p}+(1-\alpha)\slashed{k}+m)\right]}{[p^2 - m^2 + \alpha(1-\alpha)k^2 + i\epsilon]^2} \right],$$

$$(12.42)$$

where we have changed the integration variable by setting $p + \alpha k = p'$, which we continue, for simplicity of notation, to denote with p.

The trace in the numerator is calculated with the usual rules, since, in a generic number of dimensions D, the rules of the Dirac algebra still hold, i.e.

$$\{\gamma^\mu, \gamma^\nu\} = g^{\mu\nu} ,$$

$$(12.43)$$

with

$$g_{\mu\nu}g^{\mu\nu} = D .$$

$$(12.44)$$

We then set, in the space of the γ matrix indices,

$$\text{Tr}\,(\mathbf{1}) = f(D) , \quad f(4) = 4 .$$

$$(12.45)$$

As we will see, it is not necessary to know any more about the function $f(D)$.

With these rules, the trace in the numerator of (12.42) gives the result

$$\text{Tr}\left[\gamma^\mu(\slashed{p}-\alpha\slashed{k}+m)\gamma^\nu(\slashed{p}+(1-\alpha)\slashed{k}+m)\right] \qquad (12.46)$$
$$= f(D)\left[2p^\alpha p^\beta - g^{\alpha\beta}p^2 - 2\alpha(1-\alpha)k^\alpha k^\beta + g^{\alpha\beta}\left[\alpha(1-\alpha)k^2 + m^2\right]\right\}$$
$$+ (\text{terms linear in } p) .$$

Inserting the above result into equation (12.42), we note that the terms linear in p make no contribution to the integral. The non-zero integrals can be traced back to the formula (E.1) by using the relations:

$$I_2 = \int d^D p \frac{p^2}{(p^2 - s + i\epsilon)^2} = I(s, D, 1) + sI(s, D, 2) ,$$

$$(12.47)$$

$$I_2^{\alpha\beta} = \int d^D p \frac{p^\alpha p^\beta}{(p^2 - s + i\epsilon)^2} = \frac{g^{\alpha\beta}}{D} I_2 .$$

$$(12.48)$$

The latter result follows from the fact that $I_2^{\alpha\beta}$ must be proportional to $g^{\alpha\beta}$, and from (12.44).

Putting everything together we obtain

$$
\int d^D p \frac{\text{Tr}\left[\gamma^\mu(\not p - \alpha\not k + m)\gamma^\nu(\not p + (1-\alpha)\not k + m)\right]}{[p^2 - m^2 + \alpha(1-\alpha)k^2 + i\epsilon]^2} \tag{12.49}
$$

$$
= f(D)\left\{\frac{2-D}{D}g^{\alpha\beta}\left[I(s,D,1) + sI(s,D,2)\right]\right.
$$

$$
\left. +g^{\alpha\beta}\left[m^2 + \alpha(1-\alpha)k^2\right]I(s,D,2) - 2\alpha(1-\alpha)k^\alpha k^\beta I(s,D,2)\right\},
$$

and by using the relation

$$
\frac{2-D}{D}I(s,D,1) = -\frac{2s}{D}I(s,D,2) , \tag{12.50}
$$

we obtain

$$
\Pi^{\alpha\beta} = \int_0^1 d\alpha \left[\frac{i}{(2\pi)^D}\int d^D p \frac{\text{Tr}\left[\gamma^\alpha(\not p - \alpha\not k + m)\gamma^\beta(\not p + (1-\alpha)\not k + m)\right]}{[p^2 - m^2 + \alpha(1-\alpha)k^2 + i\epsilon]^2}\right]
$$

$$
= \left[-\frac{f(D)}{2^{D-1}\pi^{D/2}}\int_0^1 d\alpha\, \alpha(1-\alpha)\frac{\Gamma(2-D/2)}{s^{2-D/2}}\right]\left(g^{\alpha\beta}k^2 - k^\alpha k^\beta\right) ,
$$

$$
\tag{12.51}
$$

with $s = m^2 - \alpha(1-\alpha)k^2$. We note that we have obtained the structure required from (11.5), as a consequence of the fact that the dimensional regularisation maintains the invariance under gauge transformations.

Comparing with the definitions (11.6) and (11.7) and using the formulae (E.14) we obtain[8]

$$
e_0^2\Pi_c(k^2) = \frac{e_0^2}{2\pi^2}\int_0^1 d\alpha\, \alpha(1-\alpha)\ln\left[1 - \alpha(1-\alpha)\frac{k^2}{m^2}\right] . \tag{12.52}
$$

In conclusion, starting from (11.2)

$$
iD_{\mu\nu}(k) = \frac{-ig_{\mu\nu}}{k^2 + i\epsilon} \to \frac{-ig_{\mu\nu}}{k^2 + i\epsilon} + \frac{-ig_{\mu\alpha}}{k^2 + i\epsilon}ie_0^2\Pi^{\alpha\beta}(k)\frac{-ig_{\beta\nu}}{k^2 + i\epsilon} , \tag{12.53}
$$

with

$$
\Pi^{\alpha\beta}(k) = g^{\alpha\beta}A(k^2) + k^\alpha k^\beta B(k^2) , \tag{12.54}
$$

and

$$
A(k^2) = k^2 A'(0) + k^2\Pi_c(k^2) , \tag{12.55}
$$

we arrive at

$$
iD_{\mu\nu}(k) = \frac{-ig_{\mu\nu}Z_3}{k^2 + i\epsilon}\left[1 + e^2\Pi_c(k)\right] , \tag{12.56}
$$

---

[8]Equation (E.14) shows that the term proportional to η in the expansion of terms like $\pi^{D/2}$ contributes to the final result with a constant, which is absorbed into the value at $k = 0$ and does not influence the value of $\Pi_c(k)$. A similar argument holds for the function $f(D)$, for which only the value for $D = 4$ matters, as we expected.

with

$$e^2\Pi_c = \frac{2\alpha}{\pi} \int_0^1 dz\, z(1-z)\, \ln\left[1 - z(1-z)\frac{k^2}{m^2}\right] . \qquad (12.57)$$

We note that to this order of perturbation theory, we can identify the *bare charge*, e_0, with the *physical, or renormalised, charge*, e. In (12.57) we have highlighted, as is usual, the fine structure constant α, defined by

$$\alpha = \frac{e^2}{4\pi} \simeq \frac{1}{137} . \qquad (12.58)$$

To avoid possible confusion, the integration variable is denoted as z.

12.5 THE ANOMALOUS MAGNETIC MOMENT

The accurate calculation of the magnetic moment of the electron is certainly one of the great successes of QED. According to the Dirac theory, the gyromagnetic factor of the electron, which determines the magnetic moment in units of Bohr magnetons, is exactly equal to 2 [1]. In the context of QED, this value is modified by effects of higher orders in the fine structure constant. We will carry out the full calculation, done for the first time by Schwinger in 1949, introducing a few simplifications which make it particularly concise. We will assume also that the reader is aware of the discussion in Section 14.2 of [1].

12.5.1 Preliminaries

We begin with a premise. The most general form of the vertex function which describes the coupling between a photon and an *on-shell* electron can be written in the form

$$\bar{u}(p')\Lambda^\mu(p',p)u(p) = \bar{u}(p')\left[A_1(q^2)\gamma^\mu + A_2(q^2)p^\mu + A_3(q^2)p'^\mu \qquad (12.59)\right.$$
$$\left. + A_4(q^2)\sigma^{\mu\nu}p_\nu + A_5(q^2)\sigma^{\mu\nu}p'_\mu\right]u(p) ,$$

with $p^2 = p'^2 = m^2$ and $q = p' - p$. Conservation of the current, expressed by the relation

$$q_\mu \bar{u}(p')\Lambda^\mu(p',p)u(p) = 0 ,$$

implies that $A_2(q^2) = A_3(q^2)$ and $A_5(q^2) = -A_4(q^2)$, i.e. that

$$\bar{u}(p')\Lambda^\mu(p',p)u(p) =$$
$$\bar{u}(p')\left[A_1(q^2)\gamma^\mu + A_2(q^2)(p+p')^\mu + A_4(q^2)\sigma^{\mu\nu}q_\nu\right]u(p) . \qquad (12.60)$$

Furthermore, hermiticity requires that the functions $A_1(q^2)$ and $A_2(q^2)$ are real and that $A_4(q^2)$ is imaginary. Keeping in mind the Gordon identity [see equation (14.33) [1]]

$$2m\,\bar{u}(p')\gamma^\mu u(p) = \bar{u}(p')\left[(p+p')^\mu + i\sigma^{\mu\nu}q_\nu\right]u(p) , \qquad (12.61)$$

we can therefore write the matrix element $\bar{u}(p')\Lambda^\mu(p',p)u(p)$ in the form

$$\bar{u}(p')\Lambda^\mu(p',p)u(p) = \bar{u}(p') \left[F_1(q^2)\gamma^\mu + F_2(q^2)\frac{1}{2m}i\sigma^{\mu\nu}q_\nu \right] u(p) , \quad (12.62)$$

with $F_1(q^2)$ and $F_2(q^2)$ real. On the other hand, we know that [see equation (11.31)],

$$\Lambda^\mu(p',p) = L\gamma^\mu + \Lambda_c^\mu(p',p) \quad (12.63)$$

and since p' and p correspond to 4-momenta of physical particles ($p^2 = p'^2 = m^2$), we can use the relation (11.33), valid in the limit $q \to 0$, $p \to p'$,

$$\bar{u}(p)\Lambda_c^\mu(p,p)u(p) = 0 . \quad (12.64)$$

Comparing with (12.62), since the term in F_2 is explicitly proportional to q, it can be seen that $F_1(0) = L$. If we then calculate the expression on the right-hand side of (12.62), omitting terms of second order in q^2, we find

$$\bar{u}(p')\Lambda^\mu(p',p)u(p) = \bar{u}(p') \left[L\gamma^\mu + F_2(0)\frac{1}{2m}i\sigma^{\mu\nu}q_\nu \right] u(p) + O(q^2) . \quad (12.65)$$

Because of the Ward identity, the term L will be exactly cancelled by the correction to the external lines. Hence, the vertex function, including *all* the second-order corrections, becomes

$$ie_0\bar{u}(p')\gamma^\mu u(p) \to ie\bar{u}(p') \left[\gamma^\mu + \frac{e}{2m}F_2(0)\, i\sigma^{\mu\nu}q_\nu \right] u(p) + O(q^2) ,$$

and, again using the Gordon identity (12.61), we find

$$\bar{u}(p')\Lambda^\mu(p',p)u(p) = ie\bar{u}(p') \left[\frac{p^\mu + p'^\mu}{2m} + \frac{e}{2m}(1 + F_2(0))\, i\sigma^{\mu\nu}q_\nu \right] u(p)$$
$$+ O(q^2) . \quad (12.66)$$

In the above expression we have included the effect of the second-order corrections to the photon line which renormalises the charge, $e_0 \to e$. The first term $\propto (p+p')^\mu$ represents the interaction of the electric charge (it is the same as we would have for a spin 0 particle), the second describes the interaction of the magnetic moment. The result obtained shows that the magnetic moment of the lepton is equal to $(1 + F_2(0))$ Bohr magnetons. We can therefore write the gyromagnetic factor in the form

$$g = 2\left[1 + F_2(0)\right] = 2 + O(\alpha) .$$

To obtain the value of $F_2(0)$ we must carry out the calculation taking into account the terms linear in q in equation (12.62). The calculation is significantly simplified if we follow the following rules:

1. Omit terms which are explicitly of order q^2.

2. Omit terms which are $\propto \gamma^\mu$. As we saw, these terms determine the (divergent) constant L and are eliminated by the renormalisation.

We note that this procedure has two important advantages:

1. With the first rule, the integral simplifies considerably.

2. With the second term, not only have we omitted ultraviolet divergent terms, but we have also eliminated the infrared divergence, which is also produced by terms $\propto \gamma^\mu$.

We can therefore carry out the calculation without reverting to infrared or ultraviolet regularisation, or more strictly but in a completely equivalent manner, suppose that the regularisation is carried out and that terms $\propto \gamma^\mu$, which in the regularised theory are finite, are omitted. The terms which remain, i.e. $F_2(0)$, remain finite in the limit in which the regularisation is eliminated.

12.5.2 The calculation

We can write the quantity to be calculated in the form

$$\bar{u}(p')\Lambda^\mu(p', p)u(p) \tag{12.67}$$

$$= \frac{-i}{(2\pi)^4} \int d^4k \frac{\bar{u}(p')N^\mu(k, p, p')u(p)}{(k^2 + i\epsilon)((p' - k)^2 - m^2 + i\epsilon)((p - k)^2 - m^2 + i\epsilon)},$$

with

$$N^\mu(k, p, p') = \gamma^\alpha(\not{p}' - \not{k} + m)\gamma^\mu(\not{p} - \not{k} + m)\gamma_\alpha . \tag{12.68}$$

We begin by establishing some relationships between the variables:

$$q^2 = (p' - p)^2 = 2m^2 - 2(pp') , \quad (p' + p)^2 = 2m^2 + 2(pp') = 4m^2 - q^2 . \tag{12.69}$$

We can then combine the last two factors in the denominator by writing

$$Q = \frac{p + p'}{2} , \quad p' = Q + \frac{q}{2} , \quad p = Q - \frac{q}{2} . \tag{12.70}$$

We note also that $Q^2 = m^2 + q^2/4$, which, since we will omit terms $O(q^2)$, we can rewrite as $Q^2 = m^2 + O(q^2)$. Then we have

$$\begin{aligned}
[(p' - k)^2 &- m^2 + i\epsilon][(p - k)^2 - m^2 + i\epsilon] \\
&= (k^2 - 2(p'k) + i\epsilon)(k^2 - 2(pk) + i\epsilon) \\
&= \left[(k^2 - 2(Qk) + i\epsilon)^2 - (kq)^2\right] \\
&= (k^2 - 2(Qk) + i\epsilon)^2 + O(q^2) .
\end{aligned} \tag{12.71}$$

Hence, neglecting terms $O(q^2)$, we can recast (12.67) in the form

$$\bar{u}(p')\Lambda^\mu(p', p)u(p) = \frac{-i}{(2\pi)^4} \int d^4k \frac{\bar{u}(p') N^\mu(k, p, p') u(p)}{(k^2 + i\epsilon)(k^2 - 2(Qk) + i\epsilon)^2} + O(q^2) . \tag{12.72}$$

The next step consists of combining the denominators together, by using another parameterisation due to Feynman:

$$\frac{1}{D_1^2 D_2} = 2 \int_0^1 dz \, \frac{z}{[D_2 + (D_1 - D_2)z]^3} \, . \tag{12.73}$$

Hence, we obtain

$$
\begin{aligned}
\bar{u}(p')\Lambda^\mu(p',p)u(p) &= \frac{-i}{(2\pi)^4} \int_0^1 2z \, dz \int d^4k \frac{\bar{u}(p') \, N^\mu(k,p,p')u(p)}{(k^2 - 2z(Qk) + i\epsilon)^3} + O(q^2) \\
&= \frac{-i}{(2\pi)^4} \int_0^1 2z \, dz \int d^4k \frac{\bar{u}(p')N^\mu(k,p,p')u(p)}{[(k - zQ)^2 - z^2Q^2 + i\epsilon]^3} + O(q^2) \\
&= \frac{-i}{(2\pi)^4} \int_0^1 2z \, dz \int d^4k \frac{\bar{u}(p')N^\mu(k,p,p')u(p)}{[(k - zQ)^2 - z^2m^2 + i\epsilon]^3} + O(q^2) \, ,
\end{aligned}
\tag{12.74}
$$

where, in the last step, we have used the fact that $Q^2 = m^2 + O(q^2)$. We can now carry out the change of variables[9] $t = k - zQ$, $k = t + z(p + p')/2$, so that

$$\bar{u}\Lambda^\mu(p',p)u = \frac{-i}{(2\pi)^4} \int_0^1 2z \, dz \int d^4t \frac{\bar{u}(p') \, N^\mu(t + zQ, p, p')u(p)}{(t^2 - z^2m^2 + i\epsilon)^3} + O(q^2) \, . \tag{12.75}$$

At this point we note that in $N^\mu(t+zQ, p, p')$, defined by equation (12.68), terms of order zero in t are present, as well as terms linear and quadratic in t. The linear terms make no contribution to the integration. The terms quadratic in t give a contribution proportional to γ^μ and can be omitted in the approximation scheme we are using.

We can then write, denoting explicitly with $(\propto \gamma^\mu)$ the omitted terms,

$$
\begin{aligned}
\bar{u}(p')\Lambda^\mu(p',p)u(p) &= \frac{-i}{(2\pi)^4} \int_0^1 2z \, dz \int d^4t \frac{\bar{u}(p')N^\mu(z(p+p')/2, p, p')u(p)}{(t^2 - z^2m^2 + i\epsilon)^3} \\
&\quad + O(q^2) + (\propto \gamma^\mu) \, .
\end{aligned}
\tag{12.76}
$$

Since the numerator does not depend on t, we can immediately carry out the integral over t with the result (see Appendix E)

$$\int \frac{d^4t}{(t^2 - z^2m^2 + i\epsilon)^3} = -i\frac{\pi^2}{2z^2m^2} \, , \tag{12.77}$$

from which it follows that

$$
\begin{aligned}
\bar{u}(p')\Lambda^\mu(p',p)u(p) &= \frac{-1}{16\pi^2m^2} \int_0^1 \frac{dz}{z} \, \bar{u}(p')N^\mu(z(p+p')/2, p, p')u(p) \\
&\quad + O(q^2) + (\propto \gamma^\mu) \, . \tag{12.78}
\end{aligned}
$$

---

[9]The operation is allowed only if the integral is convergent; for this we suppose to have made a regularisation and taken the limit at the end of the calculation.

It now remains to calculate the quantity [see equation (12.68)]

$$X = \bar{u}(p') \, N^\mu \left[\frac{z(p + p')}{2}, p, p' \right] u(p)$$

$$= \bar{u}(p') \gamma^\alpha \left[p' - \frac{z(\not{p} + \not{p}')}{2} + m \right] \gamma^\mu \left[\not{p} - \frac{z(\not{p} + \not{p}')}{2} + m \right] \gamma_\alpha u(p) \,,$$

which can be rewritten in the form

$$X = \bar{u}(p') \left[-2(\not{p} - \frac{z(\not{p} + \not{p}')}{2}) \gamma^\mu (\not{p}' - \frac{z(\not{p} + \not{p}')}{2}) \right. \tag{12.79}$$

$$\left. + 4m(1 - z)(p + p')^\mu - 2m^2 \gamma^\mu \right] u(p) \,,$$

where the final term is ($\propto \gamma^\mu$) and can be omitted. In the first term we use the Dirac equation, from which it follows that on the left $\not{p} = \not{p}' - \not{q} = m - \not{q}$, while on the right $\not{p}' = \not{p} + \not{q} = m + \not{q}$. The result is

$$X = \bar{u}(p') \left[-2 \left(m(1 - z) - \not{q}(1 - \frac{z}{2}) \right) \gamma^\mu \left(m(1 - z) + \not{q}(1 - \frac{z}{2}) \right) \right.$$

$$\left. + 4m(1 - z)(p + p')^\mu \right] u(p) + (\propto \gamma^\mu)$$

$$= \bar{u}(p') \left[4m(1 - z) \left(1 - \frac{z}{2} \right) \frac{1}{2} [\not{q}, \gamma^\mu] + 4m(1 - z)(p + p')^\mu \right] u(p)$$

$$+ O(q^2) + (\propto \gamma^\mu) \,,$$

and using the relation $[\not{q}, \gamma^\mu] = 2i\sigma^{\mu\nu} q_\nu$, and the Gordon identity, we obtain

$$\bar{u}(p') N^\mu (\frac{z(p + p')}{2}, p, p') u(p)$$

$$= -2m \, z(1 - z) \bar{u}(p') i\sigma^{\mu\nu} q_\nu u(p) + O(q^2) + (\propto \gamma^\mu) \,. \tag{12.80}$$

Finally, substituting into (12.78) and carrying out the integration over z, we arrive at the result

$$\bar{u}(p') e^2 \Lambda^\mu(p', p) u(p) = \frac{e^2}{8\pi^2} \, \bar{u}(p') \frac{1}{2m} \sigma^{\mu\nu} q_\nu u(p) + O(q^2) + (\propto \gamma^\mu) \,, \tag{12.81}$$

which, comparing with (12.65), shows that

$$e^2 F_2(0) = \frac{e^2}{8\pi^2} = \frac{\alpha}{2\pi} \,, \tag{12.82}$$

i.e. that, to order α, the magnetic moment of a lepton is equal to

$$1 + \frac{\alpha}{2\pi} \quad \text{Bohr magnetons.}$$

RENORMALISATION GROUP OF QED

CONTENTS

As we saw in Chapter 11, renormalisation of QED consists essentially of carrying out, on the coupling constant of the theory, on the propagators, and on the vertices, the transformations

$$
\begin{aligned}
e^2 &\to Z_3 Z_1^{-1} Z_2 e^2 = Z_3 e^2 \ , \\
D_{\mu\nu} &\to Z_3^{-1} D_{\mu\nu} \ , \\
S_F &\to Z_2^{-1} S_F \ , \\
\Lambda_\mu &\to Z_1 \Lambda_\mu \ ,
\end{aligned}
\tag{13.1}
$$

where we have used the Ward identity, which implies that $Z_2/Z_1 = 1$.

In addition to the transformations (13.1), involving infinite coefficients, others can be considered, which correspond to different choices of the momentum scale for which the subtraction that is at the foundation of the renormalisation procedure is carried out, and which leaves the amplitudes of the physical processes invariant. In this chapter we will introduce these transformations, which are characterised by finite coefficients and form a group, the *renormalisation group of QED*.

13.1 EFFECTIVE ELECTRIC CHARGE

The expression for the photon propagator to order e_0^2 discussed in the previous sections,[1]

$$iD^{\mu\nu}(q) = \frac{-ig^{\mu\nu}}{q^2}\left[1 + e_0^2\Pi(q^2)\frac{1}{q^2}\right] , \qquad (13.2)$$

where $\Pi(q^2) = A(q^2)$, with $A(q^2)$ defined by equations (11.6) and (11.7), can be generalised to take account of higher-order diagrams which are obtained by repeating the insertion of the fermion loop into the photon line. In this way we obtain

$$\begin{aligned}
iD^{\mu\nu}(q) &= \frac{-ig^{\mu\nu}}{q^2}\left[1 + e_0^2\Pi(q^2)\frac{1}{q^2} + e_0^2\Pi(q^2)\frac{1}{q^2}e_0^2\Pi(q^2)\frac{1}{q^2} + \cdots\right]\\
&= \frac{-ig^{\mu\nu}}{q^2 - e_0^2\Pi(q^2)} \qquad (13.3)\\
&= \frac{-ig^{\mu\nu}}{q^2}\frac{1}{[1 - e_0^2\Pi(0) - e_0^2\Pi_c(q^2)]} ,
\end{aligned}$$

which can be rewritten in terms of the renormalised charge introduced in Chapter 11. To obtain the last line of (13.3) we have used the property $\Pi_c(0) = 0$. From

$$1 - e_0^2\Pi(0) \approx [1 + e_0^2\Pi(0)]^{-1} = Z_3^{-1} , \qquad (13.4)$$

it then follows that

$$iD^{\mu\nu}(q) = \frac{-ig^{\mu\nu}}{q^2}\frac{1}{Z_3^{-1} - e_0^2\Pi_c(q^2)} = \frac{-ig^{\mu\nu}}{q^2}\frac{Z_3}{1 - e^2\Pi_c(q^2)} . \qquad (13.5)$$

By substituting the photon propagator (13.5) into the expression for an amplitude, for example, that associated with the process of Mott scattering discussed in Section 12.1, we obtain

$$ie_0(\bar{u}'\gamma_\mu u) \, D^{\mu\nu}(\, e_0 j_\nu) = ie_{\text{eff}}(\bar{u}'\gamma_\mu u)\frac{-ig^{\mu\nu}}{q^2}(e_{\text{eff}} j_\nu) , \qquad (13.6)$$

which shows that the effect of the corrections due to vacuum polarisation can be described leaving the form of the propagator of the free theory unchanged and replacing the coupling constant $\alpha_0 = e_0^2/4\pi$ with an effective coupling constant $\alpha(q^2) = e_{\text{eff}}^2/4\pi$, defined by the relation

$$\alpha(q^2) = \frac{Z_3}{1 - e^2\Pi_c(q^2)} \, \alpha_0 = z_3(q^2)\alpha , \qquad (13.7)$$

---

[1]In the description of the propagation of virtual photons, we can omit the term $i\epsilon$ in the denominators.

with $\alpha = Z_3\alpha_0 = 1/137$. Therefore, besides the infinite constants which appear in the definition of Z_3, the higher-order terms of the perturbative series introduce a finite correction z_3, expressible in terms of the function $\Pi_c(q^2)$ defined by (12.52).

Using the asymptotic expression for $\Pi_c(q^2)$ (we recall that for a virtual photon $q^2 < 0$, and we define $Q^2 = -q^2$)

$$\lim_{Q^2 \to \infty} \Pi_c(q^2) = \frac{1}{12\pi^2} \ln\left(\frac{Q^2}{m^2}\right), \tag{13.8}$$

we can write the explicit form of the effective coupling constant valid at short distances:

$$\alpha(Q^2) = \frac{\alpha(0)}{1 - \frac{\alpha(0)}{3\pi} \ln \frac{Q^2}{m^2}}. \tag{13.9}$$

Equation (13.9) shows that the effective charge is an increasing function of Q^2. Consequently, as it was natural to expect, the electrostatic interaction between two charged particles decreases as their separation increases due to the effect of screening caused by vacuum polarisation. The dependence of the QED coupling constant on Q^2 has been observed experimentally at the LEP storage ring at CERN in Geneva. The results of these measurements are shown in Figure 13.1.

We note also that (13.9) predicts the appearance of a pole, called the Landau pole, for $Q^2/m^2 = \exp\left[3\pi/\alpha(0)\right]$, which from the observational point of view corresponds to an enormous value of Q^2 ($\sim m^2 \times 10^{560}$). In a theory with several charged particles, however, the Landau pole may get much closer and become relevant to physics, as we shall discuss in Chapter 17.

From the definition it follows that we can write the function $\alpha(Q^2)$ in the form

$$\frac{1}{\alpha(Q^2)} = \frac{1}{\alpha_0} - F(Q^2), \tag{13.10}$$

with $F(Q^2) = 4\pi\Pi(Q^2)$. We note also that (13.10) implies

$$\frac{1}{\alpha(0)} = \frac{1}{\alpha_0} - F(0), \tag{13.11}$$

and that combining (13.10) and (13.11) gives

$$\frac{1}{\alpha(Q^2)} = \frac{1}{\alpha(0)} - [F(Q^2) - F(0)]. \tag{13.12}$$

The choice of performing the subtraction at $Q^2 = 0$, although natural in the case of QED, is not the only one possible. The effective coupling constant can be equally obtained by choosing a different value, $Q^2 = \mu^2$, and using

$$\frac{1}{\alpha(Q^2)} = \frac{1}{\alpha(\mu^2)} - [F(Q^2) - F(\mu^2)]. \tag{13.13}$$

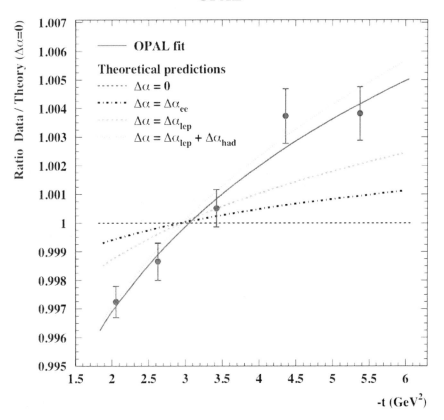

Figure 13.1 Dependence on $-t = Q^2$ of the ratio between the elastic scattering cross section for e^+–e^- (Bhabha scattering) measured at LEP [14] and the results of theoretical calculations. The horizontal line corresponds to the case in which the value of α is kept constant as Q^2 varies.

The subtraction at $\mu^2 \neq 0$ is necessary if we wish to go to the massless limit of the theory, which is useful to examine the dependence on momentum when $Q^2 >> m^2$. It is, furthermore, essential in the case of the theory of strong interactions, in which the coupling constant becomes infinite in the limit $Q^2 \to 0$.

13.2   THE GELL–MANN AND LOW EQUATION

Using the results of the previous section, we can write the renormalised photon propagator in the form

$$e^2 D_{\mu\nu}(q) = \frac{-g_{\mu\nu}}{q^2} \, d(q^2, 0, e^2) \, , \qquad (13.14)$$

where we have indicated explicitly that the subtraction is performed at the momentum scale $\mu^2 = 0$.

The explicit expression for $d(q^2, 0, e^2)$, which agrees with the square of the effective charge introduced in (13.6), is

$$d(q^2, 0, e^2) = \frac{e^2}{1 - e^2 \Pi_c(q^2, 0, e^2)} \, , \qquad (13.15)$$

with

$$\Pi_c(q^2, 0, e^2) = \Pi(q^2, 0, e^2) - \Pi(0, 0, e^2), \qquad (13.16)$$

which implies $\Pi_c(0, 0, e^2) = 0$. Using (13.16) we can also write the square of the observed electric charge as

$$e^2 = d(0, 0, e^2) \, . \qquad (13.17)$$

If we now change the renormalisation scheme by choosing a different scale, $\mu^2 \neq 0$, equations (13.15) and (13.17) become, respectively

$$d(q^2, \mu^2, e_\mu^2) = \frac{e_\mu^2}{1 - e_\mu^2 [\Pi(q^2, \mu^2, e_\mu^2) - \Pi(\mu^2, \mu^2, e_\mu^2)]} \, , \qquad (13.18)$$

and

$$e_\mu^2 = d(\mu^2, \mu^2, e^2) \, . \qquad (13.19)$$

The expressions for the propagator which are obtained from (13.15) and (13.18) must agree, because both are equal to the divergent expression, which is obtained from the perturbative expansion in powers of e_0^2. It is thus true that

$$d(q^2, 0, e^2) = d(q^2, \mu^2, e_\mu^2) \, , \qquad (13.20)$$

which for $q^2 = 0$ allows us to establish the relation between e^2 and e_μ^2

$$e^2 = d(0, 0, e^2) = d(0, \mu^2, e_\mu^2) \, . \qquad (13.21)$$

We can now repeat the steps up to this point by choosing a different scale for the subtraction, with the objective of determining a connection between the propagators obtained using two arbitrary momentum scales, μ^2 and σ^2. From the previous equations it immediately follows that

$$d(q^2, \mu^2, e_\mu^2) = d(q^2, \sigma^2, e_\sigma^2) \ , \tag{13.22}$$

and therefore that

$$e_\sigma^2 = d(\sigma^2, \sigma^2, e_\sigma^2) = d(\sigma^2, \mu^2, e_\mu^2) \ . \tag{13.23}$$

The renormalised charge at σ^2 is the evolution at $q^2 = \sigma^2$ of the renormalised charge at μ^2.

Substituting (13.23) into the right-hand side of (13.22), the functional equation originally derived by Gell–Mann and Low [15] is obtained:

$$d(q^2, \mu^2, e_\mu^2) = d[(q^2, \sigma^2, d(\sigma^2, \mu^2, e_\mu^2)] \ , \tag{13.24}$$

which relates the coupling constants corresponding to the two different scales.

13.3   THE QED β FUNCTION

The Gell–Mann and Low equation (13.24) can be solved in the region of spacelike momenta —where there are no singularities linked to the production of real particles—in which $-\mu^2, -\sigma^2$ and $-q^2 \gg m^2$.

We firstly consider equation (13.20). The dependence on m of the left-hand side, while not being specified, is singular, as seen from (13.8). Even in the $-q^2 \to +\infty$ limit we cannot get to the theory in which the electron is massless. The mass singularity disappears in the theory renormalised at μ^2; the theory at zero mass is regular and the asymptotic dependence on q^2 becomes

$$\lim_{-q^2, -\mu^2 \gg m^2} \Pi_c(q^2, \mu^2) = \frac{1}{12\pi^2} \ln \frac{q^2}{\mu^2} \ . \tag{13.25}$$

The other difference between the two sides of (13.20) is that the right-hand side exhibits an explicit dependence on μ that must be exactly cancelled by the μ-dependence of e_μ^2, since the left-hand side does not depend on μ. The invariance of physical quantities upon the choice of the renormalisation point is, in fact, the essence of the renormalisation group and requires that

$$\frac{\partial d(q^2, \mu^2, e_\mu^2)}{\partial e_\mu^2} \ \mu \frac{\partial e_\mu^2}{\partial \mu} = -\mu \frac{\partial d(q^2, \mu^2, e_\mu^2)}{\partial \mu} \ . \tag{13.26}$$

If we limit ourselves to lowest order quantities, in the left-hand side we can set $\partial d / \partial e_\mu^2 = 1$, since the derivative of e_μ^2 is of at least second order in e^2. In the right-hand side, after having differentiated d, we can set $q^2 = \mu^2$, since the further dependence on q^2 is in higher order terms. We therefore find

$$\mu \frac{\partial e_\mu^2}{\partial \mu} = -\mu \left. \frac{\partial d(q^2, \mu^2, e_\mu^2)}{\partial \mu} \right|_{q^2 = \mu^2} \ . \tag{13.27}$$

It is usual to define the β *function* by the relation

$$\mu \frac{\partial e_\mu}{\partial \mu} = \beta(e_\mu) , \tag{13.28}$$

which, using (13.18) and (13.25), leads to

$$\beta(e_\mu) = +\frac{e_\mu^3}{12\pi^2} . \tag{13.29}$$

In electrodynamics extended to include different spin $\frac{1}{2}$ particles—e.g. quarks and leptons other than the electron—with electric charge Q_f, and spin 0 particles, with electric charge Q_s, the formula for the β function is immediately generalised to

$$\beta(e_\mu) = +\frac{e_\mu^3}{12\pi^2} \left(\sum_f Q_f^2 + \frac{1}{4} \sum_s Q_s^2 \right) . \tag{13.30}$$

13.4   ASYMPTOTIC VARIATION OF THE EFFECTIVE CHARGE

In the massless theory, the function $d(q^2, \mu^2, e_\mu^2)$ depends only on the ratio q^2/μ^2. This allows use of the renormalisation group relation, (13.28), to determine its asymptotic variation with q^2.

Let us consider the effective coupling constant at a momentum scale around the renormalisation point $q^2 \sim \sigma^2$

$$e^2(q^2) = d(q^2, \sigma^2, e_\sigma) , \tag{13.31}$$

and calculate

$$
\begin{aligned}
q^2 \frac{\partial e^2(q^2)}{\partial q^2} &= \frac{\partial d(q^2, \sigma^2, e_\sigma^2)}{\partial \ln q^2} = -\frac{\partial d(q^2, \sigma^2, e_\sigma^2)}{\partial \ln \sigma^2} \\
&= -\frac{1}{2}\sigma \frac{\partial d(q^2, \sigma^2, e_\sigma^2)}{\partial \sigma}\Big|_{q^2 \sim \sigma^2} = e(q^2)\beta[e(q^2)] ,
\end{aligned}
\tag{13.32}
$$

or

$$\frac{\partial e(q^2)}{\partial \ln q^2} = \frac{1}{2}\beta[e(q^2)] . \tag{13.33}$$

We can integrate the above equation, by separation of variables, between μ^2 and a given value of q^2. Thus we obtain

$$\int_{e(\mu^2)}^{e(q^2)} \frac{de}{\beta(e)} = \frac{1}{2} \int_{\mu^2}^{q^2} d \ln q^2 = \frac{1}{2}\ln\frac{q^2}{\mu^2} . \tag{13.34}$$

If we now use (13.29), we find

$$\int_{e(\mu^2)}^{e(q^2)} \frac{de}{\beta(e)} = 12\pi^2 \int \frac{de}{e^3} = -\frac{3\pi}{2} \int d(\frac{1}{\alpha}) \,, \qquad (13.35)$$

from which it follows that

$$\frac{1}{\alpha(q^2)} = \frac{1}{\alpha(\mu^2)} - \frac{1}{3\pi} \ln \frac{q^2}{\mu^2} \,, \qquad (13.36)$$

or

$$\alpha(q^2) = \frac{\alpha(\mu^2)}{1 - \frac{\alpha(\mu^2)}{3\pi} \ln \frac{q^2}{\mu^2}} \,, \qquad (13.37)$$

which again gives equation (13.9), with $\alpha(0) \to \alpha(\mu^2)$ and $m^2 \to \mu^2$.

The reader can easily show that the expression (13.37) and the analogous expression in terms of the subtraction point σ^2 satisfy (perturbatively in the constant α which appears in the numerators) the Gell–Mann and Low equation, (13.24).

Comment.   The formula (13.2) tells us that the dominant correction to the propagator grows as $\alpha \ln(q^2/m^2)$. For larger values of the logarithm we can no longer state that the higher-order terms are negligible. Actually the nth term of the series which we have conjectured in (13.3) is of order $[\alpha \ln(q^2/m^2)]^n$ and when the lowest-order term becomes of order one, to obtain a meaningful result, it is necessary to resum all the terms of the series, which are known as the *leading logarithmic terms* or *leading logs*.

The renormalisation group equations are based on the β function, which is obtained from a legitimate perturbative calculation, not invalidated by large logarithms; the calculation of the right-hand side in equation (13.27) is carried out by setting $\ln(q^2/\mu^2) = 1$ at the end. The result found in (13.37) shows that the renormalisation group equation (13.33) sums all the dominant logarithms in a rigorous way.

The calculation of the β function can be carried out to higher orders, to still higher orders, and so on (naturally with rapidly increasing difficulty!). Applying these results, the logarithms of order $\alpha^n [\ln(q^2/m^2)]^{n-1}$, $\alpha^n [\ln(q^2/m^2)]^{n-2}$, etc. are summed, and denoted as Next to Leading Logs (NLL), Next to Next to Leading Logs (NNLL), etc. The calculations of various variables of QCD, in which the effective constant is not small like that in electrodynamics, must be performed at least up to NNLL terms.

QUANTISING A NON-ABELIAN THEORY

CONTENTS

The gauge transformations considered in Chapter 5 for the electromagnetic field were extended by Yang and Mills [16] to local transformations based on a non-Abelian group. The resulting theories, known as Yang–Mills or gauge theories, provide the basis for a description of electroweak interactions and quantum chromodynamics.

The fundamental elements and the relevant aspects of Yang–Mills theories, at the level of the classical Lagrangian, are summarised in [2]. In this chapter we discuss the procedure for quantisation of the gauge fields in the formulation based on the Feynman path integral, following the method introduced by Faddeev and Popov in 1967 [17].

14.1   FUNDAMENTALS

We recall briefly the fundamental concepts and the notation.

Gauge transformations.   A gauge transformation is a correspondence between the points of space-time, x, and the elements, g, of a continuous group G: $x \to g(x)$. We consider the unitary matrices, $U(g)$, which represent G in a vector space. The following composition law must apply

$$U(g_2 g_1) = U(g_2)U(g_1) . \tag{14.1}$$

The gauge transformations act on the vectors, $\psi(x)$, of the representation as

$$\psi(x) \to U(g(x))\psi(x) = U(x)\psi(x) . \tag{14.2}$$

In applications, $\psi(x)$ are the quark and lepton fields and the Higgs field (matter fields) and the groups are $G = SU(2) \otimes U(1)$ or $SU(3)_{colour}$ for electroweak interactions or QCD, respectively (see [2]).

For infinitesimal transformations

$$U(x) = 1 + iT^A \alpha^A(x) \quad , \quad A = 1, \cdots N \; , \tag{14.3}$$

where T^A are the generators of the Lie algebra of G, N is the dimension of the algebra and $\alpha^A(x)$ are infinitesimal functions of the coordinates; the sum over repeated indices is understood. The generators satisfy commutation rules which characterise the algebra

$$\left[T^A, T^B\right] = if_{ABC}T^C \; , \tag{14.4}$$

with the *structure constants* of the group, f_{ABC}, completely antisymmetric in the indices.

The gauge fields are represented by matrices of the form

$$A_\mu(x) = T^A A_\mu^A(x) \; , \tag{14.5}$$

which transform as

$$A_\mu(x) \to (A_\mu)^U = U(x)A_\mu(x)U^\dagger(x) - iU(x)\partial_\mu U^\dagger(x) \; , \tag{14.6}$$

or, for infinitesimal transformations,

$$A_\mu^A \to (A_\mu^A)' = A_\mu^A - f_{ABC}\alpha^B A_\mu^C - \partial_\mu \alpha^A \; . \tag{14.7}$$

Applying (14.6) successively with two gauge matrices, U and V, we find the composition law of two gauge transformations

$$\begin{aligned}
A_\mu(x) \to V &\left[U A_\mu U^\dagger - iU\partial_\mu U^\dagger\right] V^\dagger - iV\partial_\mu V^\dagger \\
&= VU A_\mu U^\dagger V^\dagger - iVU\partial_\mu U^\dagger V^\dagger - iVUU^\dagger\partial_\mu V^\dagger \\
&= (A_\mu)^{VU} \; ,
\end{aligned} \tag{14.8}$$

in agreement with (14.1). Equation (14.7) clearly shows that the transformation law of gauge fields does not depend on the specific representation that we have chosen for ψ, but only on the algebra of G, characterised by the structure constants.

Starting from (14.7), we can define the covariant derivative of ψ, $D_\mu(A)\psi$, which transforms like ψ

$$\begin{aligned}
D_\mu(A)\psi &= \partial_\mu \psi + iA_\mu \psi \; , \\
D_\mu(A)\psi &\to \partial_\mu(U\psi) + iU A_\mu(x)U^\dagger U\psi + U(\partial_\mu U^\dagger)U\psi \\
&= U\left[\partial_\mu \psi + iA_\mu \psi\right] \\
&= U D_\mu(A)\psi \; .
\end{aligned} \tag{14.9}$$

The gauge transformations of QED correspond to the Abelian case and are obtained from the previous formulae by setting

$$f_{ABC} = 0 \quad , \quad T^A \to Q \; . \tag{14.10}$$

The Yang–Mills tensor. The final element necessary for the theory is the Yang–Mills tensor, the analogue of the Maxwell tensor of electromagnetism. In matrix form, starting from (14.5), we define

$$G_{\mu\nu} = T^A G^A_{\mu\nu} = \partial_\nu A_\mu - \partial_\mu A_\nu + i\,[A_\nu, A_\mu]\;,$$
$$G^A_{\mu\nu} = \partial_\nu A^A_\mu - \partial_\mu A^A_\nu + f_{ABC} A^B_\mu A^C_\nu\;. \tag{14.11}$$

The Yang–Mills tensor transforms linearly (as matter fields do) according to a group representation

$$G_{\mu\nu} \to (G_{\mu\nu})^U = U G_{\mu\nu} U^\dagger, \tag{14.12}$$

or

$$G^A_{\mu\nu} \to (G^A_{\mu\nu})' = G^A_{\mu\nu} - f_{ABC}\alpha^B G^C_{\mu\nu}\;. \tag{14.13}$$

If we define

$$(T^C_{reg})_{AB} = i f_{ACB}\;, \tag{14.14}$$

it can be seen that these matrices satisfy the commutation rules of the algebra and therefore define an irreducible representation of the algebra, called the regular or adjoint representation (see [2], Appendix C). In terms of these matrices, (14.13) takes the form

$$G^A_{\mu\nu} \to (G^A_{\mu\nu})' = \left[(1 + i T^C_{reg}\alpha^C) G_{\mu\nu}\right]^A\;. \tag{14.15}$$

The Yang–Mills Lagrangian (classical limit). The starting point is a Lagrangian of matter fields, $\mathcal{L}_0(\psi, \partial^\mu \psi)$, which should be *invariant under global transformations* of the group G. From (14.9) it follows that a Lagrangian *invariant for local transformations* of G is obtained with the minimal substitution: $\partial^\mu \to D(A)^\mu$, or that

$$\mathcal{L}_{mat} = \mathcal{L}_0[\psi, D(A)^\mu \psi] \tag{14.16}$$

is gauge invariant.

From (14.15) it follows that the function of the gauge fields represented by

$$\mathcal{L}_{gauge} = -\frac{1}{4g^2}\sum_A (G^A_{\mu\nu} G^{A,\mu\nu})\;, \tag{14.17}$$

is an invariant function and can be chosen as the Lagrangian of the gauge fields.

For a *simple* group, i.e. one that cannot be decomposed into products of commuting factors (for example $SU(3)_{colour}$), (14.17) is unique, but for *semi-simple* groups (for example $SU(2) \otimes U(1)$), the matrices $(T^C_{reg})_{AB}$ are block-diagonal and the sum in (14.17) limited to each block is itself invariant. In this case we have as many constants g as there are simple factors

$$\mathcal{L}_{gauge} = -\sum_i \left[\frac{1}{4g_i^2} \sum (G_{\mu\nu}^A G^{A,\mu\nu})_i \right] , \tag{14.18}$$

where the internal sum extends over each block. The constants g or g_i in (14.17) and (14.18) are the coupling constants of the gauge fields, as is immediately seen by rescaling the fields A_μ^A in the previous equations:

$$A_\mu^A \to g A_\mu^A ,$$
$$D_\mu(A)\psi \to \partial_\mu \psi + ig T^A A_\mu^A \psi ,$$
$$G_{\mu\nu}^A \to g \left[\partial_\nu A_\mu^A - \partial_\mu A_\nu^A + g f_{ABC} A_\mu^B A_\nu^C \right] . \tag{14.19}$$

In the limit $g \to 0$, the Lagrangian of the gauge fields reduces to a sum of Maxwell Lagrangians, one for each index A.

In the same limit, the gauge fields describe N massless, free particles with spin 1, similar to the photon, while matter is described by the Lagrangian $\mathcal{L}_0$. The Standard Theory, in this limit, contains quarks and leptons interacting with the Higgs field (Yukawa interactions).

In conclusion, the Lagrangian of a Yang–Mills theory is given by

$$\mathcal{L}_{YM} = \mathcal{L}_0[\psi, D(A)^\mu \psi] + \mathcal{L}_{gauge} . \tag{14.20}$$

14.2   QUARKS IN QUANTUM CHROMODYNAMICS

In this section and the following we narrow our discussion to the case of quantum chromodynamics (QCD), the gauge theory which is invariant under the $SU(3)$ group associated with the *colour* quark symmetry, both for its physical importance and because it contains all the relevant elements for a general gauge theory with exact symmetry.[1]

QCD describes the fundamental strong interactions, due to interactions between $SU(3)$ gauge fields, the *gluons*, described by the regular **8** representation, and quarks, spin $\frac{1}{2}$ particles attributed to the fundamental **3** representation of the group (correspondingly, antiquarks transform according to the conjugate representation **3̄**).

At present we know of six different types of quark, characterised by quantum numbers denoted by the generic name *flavour*, which identify their properties under electromagnetic and weak interactions, as shown in Figure 1.1. For weak interactions, the six quarks can be classified into three doublet-singlet families (or generations), with identical interactions, distinguished only by

---

[1]The case of a gauge theory with spontaneously broken symmetry will be considered in Chapters 18 and 19 and is introduced in more detail in Ref. [2].

their different masses:

$$\begin{pmatrix} u \\ d \end{pmatrix}_L, \; u_R, \; d_R \quad \text{(generation one)} ,$$

$$\begin{pmatrix} c \\ s \end{pmatrix}_L, \; c_R, \; s_R \quad \text{(generation two)} ,$$

$$\begin{pmatrix} t \\ b \end{pmatrix}_L, \; t_R, \; b_R \quad \text{(generation three)} . \tag{14.21}$$

The suffixes L and R in (14.21) denote *left-handed* and *right-handed* components of the fields. In each doublet, the electric charge has the value $\frac{2}{3}$ and $-\frac{1}{3}$ for the upper and lower component, respectively. To the previous properties, we must add the colour charge.

The lightest baryons (proton, neutron, hyperons) are made of three spin $\frac{1}{2}$ quarks in an overall symmetric state for exchange of the quantum numbers of space, spin and flavour. Colour was originally introduced to reestablish the antisymmetry required by the half-integer spin of the quarks. The choice of $SU(3)$ is motivated by the possibility to attribute the quarks to the **3** representation, which allows us to describe baryons as completely antisymmetric, colour singlet states.

The three colour states of each quark are traditionally denoted with the names of three different colours, for example *blue, green, red,* or by assigning an index with three values to the field of a quark of a given flavour, e.g. $u_{L,R} \to u_{L,R}^a$, with $a = 1, 2, 3$.

The colour interactions conserve parity, hence *left-* and *right-handed* components enter on the same level into the Dirac combination:

$$u_L^a + u_R^a = \frac{1 - \gamma_5}{2} u^a + \frac{1 + \gamma_5}{2} u^a = u^a . \tag{14.22}$$

A generation structure similar to (14.21) is encountered in the *leptons*, spin $\frac{1}{2}$ particles sensitive only to electromagnetic and weak interactions:

$$\begin{pmatrix} \nu_e \\ e \end{pmatrix}_L, \; e_R \quad \text{(generation one)} ,$$

$$\begin{pmatrix} \nu_\mu \\ \mu \end{pmatrix}_L, \; \mu_R \quad \text{(generation two)} ,$$

$$\begin{pmatrix} \nu_\tau \\ \tau \end{pmatrix}_L, \; \tau_R \quad \text{(generation three)} , \tag{14.23}$$

where we have assumed that the observed neutrinos exist only in *left-handed* form.

The assignment of zero colour charge to the leptons implies in a natural way that these particles are not affected by strong interactions.

14.3 THE FADDEEV–POPOV DETERMINANT

Using the notation of Section 5.1 adapted to the case of a general gauge group, we denote with $f^A[A(x)]$, or simply $f^A(A)$, the function of A_μ that fixes the gauge for all components of the gauge field, for example

$$f^A[A(x)] = \partial^\mu A_\mu^A(x) , \quad \text{(Lorenz gauge)} , \tag{14.24}$$

and by $B[f(A)]$ a functional of f which introduces a convergence factor along the gauge orbits, which are characterised by $(A_\mu^A)^U = (A_\mu^A)_0 + f_{ABC}(A_\mu^B)_0 \alpha^C - \partial_\mu \alpha^A$, with $f(A_0) = 0$, and illustrated in Figure 14.1.

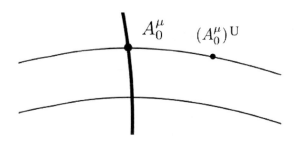

Figure 14.1 Space of the functions $A^\mu(x)$. Trajectories composed of paths connected by gauge transformations are indicated by the horizontal lines. Along coordinates orthogonal to the gauge trajectories (shown by the vertical line) we find physically distinct paths.

To restrict the integration to the vertical line, we can use a δ-function

$$B[f(A)] = \delta[f(A)] . \tag{14.25}$$

In what follows we will use an alternative, the Gaussian functional:

$$B[f] = e^{\frac{-i}{2\alpha} \int d^4x \, f(x)^2} = e^{\frac{-i}{2\alpha} \int d^4x \, (\partial_\mu A^\mu)^2} , \tag{14.26}$$

from which the δ-function can be obtained by taking the limit $\alpha \to 0$.

We now consider the "naive" generating functional

$$W = \int d[A_\mu] \, e^{iS(A)} , \tag{14.27}$$

where $S(A)$ is the Yang–Mills action. To restrict the integration only to non-equivalent configurations, we could insert the function $\delta[f(A^U)]$ and integrate over the gauge transformations, so as to extract the measure of the gauge orbit. However, this would not be correct, as seen from a very simple case, from the fact that

$$\int dx \, \delta[f(x)] = \frac{1}{f'(x_0)} \neq 1 . \tag{14.28}$$

The correct factor to insert in (14.27) is, rather, the analogue of

$$\int dx f'(x)\, \delta[f(x)] = \int df\, \delta(f) = 1 \ . \tag{14.29}$$

In the case of the functional integration, this means inserting into (14.27) a factor 1 written in the form

$$1 = \Delta[A] \int d[U(x)]\, \delta[f(A^U(x))] \ , \tag{14.30}$$

where

$$U(x) = e^{i\alpha^A(x)T^A} \ , \quad d[U(x)] = \Pi_A d[\alpha^A(x)]; \ , \tag{14.31}$$

and $\Delta[A]$ is the Faddeev–Popov determinant:

$$\Delta[A] = \mathrm{Det}[K_{FP}] = \mathrm{Det}\left[\frac{\delta f^A[A(x)^U]}{\delta[\alpha^B(y)]}\right] \ . \tag{14.32}$$

Equation (14.32) shows that Δ is simply the Jacobian of the transformation which changes the variables of the gauge transformation to the coordinates of the gauge orbit in Figure 14.1, i.e. the functions $f^A(A^U)$.

As a first step, we will show that the Faddeev–Popov determinant is gauge invariant. To see this, we make the substitution $A \to A^V$ in (14.30), where V is a fixed gauge transformation. Equation (14.30) becomes

$$1 = \Delta[A^V] \int d[U(x)]\delta[f((A^V)^U(x))]$$

$$= \Delta[A^V] \int d[U(x)]\delta[f(A^{(UV)(x)})] \ . \tag{14.33}$$

Combining the gauge transformation U with a fixed one V, we obtain a new gauge transformation which, as U varies, covers the same functional space as U; multiplication by V is the equivalent of a translation in the manifold of gauge transformations, under which the measure $d[U]$ is invariant. Therefore we find

$$1 = \Delta[A^V] \int d[U(x)]\, \delta[f(A^{(UV)(x)})]$$

$$= \Delta[A^V] \int d[U(x)]\, \delta[f(A^U)] = \frac{\Delta[A^V]}{\Delta[A]} \ . \tag{14.34}$$

We now substitute both sides of (14.30) into (14.27). Using the gauge invariance of the integration measure over A_μ, from the Yang–Mills action and from K_{FP}, we find

$$
\begin{aligned}
W &= \int d[A_\mu]\, e^{iS(A)} = \int d[A_\mu]\, \Delta[A] \int d[U(x)] \delta[f(A^U(x))]\, e^{iS(A)} \\
&= \int d[A_\mu^U] \int d[U(x)]\, \Delta[A^U]\, \delta[f(A^U(x))]\, e^{iS(A^U)} \\
&= \left[\int d[U(x)] \right] \times \int d[A_\mu]\, \Delta[A]\, \delta[f(A)]\, e^{iS(A)} \, .
\end{aligned}
\tag{14.35}
$$

The factor which multiplies the final term of (14.35) is a multiplicative constant independent of the fields, which we can omit from the generating functional. Furthermore, we can replace the Dirac δ-function with a more regular functional, such as the Gaussian functional in (14.26), to obtain

$$
W[A] = \int d[A_\mu]\, \Delta[A]\, e^{i[S(A) + S_{gf}(A)]} \, ,
\tag{14.36}
$$

with

$$
S_{gf}(A) = -\frac{1}{2\alpha} \int d^4x\, (\partial^\mu A_\mu^A)^2 \, .
\tag{14.37}
$$

The same conclusion is reached if we consider the vacuum expectation value of a gauge-invariant observable $\mathcal{O}[A]$. Starting from the formula analogous to (14.27) and normalising the matrix element leads to

$$
\begin{aligned}
\langle 0|\mathcal{O}[A]|0\rangle &= \frac{\int d[A_\mu]\, e^{iS[A]}\, \mathcal{O}[A]}{\int d[A_\mu]\, e^{iS[A]}} = \frac{\int d[A_\mu]\, e^{iS[A]}\, \mathcal{O}[A]\, \Delta[A]\, B[f(A)]}{\int d[A_\mu]\, e^{iS[A]}\, \Delta[A]\, B[f(A)]} \\
&= \frac{\int d[A_\mu]\, \Delta[A]\, e^{iS_{new}[A]}\, \mathcal{O}[A]}{\int d[A]_\mu\, \Delta[A]\, e^{iS_{new}[A]}} \, .
\end{aligned}
\tag{14.38}
$$

The first fraction corresponds to the naive prescription [cf. equation (5.5)] derived by analogy with the case of the scalar field, and which formally satisfies the gauge invariance if $\mathcal{O}[A]$ is gauge invariant. However, the numerator and the denominator are both separately divergent because of the volume of the gauge orbits. In the second fraction, the functional $B[f]$ provides a convergence factor along the gauge orbit and both the numerator and denominator are separately convergent. In the third fraction we have absorbed, for convenience, the functions $B[f]$ into an overall action, S_{new}

$$
e^{iS_{new}[A]} = e^{iS[A]}\, B[f(A)] = e^{i(S[A] + S_{gf}[A])} \, .
\tag{14.39}
$$

The overall action S_{new} is different from the Yang–Mills action and *is not gauge invariant*. However, the equality (14.38) which we just derived shows that *for gauge-invariant quantities* the third fraction gives the gauge-invariant result required.

Note. In order that the change of variables from the gauge transformations to the coordinates of the gauge orbit in Figure 14.1 [i.e. the functions $f^A(A^U)$] should be valid, it is necessary that the relation between the f^A and the α^A should be unique. This implies that the point on the gauge orbit in Figure 14.1 must not return to the same value for different values of the α^A, or, equivalently, that there are not multiple solutions of the equation

$$f^A(A) = (\partial^\mu A_\mu^A) = 0 \ . \tag{14.40}$$

The possibility of multiple solutions of the gauge conditions, for non-Abelian gauge theories, was considered by Gribov [18] and the different possible solutions of (14.40) are called *Gribov copies*. It is thought that Gribov copies can play a role in the theory of quark confinement. We will remain in the framework of perturbation theory in which we move in an infinitesimal neighbourhood of A_0 where Gribov copies are not encountered, and we can therefore ignore the problem.

We conclude with the calculation of the Faddeev–Popov determinant in the case in which the function that fixes the gauge is given by the Lorenz condition (14.40). Using (14.32) and (14.7), we find

$$\begin{aligned} K_{FP} &= \frac{\delta[\partial^\mu (A_\mu^A(x))^U]}{\delta \alpha^B(y)} \\ &= \partial^\mu \left[f_{ACB}\delta(x-y)A_\mu^C(x) - \delta(x-y)\delta_{AB}\partial_\mu \right] \\ &= -\partial^\mu \delta(x-y)(D(A)_\mu)_{AB} \ , \end{aligned} \tag{14.41}$$

where the covariant derivative is that appropriate for the regular representation.

In the case of QED ($f_{ABC} = 0$)

$$\Delta[A] = \det(-\Box) \ , \tag{14.42}$$

independent of A_μ. In this case $\Delta[A]$ is a constant, which is simplified in equation (14.38), and adding the gauge-fixing term to the action is sufficient, as anticipated in Chapter 5.

A suggestive and useful representation of the Faddeev–Popov determinant can be obtained by noting that the determinant can itself be expressed as a Gaussian functional integral of *anticommuting functions* as we saw in Section 6.1.3. If $\phi(x)$ is a complex function of this type, we may write

$$\Delta[A] = \int d[\phi]d[\phi^\dagger]e^{iS_{ghost}} \ ,$$

$$S_{ghost} = \int d^4x \left[(\partial^\mu \phi^\dagger)D_\mu(A)\phi \right] \ . \tag{14.43}$$

Equation (14.43) describes massless scalar particles interacting with the gauge fields. However, these particles have the wrong statistics to represent spin 0 physical particles. The spin–statistics theorem can be violated by allowing the states of these particles to have a negative norm. States of this kind are not allowed among the physical states of quantum mechanics, but can appear as intermediate states of the S-matrix between physical states.[2] The need for diagrams with internal ghost loops, in addition to diagrams which contain gauge field loops, was derived in 1963 by Feynman [19] and subsequently by deWitt [20].

In conclusion, the overall action to insert into the functional integral over the gauge fields is

$$S_{tot} = \int d^4x \, \mathcal{L}_{tot} , \qquad (14.44)$$

$$\mathcal{L}_{tot} = \mathcal{L}_{YM} - \frac{1}{2\alpha}(\partial^\mu A_\mu^A)^2 + (\partial^\mu \phi^\dagger)D_\mu(A)\phi , \qquad (14.45)$$

with $\mathcal{L}_{YM}$ given by (14.20).

14.4 FEYNMAN RULES

We can expand the Lagrangian (14.45) in powers of the fields, obtaining quadratic, cubic and quartic terms:

$$\mathcal{L}_{tot} = \mathcal{L}^{(2)} + \mathcal{L}^{(3)} + \mathcal{L}^{(4)} . \qquad (14.46)$$

The expansion corresponds to an expansion in powers of the coupling constant g:

$$\mathcal{L}^{(2)} = \mathcal{O}(g^0) , \quad \mathcal{L}^{(3)} = \mathcal{O}(g) , \quad \mathcal{L}^{(4)} = \mathcal{O}(g^2) . \qquad (14.47)$$

The quadratic terms give rise to the free propagators. Referring to the first line of Figure 14.2, for the quarks (spin $\frac{1}{2}$ and mass m) and for the FP ghosts (spin 0 and zero mass), we find [cf. equations (3.62) and (6.64)]

$$iS(p) = \frac{i}{\not{p} - m + i\epsilon}\, \delta_{ab} , \qquad \text{(spin } \frac{1}{2}; \, a,b = 1,2,3) , \qquad (14.48)$$

$$i\Delta(p) = \frac{i}{p^2 + i\epsilon}\, \delta_{AB} , \qquad \text{(FP ghost; } A,B = 1,\cdots,8) . \qquad (14.49)$$

For the gauge fields, $\mathcal{L}^{(2)}$ coincides, for all values of A, with the Lagrangian of the electromagnetic field considered in Chapter 5. In the general gauge, identified by the parameter α of equation (14.26), we find [cf. equation (5.19)],

$$i\Delta_{\mu\nu}^{AB}(p) = \frac{i}{p^2 + i\epsilon}\left[-g_{\mu\nu} + (1-\alpha)\frac{p_\mu p_\nu}{p^2}\right]\delta_{AB} ,$$

$$\text{(gauge fields; } A,B = 1,\cdots,8) . \qquad (14.50)$$

---

[2]Confirming our previous statement, (14.43) shows that in the abelian case, $f_{ABC} = 0$, ghosts are free particles that can simply be omitted.

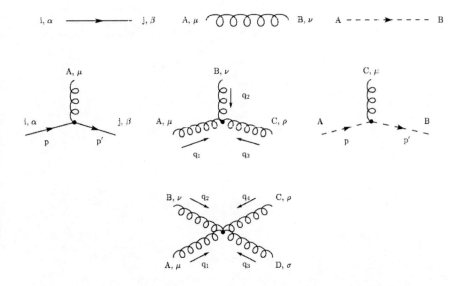

Figure 14.2 The relevant Feynman diagrams for propagators and vertices of a Yang–Mills theory with exact gauge symmetry, for example QCD. Codes for the lines: fermion matter fields, continuous arrowed lines; gauge fields, wavy lines; ghosts, dashed arrowed lines. The indices present on the figure refer to QCD. Quarks (representation 3 and $\bar{3}$), i, j: colour indices; α, β: Dirac spin indices. Gauge and ghost fields (representation 8), $A, B, \cdots$: colour indices; $\mu, \nu, \cdots$: Lorentz indices. Conservation of 4-momentum at every vertex is to be understood; the 4-momenta follow the directions of the lines for quarks and ghosts, or the direction of the small arrows for the gauge fields (thus, for example, $q_1 + q_2 + q_3 = 0$).

The cubic terms give rise to the vertices illustrated in the second line of Figure 14.2. Once again in terms of the Fourier transform we find, in order[3]

$$iV_{ab}^{A\mu}(p', p) = ig(\frac{\lambda^A}{2})_{ab}\gamma^\mu , \tag{14.51}$$

$$i\Gamma(q_1, q_2, q_3)_{ABC}^{\mu\nu\rho} = -gf_{ABC} \times V^{\mu\nu\rho}(q_1, q_2, q_3) =$$
$$-gf_{ABC} \times [g_{\mu\nu}(q_1)_\rho - g_{\mu\rho}(q_1)_\nu + g_{\nu\rho}(q_2)_\mu - g_{\nu\mu}(q_2)_\rho + g_{\rho\mu}(q_3)_\nu - g_{\rho\nu}(q_3)_\mu]$$
$$= -gf_{ABC} [g_{\mu\nu}(q_1 - q_2)_\rho + g_{\nu\rho}(q_2 - q_3)_\mu + g_{\rho\mu}(q_3 - q_1)_\nu] , \tag{14.52}$$

$$i(G_{C\mu})_{BA}(p', p) = -gf_{BCA}p'_\mu . \tag{14.53}$$

---

[3]The first term corresponds to the vertex $ie\gamma^\mu$ of electrodynamics (compare with the Feynman rules of Chapter 10).

Finally, for the vertex in the last line of Figure 14.2, we find

$$i\Gamma^{\mu\nu\rho\sigma}_{ABCD} = -ig^2 \left[f_{ABN}f_{CDN}(g_{\mu\rho}g_{\nu\sigma} - g_{\mu\sigma}g_{\nu\rho}) \right. \tag{14.54}$$
$$\left. + f_{ACN}f_{BDN}(g_{\mu\nu}g_{\rho\sigma} - g_{\mu\sigma}g_{\rho\nu}) + f_{ADN}f_{CBN}(g_{\mu\rho}g_{\sigma\nu} - g_{\mu\nu}g_{\rho\sigma}) \right].$$

To the usual Feynman rules which associate diagrams with amplitudes as in electrodynamics, we must add the following:

- Ghost lines form closed loops; they cannot appear as external lines;

- Each ghost loop contributes a factor (-1).

THE β FUNCTION IN QCD

CONTENTS

As applications of the concepts developed in the previous chapter, we derive the β function and the proof of asymptotic freedom of QCD.

15.1   VACUUM POLARISATION

The diagrams associated with the renormalisation of the gluon propagator are shown in Figure 15.1. In preparation for the proof of the asymptotic freedom of QCD, we calculate the corresponding amplitudes. Aiming for the high energy limit, we consider the limit of massless fermions and we renormalise the amplitudes to a momentum transfer value of $q^2 = \mu^2 < 0$. For each diagram, we calculate the amplitude resulting from the Feynman rules. The vacuum polarisation tensor, as seen from (11.2), is obtained from

$$\text{Feynman amplitude} = i\Pi_{\mu\nu} \ . \tag{15.1}$$

We can restrict the calculation to the first three diagrams, since the diagram in Figure 15.1(d) contributes a constant, that is independent of q, which disappears in the renormalisation procedure.

Fermions.   The first diagram of Figure 15.1 describes the creation and annihilation of a fermion–antifermion pair, and is obtained directly from the calculation of the QED vacuum polarisation, Chapter 12, with the substitution

$$e^2 \to g^2 \text{Tr}(\frac{\lambda^A}{2} \frac{\lambda^B}{2}) \ . \tag{15.2}$$

The Gell–Mann $SU(3)$ matrices are normalised according to

$$\text{Tr}(\lambda^A \lambda^B) = 2\delta^{AB} , \qquad (15.3)$$

and we define, for each representation R, the quantity $T(R)$ by

$$\text{Tr}(T^A T^B) = \delta^{AB} T(R) . \qquad (15.4)$$

$T(R)$ is proportional to the value of the operator $C(R) = \sum_A (T^A T^A)$, known as the Casimir operator.

The vacuum polarisation due to fermions, represented by diagram (a) of Figure 15.1, is therefore obtained by the substitution

$$e^2 \rightarrow \frac{1}{2} g^2 \delta^{AB} = g^2 T(R) \delta^{AB} , \qquad (15.5)$$

where R denotes the fermion representation, the **3** representation in the case of quarks.[1]

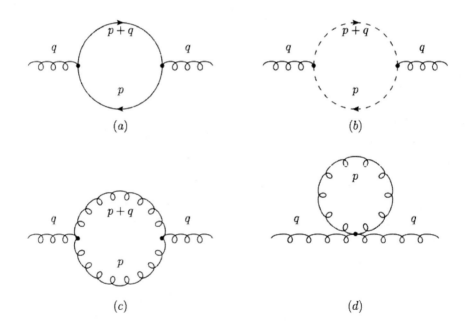

Figure 15.1   Vacuum polarisation corrections in QCD.

From the definitions given in Chapter 11 and from (12.52) we find, in the

---

[1]On the representation **3**, $T(\mathbf{3}) = 1/2$ from which, multiplying (15.3) by δ^{AB}, we obtain $8T(\mathbf{3}) = 3C(\mathbf{3})$ or $C(\mathbf{3}) = 4/3$.

$m \to 0$ limit

$$i(\Pi^{(a)})_{\mu\nu}^{AB}(k) = \delta^{AB}(g^{\mu\nu}q^2 - q^\mu q^\nu)i\Pi^{(a)}(q^2) ,$$

with

$$i\Pi^{(a)}(q^2) = i\frac{g^2}{2\pi^2}T(R)\int_0^1 dx\; x(1-x)\log\left[\frac{m^2 - x(1-x)q^2}{m^2 + x(1-x)|\mu^2|}\right]$$

$$= i\frac{g^2}{12\pi^2}T(R)\;\log\left[\frac{-q^2}{|\mu^2|}\right] . \tag{15.6}$$

Gauge fields. We now calculate the amplitude for the diagram of Figure 15.1(c) in the Feynman gauge, corresponding to $\alpha = 1$. From (14.52) we find

$$i(\Pi^{(c)})_{\mu\nu}^{AB} = \frac{1}{2}g^2 f_{ALC}f_{LBC}\int \frac{d^D p}{(2\pi)^4}\; V_{\mu\rho\sigma}(q, -p - q, p)(-g^{\rho\alpha})$$

$$\times V_{\alpha\nu\delta}(p + q, -q, -p)(-g^{\delta\sigma})\frac{(i)^2}{[(p+q)^2 + i\epsilon][(p)^2 + i\epsilon]} . \tag{15.7}$$

Proceeding as in Appendix E, we integrate over the momentum p in D dimensions. V is the colour-independent part of the Yang–Mills vertex in (14.52), and we recall that by convention the momenta are all taken as incoming (this explains the various minus signs on the indicated momenta). The factor $\frac{1}{2}$ in the formula, which does not follow directly from the Feynman rules, will be discussed in a note at the end of this section.

Recalling the definition of the generators in the adjoint representation, see equation (14.15), the colour factors lead to the constant $T(R)$ where now R is the adjoint representation, which we denote with A. Thus

$$T(A)\delta^{AB} = \text{Tr}[T_{reg}^A T_{reg}^B] =$$

$$= if_{CALi}f_{LBC} = f_{ALC}f_{LCB} = -f_{ALC}f_{LBC} . \tag{15.8}$$

Carrying out the multiplications shown, we cast $(\Pi^{(c)})_{\mu\nu}^{AB}$ in the form

$$i(\Pi^{(c)})_{\mu\nu}^{AB} = g^2\delta^{AB}T(A)\int \frac{d^D p}{(2\pi)^4}\; \frac{N_{\mu\nu}(p,q)}{[(p+q)^2 + i\epsilon][(p)^2 + i\epsilon]} , \tag{15.9}$$

with

$$N_{\mu\nu}(p,q) = -q_\mu q_\nu + \frac{5}{2}(q_\mu p_\nu + p_\mu q_\nu) + 5p_\mu p_\nu + g_{\mu\nu}(\frac{5}{2}q^2 + (qp) + p^2) . \tag{15.10}$$

We now introduce the integration over the Feynman parameter, equation (E.8), to obtain

$$i(\Pi^{(c)})_{\mu\nu}^{AB}(q) = g^2\delta^{AB}T(A)\int_0^1 dx \int \frac{d^D p}{(2\pi)^4}\; \frac{N_{\mu\nu}(p,q)}{[(p+xq)^2 - s + i\epsilon]^2} , \tag{15.11}$$

where

$$s = -x(1-x)q^2 , \qquad (15.12)$$

and changing the integration variable to $p = k - xq$, we arrive at

$$N_{\mu\nu} = -q_\mu q_\nu[1+5x(1-x)]+g_{\mu\nu}q^2[\tfrac{5}{2}-x(1-x)]+5k_\mu k_\nu+g_{\mu\nu}k^2+\cdots . \quad (15.13)$$

The ellipsis stands for terms linear in k which vanish in the integration. Furthermore, we can make the substitution $k_\mu k_\nu \to g_{\mu\nu}k^2 D^{-1}$ in the integral and also directly set $D = 4$ in the numerator.

In conclusion, we must calculate the two integrals:

$$I_1 = \int d^D k \; \frac{1}{[k^2 - s + i\epsilon]^2} , \qquad (15.14)$$

$$I_2 = \int d^D k \; \frac{k^2}{[k^2 - s + i\epsilon]^2} . \qquad (15.15)$$

In the notation of Appendix E, having set $D = 4 - \eta$,

$$I_1 = I(s, D, 2) = i\pi^2\Gamma(\eta/2)\frac{1}{s^{\eta/2}} , \qquad (15.16)$$

$$I_2 = I(s, D, 1) + sI(s, D, 2) = 2sI(s, D, 2) . \qquad (15.17)$$

Taking the limit $\eta \to 0$, we find

$$i(\Pi^{(c)})_{\mu\nu}^{AB}(q) = i\frac{g^2}{16\pi^2}\delta^{AB}T(A)\int_0^1 dx \; P_{\mu\nu}(q,x)\left[\frac{2}{\eta} - \log[-x(1-x)q^2]\right]$$

$$= -i\frac{g^2}{16\pi^2}\delta^{AB}T(A) \; \log[-q^2]\int_0^1 dx \; P_{\mu\nu}(q,x) + \text{Pol}(q) ,$$

$$(15.18)$$

with

$$P_{\mu\nu}(q, x) = -q_\mu q_\nu[1 + 5x(1 - x)] + \tfrac{1}{2}g_{\mu\nu}q^2[5 - 11x(1 - x)] , \qquad (15.19)$$

where $\text{Pol}(q)$ is a polynomial in q, which includes the infinite constant $2/\eta$ and that is determined by the condition that $\Pi_{\mu\nu}$ vanishes at the renormalisation point $q^2 = \mu^2 < 0$.

We can now integrate directly over x, with the result

$$i(\Pi^{(c)})_{\mu\nu}^{AB}(q) = i\frac{g^2}{32\pi^2}\delta^{AB}T(A)\log\left(\frac{-q^2}{|\mu|^2}\right)(-\frac{19}{6}g_{\mu\nu}q^2 + \frac{11}{3}q_\mu q_\nu) . \quad (15.20)$$

Ghost fields. For the diagram of Figure 15.1(b), by using the vertex in (14.51), we find

$$i(\Pi^{(b)})_{\mu\nu}^{AB}(q) = (-1)f_{CAL}f_{LBC}\int\frac{d^D p}{(2\pi)^4}\,p_\mu(p+q)_\nu\frac{(i)^2}{(p^2+i\epsilon)[(p+q)^2+i\epsilon]}\,,$$

(15.21)

where we have made explicit the minus sign associated with the ghost loop.

Recalling equation (15.8) and introducing the integration over the Feynman parameter, we find

$$i(\Pi^{(b)})_{\mu\nu}^{AB}(q) = -\delta^{AB}T(A)\int_0^1 dx\int\frac{d^D k}{(2\pi)^4}\frac{(k-xq)_\mu(k+(1-x)q)_\nu}{(k^2-s+i\epsilon)^2}\,.$$

(15.22)

Proceeding as before leads to

$$i(\Pi^{(b)})_{\mu\nu}^{AB}(q) = i\frac{g^2}{32\pi^2}\delta^{AB}T(A)\left(-\frac{1}{3}q_\mu q_\nu - \frac{1}{6}g_{\mu\nu}q^2\right)\log\left(\frac{-q^2}{|\mu^2|}\right)\,,$$

(15.23)

and combining this result with (15.20), we find the gauge-invariant result

$$i(\Pi^{(b+c)})_{\mu\nu}^{AB}(q) = i\frac{g^2}{32\pi^2}\delta^{AB}T(A)\frac{10}{3}\,\log\left(\frac{-q^2}{|\mu^2|}\right)(-g_{\mu\nu}q^2 + q_\mu q_\nu)\,.$$

(15.24)

Finally, adding the contribution of n_f quark triplets, we arrive at the overall result

$$i(\Pi^{(a+b+c)})_{\mu\nu}^{AB}(q) = i\frac{g^2}{16\pi^2}\delta^{AB}\left[\frac{5}{3}T(A) - \frac{4}{3}n_f T(R)\right]$$

$$\times \log\left(\frac{-q^2}{|\mu^2|}\right)(-g_{\mu\nu}q^2 + q_\mu q_\nu)\,.$$

(15.25)

Note. The factor $\frac{1}{2}$ which appears in (15.7) requires explanation. We consider the case in which $A = B = 1$ and the intermediate particles are gluons of type 2 and 3. The factor in front of the integral is

$$g^2\,f_{123}f_{213}\,.$$

(15.26)

There are no further multiplicity factors because the vertices are symmetric for the exchange of particles 2 and 3. If we sum the amplitudes corresponding to the other pairs of intermediate particles, we obtain a factor

$$g^2\sum_{L>C}f_{1LC}f_{L1C} = \frac{1}{2}g^2\sum_{L,C}f_{1LC}f_{L1C} = \frac{1}{2}g^2 f_{1LC}f_{L1C}\,.$$

(15.27)

Note that the factor $\frac{1}{2}$ does not appear in the vacuum polarisation due to ghosts, equation (15.23). This is because the ghost fields are complex and the type 1 gluon can transform into two distinct pairs: ghost 2–antighost 3 or antighost 2–ghost 3, thus giving rise to two equal amplitudes corresponding to the 2–3 pair.

15.2 CORRECTIONS TO QUARK PROPAGATOR AND VERTEX

The QCD corrections to the vertex and to the fermion propagator in the Feynman gauge are not equal to each other, in contrast to what happens in QED. Consequently, it is necessary to calculate them individually to obtain the charge renormalisation. In this section we demonstrate the calculation on the basis of the Feynman diagrams shown in Figure 15.2, beginning with the two-point function.

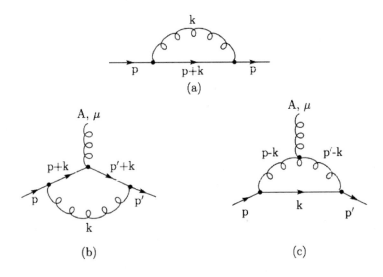

Figure 15.2 Renormalisation of the propagator and vertex in QCD.

Propagator correction. We denote with $i\Sigma(p)$ the amplitude obtained with the Feynman rules from Figure 15.2(a):

$$i\Sigma(p) = -(g^2 \sum_A \frac{\lambda^A}{2} \frac{\lambda^A}{2}) \int \frac{d^D k}{(2\pi)^4} \frac{[\gamma_\mu(\not{p} + \not{k} + m)\gamma^\mu]}{(k^2 + i\epsilon)[(p + k)^2 - m + i\epsilon]}$$

$$= \frac{g^2}{16\pi^4} C(R) \int_0^1 dx \left[(D - 2)(1 - x)\not{p} - mD\right] I(s, D, 2) ,$$

with

$$s = m^2 - x(1 - x)p^2 . \tag{15.28}$$

Renormalising as before to a momentum value $-p^2 = |\mu^2| >> m^2$ and neglecting m, we find

$$i\Sigma(p) = -i\frac{g^2}{16\pi^2}C(R)\,\log\left(\frac{-p^2}{|\mu^2|}\right)\,\not{p}\,. \qquad (15.29)$$

Vertex correction. We denote with $ig\Lambda_\mu^A$ the amplitude obtained with the Feynman rules from Figures 15.2(b) and (c). In the first case, we obtain

$$ig(\Lambda^{(b)})_\mu^A = (-ig)^3\left(\frac{\lambda^B}{2}\frac{\lambda^A}{2}\frac{\lambda^B}{2}\right)$$
$$\times \int \frac{d^Dk}{(2\pi)^4}i\frac{\gamma^\nu(\not{p}-\not{k}+m)\gamma_\mu(\not{p}'-\not{k}+m)\gamma^\nu}{[(p-k)^2-m^2+i\epsilon][(p'-k)^2-m^2+i\epsilon]}\frac{-ig_{\mu\nu}}{k^2+i\epsilon}\,. \qquad (15.30)$$

We can reduce the product of the three λ matrices as follows. We set

$$\frac{\lambda^A}{2} = T^A\,, \qquad (15.31)$$

and use the commutation rules (14.4) and the definition (14.14) to find

$$T^BT^AT^B = (T^BT^B)T^A + T^B(if_{ABC}T^C) = C(R)T^A + \frac{1}{2}if_{ABC}[T^B,T^C]$$
$$= C(R)T^A + \frac{1}{2}(if_{ABC}if_{BCL})T^L = C(R)T^A - \frac{1}{2}\text{Tr}(T^A_{reg}T^L_{reg})T^L$$
$$= [C(R) - \frac{1}{2}T(A)]\frac{\lambda^A}{2}\,. \qquad (15.32)$$

To determine the asymptotic behaviour of QCD, we are interested in the terms which depend on the logarithm of the renormalisation point. It is easy to confirm that, in the expansion of (15.30), a $\log\mu^2$ factor appears only from the renormalisation of formally divergent terms in the integral of (15.30), which in its turn is associated with the term proportional to $k_\lambda k_\sigma$ in the numerator of the fermion line. Therefore, in the numerator of (15.30), we can neglect m as well as p and p'. Keeping in mind also that for symmetric integration $k_\lambda k_\sigma \to 1/4k^2g_{\lambda\sigma}$, we find

$$ig(\Lambda^{(b)})_\mu^A = [C(R) - \frac{1}{2}T(A)](-ig\frac{\lambda^A}{2})\frac{-ig^2}{16\pi^4}\gamma_\mu$$
$$\times \int d^Dk\frac{1}{[k^2-m^2+i\epsilon][(k-q)^2-m^2+i\epsilon]}\,, \qquad (15.33)$$

where, in the denominator, we have redefined $k-p \to k$ and $q = p'-p$.

Proceeding as before and renormalising at the point $q^2 = \mu^2 < 0$, we obtain

$$ig(\Lambda^{(b)})_\mu^A = \frac{-g^2}{16\pi^2}[C(R) - \frac{1}{2}T(A)]\,\log\left(\frac{-q^2}{|\mu^2|}\right)\,(-ig\frac{\lambda^A}{2}\gamma_\mu)\,. \qquad (15.34)$$

Finally, we calculate the amplitude for the diagram of Figure 15.2(c). We find:

$$ig(\Lambda^{(c)})^A_\mu = (-ig)^2(\frac{\lambda^B}{2}\frac{\lambda^C}{2})\int \frac{d^D k}{(2\pi)^4} i\frac{\gamma_\rho k\gamma_\lambda}{(k^2+i\epsilon)}\frac{(-ig_{\rho\nu})(-ig_{\lambda\sigma})}{[(p-k)^2+i\epsilon][(p'-k)^2+i\epsilon]}$$
$$\times(-gf_{ABC})[g_{\mu\nu}(q+p'-k)_\sigma+g_{\nu\sigma}(-p'+k-p+k)_\mu+g_{\sigma\mu}(p-k-q)_\nu] . \tag{15.35}$$

The colour factors are treated as before:

$$if_{ABC}(\frac{\lambda^B}{2}\frac{\lambda^C}{2}) = \frac{1}{2}if_{ABC}\,if_{BCL}\frac{\lambda^L}{2} = -\frac{1}{2}T(A)\frac{\lambda^A}{2} . \tag{15.36}$$

In the numerator we again set $D = 4$ and consider only terms quadratic in k. With this the numerator reduces to

$$\gamma_\nu k\gamma_\sigma\,[-k_\sigma g_{\mu\nu} + 2k_\mu g_{\nu\sigma} - k_\nu g_{\sigma\mu}] \to -3\gamma_\mu k^2 , \tag{15.37}$$

for symmetric integration. Introducing the Feynman parameters and putting everything together, we find

$$ig(\Lambda^{(c)})^A_\mu = -\frac{g^3}{16\pi^4}\frac{3}{2}T(A)(\frac{\lambda^A}{2})\int_0^1 dx \int d^D k\frac{1}{(k^2-s+i\epsilon)^2}, \tag{15.38}$$

with

$$s = -x(1-x)q^2 . \tag{15.39}$$

Renormalising at the point $q^2 = \mu^2 < 0$, we obtain

$$ig(\Lambda^{(c)})^A_\mu = -\frac{g^2}{16\pi^2}\frac{3}{2}T(A)\,\log(\frac{-q^2}{|\mu^2|})\,(-ig\frac{\lambda^A}{2}\gamma_\mu) , \tag{15.40}$$

or

$$ig(\Lambda^{(b+c)})^A_\mu = \frac{-g^2}{16\pi^2}[C(R) + T(A)]\,\log(\frac{-q^2}{|\mu^2|})\,(-ig\frac{\lambda^A}{2}\gamma_\mu) . \tag{15.41}$$

15.3 ASYMPTOTIC FREEDOM

We rewrite the results found in Sections 15.1 and 15.2:

$$i\Pi^{AB}_{\mu\nu} = iC_3\log(\frac{q^2}{\mu^2})\,\delta_{AB}(g_{\mu\nu}q^2 - q_\mu q_\nu) = i\delta_{AB}(g_{\mu\nu}q^2 - q_\mu q_\nu)\,(z_3 - 1) , \tag{15.42}$$

$$i\Sigma = -iC_2\log(\frac{p^2}{\mu^2})\,p = -ip\,(z_2 - 1) , \tag{15.43}$$

$$ig\Lambda^A_\mu = -C_1\log(\frac{p^2}{\mu^2})\,(-ig\frac{\lambda^A}{2}\gamma_\mu) = (-ig\frac{\lambda^A}{2}\gamma_\mu)\,(z_1 - 1) , \tag{15.44}$$

with

$$C_3 = -\frac{g^2}{16\pi^2}[\frac{5}{3}T(A) - \frac{4}{3}n_f T(R)] ,$$ (15.45)

$$C_2 = \frac{g^2}{16\pi^2}C(R) ,$$ (15.46)

$$C_1 = \frac{g^2}{16\pi^2}[T(A) + C(R)] .$$ (15.47)

The $z_{1,2,3}$ are the finite renormalisation constants of the vertex, the fermion propagator and the gluon propagator, respectively. From the considerations of Chapter 11 we see that the effective constant which determines the renormalised vertex is

$$g_{eff} = g(\mu) \left(\sqrt{z_3}\frac{z_2}{z_1} \right) .$$ (15.48)

The effective constant depends on q^2, p^2, $(p')^2$, i.e. on the momentum scale at which the process occurs, because the z_i depend on these physical variables. But it should not depend on μ, which is an arbitrarily chosen value. For this reason we have shown the renormalisation constant as a function of μ; the dependence of $g(\mu)$ on μ must balance the dependence of μ on the z_i, which is made explicit in (15.44) and (15.47).[2]

Differentiating both sides of (15.48) with respect to $\log(\mu)$, we find (the equation which follows shows that the derivative of $g(\mu)$ is of order g^3, hence we can set $z_i = 1$ in the first term of the right-hand side)

$$0 = \frac{\partial g(\mu)}{\partial \log \mu} + g(\mu)\frac{\partial}{\log \mu}(\sqrt{z_3}\frac{z_2}{z_1}) ,$$ (15.49)

from which it follows that:

$$\beta(g) = -g(\mu)\frac{\partial}{\log \mu}(\sqrt{z_3}\frac{z_2}{z_1}) = g[\frac{1}{2}C_3 + C_2 - C_1]$$

$$= -\frac{g^3}{16\pi^2}[\frac{11}{3}T(A) - \frac{4}{3}n_f T(R)] .$$ (15.50)

We have obtained a very important result. In the first place, the β function found is *universal*, i.e. it does not depend on the specific nature of the particle whose interaction we are observing. This property arises because even if the ratio z_2/z_1 is not equal to unity as in QED, it is nevertheless universal: in the $C_2 - C_1$ difference the part that depends on R cancels.

Furthermore, the z can depend only on the ratio q^2/μ^2, hence from (15.50)

---

[2]We are repeating with different notation the arguments of Chapter 13, which lead to the Gell–Mann and Low equation of electrodynamics.

we can obtain the behaviour of g_{eff} as a function of the momentum scale

$$q^2 \frac{\partial g_{eff}}{\partial q^2} = \frac{1}{2} q \frac{\partial g_{eff}}{\partial q} = g(\mu) q \frac{\partial}{\partial q} (\sqrt{z_3} \frac{z_2}{z_1}) = -\frac{1}{2} g(\mu) \mu \frac{\partial}{\partial \mu} (\sqrt{z_3} \frac{z_2}{z_1})$$

$$= \frac{1}{2} \beta(g_{eff}) \ . \tag{15.51}$$

From (15.50) we see—and this is the real surprise—that the QCD β func-tion is negative, at least for not too large values of n_f; the QCD coupling constant decreases as the momentum scale involved in the process increases. In the limit of infinite momentum, the strong interactions transmitted by glu-ons are switched off. The name asymptotic freedom originates here, in view of this truly exceptional property.

In processes at very high momentum transfer, therefore, quarks and gluons, the constituents of the proton, behave as if they were free, a characteristic which explains the scaling laws of the deep inelastic cross sections. These properties were experimentally observed before the development of QCD, and until the discovery of asymptotic freedom, have been considered to be truly mysterious.

UNITARITY AND GHOSTS

CONTENTS

The unitarity condition on the scattering matrix, derived in Chapter 7, leads to a relation known as the *optical theorem* which connects the amplitudes of real and virtual processes. If we write

$$S = \mathbf{1} + iT \ , \tag{16.1}$$

the condition that S should be a unitary matrix, equation (7.22), implies the relation

$$-i(T - T^\dagger) = TT^\dagger \ . \tag{16.2}$$

We take a diagonal matrix element and introduce a complete system of states into the product of the operators. Thus we obtain

$$-i(< \alpha|T|\alpha > - < \alpha|T^\dagger|\alpha >) = 2\mathrm{Im}\, T_{\alpha\alpha} = \sum_\beta |T_{\alpha\beta}|^2 \ . \tag{16.3}$$

The left-hand side of the equation is called the *absorptive part* of the amplitude. In optics, the forward photon scattering amplitude determines the refractive index of the medium and its imaginary component determines the absorption length of the photon in the same material. The optical theorem expresses the fact that absorption is determined by the inelastic production of states different from the initial state, characterised by the set of quantum numbers summarised by the index β in (16.3).

Representation in terms of Feynman diagrams allows us to analyse the unitarity relation in a simple and direct way.

16.1 THE CUTKOSKY RULE

We consider a particular connected Feynman diagram, of a certain perturbative order, which describes the elastic amplitude $\alpha \to \alpha$. It can be possible to *cut* a certain number of internal lines so as to disconnect the diagram and produce two connected diagrams which describe the process $\alpha \to \beta$, where β denotes the set of particles associated with the cut lines. If the centre of mass energy of α allows the transition $\alpha \to \beta$, the latter represents the possible result of an inelastic reaction which begins from the state α. Each β state of this type corresponds to a line of singularity of the amplitude in the complex total energy plane and the discontinuity of the elastic amplitude across the cut is obtained with the *Cutkosky rule* [21].

We consider the Feynman propagator

$$D_F(q) = \frac{i}{q^2 - m^2 + i\epsilon} Np(q) , \qquad (16.4)$$

where p is a polynomial in q, which *on the mass-shell* reduces to the projector of the positive energy states of the particle

$$p(q) = \sum_{spin(\beta)} |\beta><\beta| = \begin{cases} 1 \\ (\not{q} + m)/2m \\ (-g_{\mu\nu} + q_\mu q_\nu/m^2) \end{cases} , \qquad (16.5)$$

for the cases of spin 0, spin $\frac{1}{2}$ and spin 1 with mass, respectively. $N = 2m$ or 1 for half-integer or integer spin.

According to the Cutkosky rule:

- The discontinuity of $T_{\alpha\alpha}$, defined as

$$\mathrm{Disc}T_{\alpha\alpha} = T_{\alpha\alpha}(m^2 - i\epsilon) - T_{\alpha\alpha}(m^2 + i\epsilon) = 2\,i\mathrm{Im}T_{\alpha\alpha} , \qquad (16.6)$$

  is obtained with the substitution

$$D_F(q) \to 2\pi\delta^{(+)}(q^2 - m^2)\,Np(q) , \quad \text{(Cutkosky rule)} , \qquad (16.7)$$

  for each cut line. The superscript $(+)$ indicates that we must take the positive energy solution for the argument of the δ-function.

The starting point is the well known relation[1]

$$\frac{1}{x \pm i\epsilon} = P(\frac{1}{x}) \mp i\pi\delta(x) . \qquad (16.8)$$

Applied to the propagator, the above relation translates to

$$\frac{d^4q}{(2\pi)^4} D_F(q, m^2 \mp i\epsilon) = \frac{i\,d^4q}{(2\pi)^4} P\left(\frac{Np}{q^2 - m^2}\right) \pm (1/2)\frac{d^3q}{(2\pi)^3}\frac{Np}{2q^0} . \qquad (16.9)$$

---

[1] P denotes the principal part of the integration in x, and the prescription $\pm i\epsilon$ indicates that the pole is below or above the path of integration.

The first term, after the Wick rotation in the complex plane of q^0 which eliminates the factor i, provides the part of the amplitude that is continuous across the line of singularity. The second term determines the discontinuity of the amplitude for each cut line (for a proof, see [21]).

Using this rule, it is easy to show that the optical theorem is satisfied in the case of the internal lines listed in (16.5). In fact, given a diagram $\mathcal{G}$, the absorptive part that is obtained by cutting it according to a given class of β states takes the form

$$-i\text{Disc}[T^{\mathcal{G}}] = \sum_{\beta} \int \Pi_i \frac{d^3 q_i}{q_i^0 (2\pi)^3} N_i p(q_i)\, T_1^{\mathcal{G}}(\alpha \to \beta)[T_2^{\mathcal{G}}(\alpha \to \beta)]^{\star}\ , \quad (16.10)$$

where the index i goes on the cut lines and β denotes the other quantum numbers of the intermediate state.

$T_1^{\mathcal{G}}$ and $T_2^{\mathcal{G}}$ are the amplitudes of two specific diagrams which give rise to the $\alpha \to \beta$ transition and it is not difficult to confirm that, summing all the diagrams $\mathcal{G}$ which permit the same intermediate state, leads to the modulus squared of the total amplitude for the transition $\alpha \to \beta$, or that

$$-i\text{Disc}[T_{\alpha\alpha}] = \sum_{\beta} |T_{\alpha\beta}|^2\ , \quad (16.11)$$

as required.

Note. In carrying out the calculation, it is necessary to remember that, if we denote with F the amplitude that is obtained by applying the Feynman rules to a given diagram, we have

$$F_{\alpha\beta} = iT_{\alpha\beta}\ . \quad (16.12)$$

Therefore, if we denote with $\tilde{F}$ the result that is obtained by applying the Cutkosky rule to F, we obtain

$$\tilde{F} = i\text{Disc}\, T = -2\,\text{Im}\, T\ , \quad (16.13)$$

and the optical theorem takes the form:

$$-\tilde{F}_{\alpha\alpha} = \sum_{\beta} |F_{\alpha\beta}|^2\ . \quad (16.14)$$

16.2   THE INELASTIC REACTION $u + \bar{u} \to d + \bar{d}$

As a first check, we can consider the fermion–antifermion annihilation process in QED to order e^4, applying the optical theorem to the discontinuity related to the production of a fermion–antifermion pair with a flavour different from the initial flavour, for example $u + \bar{u} \to d + \bar{d}$. In terms of Feynman diagrams, we must verify what is shown symbolically in Figure 16.1, where the vertical dashed line in the diagram on the left-hand side denotes the application of the substitution (16.7) to the cut lines in the Feynman amplitude.

Left-hand side. The Feynman rules for the left-hand side give

$$F = \frac{e^4}{q^4}(Q_u Q_d)^2 \, j_\mu(i\Pi^{\mu\nu})j_\nu^\star \, ,$$

with

$$j_\mu = \bar{u}(p^-)\gamma^\mu v(p^+) \, ,$$

and $p^- + p^+ = q$. $\Pi^{\mu\nu}$ is the QED vacuum polarisation tensor, calculated in Section 12.4 starting from equation (12.41), which we recall for convenience:

$$i\Pi^{\mu\nu}(q) = \frac{1}{(2\pi)^D}\int d^D p \frac{i}{p^2 - m^2 + i\epsilon}\frac{i}{(p-q)^2 - m^2 + i\epsilon}$$
$$\times \operatorname{Tr}\left[\gamma^\mu(\not{p}+m)\gamma^\nu(\not{p}-\not{q}+m)\right] \, . \quad (16.15)$$

We apply the substitution (16.7) and set $D = 4$ to obtain

$$i\tilde{\Pi}^{\mu\nu} = \frac{1}{(2\pi)^4}\int d^4 p \, (2\pi)^2 \delta^{(+)}(p^2 - m^2)\, \delta^{(+)}\left[(p-q)^2 - m^2\right]$$
$$\times \operatorname{Tr}\left[\gamma^\mu(\not{p}+m)\gamma^\nu(\not{p}-\not{q}+m)\right] \, .$$

The integral is manifestly Lorentz invariant, and furthermore, because of the δ-function,

$$q_\mu \tilde{\Pi}^{\mu\nu} = 0 \, . \quad (16.16)$$

As a consequence, we set

$$\tilde{\Pi}^{\mu\nu} = (g^{\mu\nu}q^2 - q^\mu q^\nu)\tilde{\Pi}(q^2) \, , \quad (16.17)$$

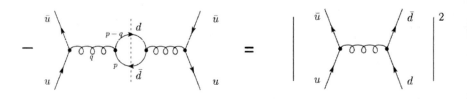

Figure 16.1 Contribution of the intermediate $d\bar{d}$ state to the unitarity relation for the amplitude $u + \bar{u} \to u + \bar{u}$.

from which it follows that

$$
\begin{aligned}
i\tilde{\Pi}(q^2) &= \frac{1}{3q^2} \, g_{\mu\nu} i\tilde{\Pi}^{\mu\nu} \\
&= \frac{1}{(2\pi)^2} \frac{4(q^2 + 2m^2)}{3q^2} \int d^4p \; \delta^{(+)}(p^2 - m^2) \; \delta^{(+)} \left[(p - q)^2 - m^2 \right] \\
&= \frac{1}{6\pi} \beta \frac{q^2 + 2m^2}{q^2} \;,
\end{aligned}
\tag{16.18}
$$

where $\beta = \sqrt{1 - q^2/4m^2}$ is the velocity of the final fermion in the centre of mass frame.

In conclusion, taking the limit $m = 0$, as we have done in the calculations up to now, and using conservation of the current j_μ we find[2]

$$
\tilde{F} = \frac{e^4}{q^4}(Q_u Q_d)^2 (i\tilde{\Pi})(-q^2 \mathbf{j} \cdot \mathbf{j}^\star) = -\frac{e^4}{6\pi q^2}(Q_u Q_d)^2 \, \mathbf{j} \cdot \mathbf{j}^\star \;.
\tag{16.19}
$$

We obtain the same result starting from the outcome of the full calculation of $\Pi^{\mu\nu}$, equation (15.6). With obvious substitutions we find

$$
i \, \Pi^{\mu\nu}(q) = (g^{\mu\nu} q^2 - q^\mu q^\nu) i\Pi(q^2) \;,
$$

with

$$
\begin{aligned}
i\Pi(q^2) = i\frac{1}{2\pi^2} \int_0^1 dx \; x(1 - x) \log &\left[\frac{m^2 - x(1 - x)q^2 - i\epsilon}{m^2 + x(1 - x)|\mu|^2} \right] \\
&\to i\frac{1}{12\pi^2} \, \log \left[\frac{-q^2 - i\epsilon}{|\mu|^2} \right] \;,
\end{aligned}
\tag{16.20}
$$

and

$$
T_{\alpha\alpha} = -iF = \frac{e^4}{q^4} \, (Q_u Q_d)^2 j_\mu \Pi^{\mu\nu} j_\nu^\star = -\frac{e^4}{q^2}(Q_u Q_d)^2 \Pi(q^2) \mathbf{j} \cdot \mathbf{j}^\star \;.
\tag{16.21}
$$

The logarithm in (16.20) is an analytic function of q^2, with a cut along the real axis of q^2 for $q^2 > 4m^2$. The discontinuity is across the cut and given by

$$
-i\mathrm{Disc}\,\Pi(q^2) = -i \left[\Pi^{(a)}(q^2 + i\epsilon) - \Pi^{(a)}(q^2 - i\epsilon) \right] \;.
\tag{16.22}
$$

Since $\ln(z) = \ln(|z|) + i\,Arg(z)$ e $Arg[-q^2 - i\epsilon] = -\pi$, we find

$$
-i\mathrm{Disc}\,\Pi = -2\pi \;,
$$

$$
\begin{aligned}
2\,\mathrm{Im}\,T_{\alpha\alpha} = -i\mathrm{Disc}\,T_{\alpha\alpha} &= -(-2\pi)\frac{e^4}{12\pi^2 q^2}(Q_u Q_d)^2 \, \mathbf{j} \cdot \mathbf{j}^\star \\
&= \frac{e^4}{6\pi q^2}(Q_u Q_d)^2 \, \mathbf{j} \cdot \mathbf{j}^\star = -\tilde{F} \;,
\end{aligned}
\tag{16.23}
$$

in agreement with (16.13), providing a nice check of the Cutkosky rule.

---

[2]Note that in the centre-of-mass frame, $\partial^\mu j_\mu = 0$ implies $j_0 = 0$.

The right-hand side. We denote the momenta of the outgoing quark and antiquark as $p_{1,2}$, and then

$$\sum_{fin} |T(\alpha \to \beta)|^2 = \int \frac{d^3 p_1 d^3 p_2}{(2\pi)^6} (2\pi)^4 \delta^{(4)}(P_{in} - P_{fin}) \frac{m^2}{p_1^0 p_2^0} |F(\alpha \to \beta)|^2 ,$$

$$|F(\alpha \to \beta)|^2 = \frac{e^4 (Q_u Q_d)^2}{q^4} (j_\mu j_\nu^\star) \text{Tr} \left[\frac{\not{p}_2 - m}{2m} \gamma^\mu \frac{\not{p}_1 + m}{2m} \gamma^\nu \right] . \tag{16.24}$$

Easy steps (see for example [1]) lead, for $m = 0$, to

$$\sum_{fin} |T(\alpha \to \beta)|^2 = \frac{e^4}{6\pi q^2} (Q_u Q_d)^2 \mathbf{j} \cdot \mathbf{j}^\star , \tag{16.25}$$

which agrees with (16.23).

16.3   THE CASE OF QED

The Cutkosky rule is not sufficient to prove the unitarity of theories which contain gauge fields, because the numerator $p(q)$ in (16.4) depends on the gauge. We consider here the case of QED in the Feynman gauge, in which the propagator is given by

$$D_F(q)^{\mu\nu} = \frac{-i g^{\mu\nu}}{q^2 + i\epsilon} . \tag{16.26}$$

The relationship between $-g^{\mu\nu}$ and the projector onto the two photon states with helicity $-g^{\mu\nu}$ was discussed in Chapter 5.

We denote the spatial part of q^μ as $\mathbf{q}$ and introduce the two purely spatial polarisation vectors, $\epsilon_{1,2}(q)^\mu$, orthogonal to $\mathbf{q}$. The projectors of the physical states

$$P^{\mu\nu} = \sum_{i=1,2} \epsilon_i(q)^\mu (\epsilon_i(q)^\nu)^\star , \tag{16.27}$$

project onto the plane orthogonal to $\mathbf{q}$. To obtain a complete system we introduce a purely timelike vector, $n^\mu = (\mathbf{0}, 1)$, and the unit vector $\mathbf{q}/|\mathbf{q}|$, defining

$$v^\mu = \frac{q^\mu - q^0 n^\mu}{|\mathbf{q}|} . \tag{16.28}$$

The completeness relation can therefore be written as

$$-g^{\mu\nu} = \sum_{i=1,2} \epsilon_i(q)^\mu (\epsilon_i(q)^\mu)^\star + \frac{v^\mu v^\nu}{|\mathbf{q}|^2} - n^\mu n^\nu$$

$$= P^{\mu\nu} + \frac{q^\mu q^\nu}{|\mathbf{q}|^2} - q^0 \frac{q^\mu n^\nu + n^\mu q^\nu}{|\mathbf{q}|^2} + q^2 \frac{n^\mu n^\nu}{|\mathbf{q}|^2} , \tag{16.29}$$

where $q^2 = (q^0)^2 - |\mathbf{q}|^2$.

If we substitute this relation into the Cutkosky rule, (16.7), the fourth term vanishes, because $q^2\delta(q^2) = 0$, but the second and third terms in (16.29) appear to disrupt the unitarity relation, which is saturated by the transverse projector, $P^{\mu\nu}$, alone. However, we can prove that these extra terms also make zero contribution since

- the photon is coupled to a conserved current,

and

- QED is a commutative (i.e. Abelian) gauge theory.

Proof. We consider the simple case of the intermediate state of two photons. Figure 16.2 shows the two diagrams which correspond to the two possible ways in which a photon with 4-momentum q coupled to the charge Q_A can end on a fermion line, together with a second photon with fixed 4-momentum coupled to Q_B. If in place of the photon polarisation we set q^μ, as occurs in the second or third term of (16.29), we find, associated with the second fermion line, the amplitude

$$A = \bar{u}(p')Q_B\!\!\not{\epsilon}\frac{i}{\not{p}+\not{q}-m}Q_A\!\!\not{q}u(p) + \bar{u}(p')Q_A\!\!\not{q}\frac{i}{\not{p}'-\not{q}-m}Q_B\!\!\not{\epsilon} \ . \quad (16.30)$$

Alternatively, we write $\not{q} = (\not{p}+\not{q}-m)-(\not{p}-m) = (\not{p}'-m)-(\not{p}'-\not{q}-m)$ and use the equation of motion on the external spinors to find, for any polarisation ϵ, of the second photon

$$A = i\bar{u}(p')Q_BQ_A\!\!\not{\epsilon}u(p) - i\bar{u}(p')Q_AQ_B\!\!\not{\epsilon}u(p) = 0 \ , \quad (16.31)$$

since $Q_A = Q_B = e$.

This simple example is easily generalised and shows that, because the

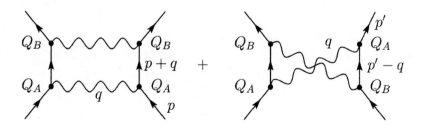

Figure 16.2 The two diagrams correspond to the ways in which a photon with 4-momentum q coupled to the charge Q_A can end on a fermion line, together with a second photon with fixed 4-momentum coupled to Q_B.

second and third terms in (16.29) contain at least one factor q^μ or q^ν, we can substitute $-g_{\mu\nu} \to P^{\mu\nu}$ in the Cutkosky rule and unitarity is satisfied in QED *after having summed over the ways in which a given photon is coupled to the lines representing the charged particles.*

16.4   NON-ABELIAN GAUGE THEORIES

The argument found in QED obviously does not apply to Yang–Mills theories, in particular QCD, because the charges do not commute with each other. The previous result shows, however, in which direction to go; the result found in (16.31) could be cancelled by a contribution connected to the commutator of the two charges. This is actually required by the fact that the gluons themselves are charged, in contrast to the photon. In addition to diagrams in which the gluon ends on a fermion line, we must consider those in which the gluon ends on all possible gluon lines. The question of unitarity of the one-loop amplitudes in an exact Yang–Mills theory was considered for the first time by Feynman [19]. The surprising result obtained by Feynman is that the additional terms in the completeness relation (16.29) do not vanish even when including the diagrams related to the couplings with three gluons, when one considers loops made by gluons only. However Feynman proved that unitarity could be satisfied by introducing, for each gluon loop, a loop of anticommuting scalar fields, also massless. The justification given by Feynman is that the additional contribution is necessary to completely cancel the contribution of the spurious polarisations introduced by the gauge. These scalar fields do not satisfy the spin-statistics theorem and cannot appear in the external states since they should be associated with a negative metric, denoted in quantum mechanics by the name *ghost*[3].

As we saw in Section 14.3, the need for ghosts arises independently from the argument of Faddeev and Popov, as a necessary step to fix the gauge in the formulation of QCD by means of the Feynman integral. In the Feynman gauge, the Faddeev–Popov ghosts are in fact the scalar fields hypothesised by Feynman to restore the unitarity of the amplitudes. The argument of Feynman provides an illuminating physical basis for the abstract argument of Faddeev and Popov.

To analyse the situation, we consider the simple case of the discontinuity linked to two gluon intermediate states in the fermion–antifermion annihilation $u + \bar{u} \to u + \bar{u}$.

To calculate the discontinuity in the elastic amplitude, it is helpful first to assemble the terms of the total amplitudes $u + \bar{u} \to$ gluon + gluon and gluon + gluon $\to u + \bar{u}$, and then calculate the discontinuity (see, for example, [22]).

The amplitude $u + \bar{u} \to$ gluon + gluon is described by the diagrams (A),

---

[3]Feynman was interested in the Yang–Mills theory as a simple prototype for quantum gravity, later studied by De Witt [20], who also demonstrated the need of ghost fields in the quantisation of gravity.

(B) and (C) in Figure 16.3. The resulting amplitude is

$$\mathcal{M}_{AB}^{\mu\nu}(q_1,q_2) = \bar{v}(p^+)(-ig)\frac{\lambda^A}{2}\gamma^\mu \frac{i}{\not{p}^- - \not{q}_2}(-ig)\frac{\lambda^B}{2}\gamma^\nu u(p^-)$$

$$+ \bar{v}(p^+)(-ig)\frac{\lambda^B}{2}\gamma^\mu \frac{i}{\not{p}^- - \not{q}_1}(-ig)\frac{\lambda^A}{2}\gamma^\nu u(p^-)$$

$$+ \bar{v}(p^+)(-ig)\frac{\lambda^C}{2}\gamma^\rho u(p^-)\frac{-ig_{\rho\sigma}}{q^2}i\Gamma_{CAB}^{\sigma\mu\nu}(q,-q_1,-q_2)\ , \quad (16.32)$$

where Γ is the three-gluon vertex given in (14.51). We use the centre of mass of the reaction, and $p^\mp$ are the momenta of the quark and antiquark, with

$$p^- + p^+ = q_1 + q_2 = q\ . \quad (16.33)$$

In what follows, we also set $j_\mu^A = \bar{\psi}(-ig)(\lambda^A/2)\gamma_\mu\psi$ and recall that, in this reference frame, two further relations hold

$$< 0|j_0^A|u\bar{u}> = 0\ , \quad \text{(conservation of the current)}\ , \quad (16.34)$$

$$< 0|j_3^A|u\bar{u}> = 0\ , \quad \text{(massless fermions)}\ . \quad (16.35)$$

Subsequently, denoting the amplitude for gluon + gluon $\to u + \bar{u}$ as $\mathcal{N}_{ED}^{\rho\sigma}$, leads to

$$\mathcal{N}_{ED}^{\rho\sigma}(q_1,q_2) = \bar{u}(p^-)(-ig)\frac{\lambda^E}{2}\gamma^\mu \frac{i}{\not{p}^- - \not{q}_2}(-ig)\frac{\lambda^D}{2}\gamma^\nu v(p^+)$$

$$+ \bar{u}(p^-)(-ig)\frac{\lambda^D}{2}\gamma^\mu \frac{i}{\not{p}^- - \not{q}_1}(-ig)\frac{\lambda^E}{2}\gamma^\nu v(p^+)$$

$$+ i\Gamma_{DLE}^{\rho\alpha\sigma}(q_1,-q,-q2)\frac{-ig_{\alpha\beta}}{q^2}\bar{u}(p^-)(-ig)\frac{\lambda^L}{2}\gamma^\beta v(p^+)\ . \quad (16.36)$$

The following relations are immediately obtained

$$\mathcal{M}_{AB}^{\mu\nu}(q_1,q_2)(q_1)_\mu = -(-ig)^2 f_{ABC}\frac{1}{q^2}(q_2)_\nu\bar{v}(p^+)\frac{\lambda^C}{2}\not{q}_1 u(p^-)\ , \quad (16.37)$$

$$\mathcal{M}_{AB}^{\mu\nu}(q_1,q_2)(q_2)_\nu = +(-ig)^2 f_{ABC}\frac{1}{q^2}(q_1)_\mu\bar{v}(p^+)\frac{\lambda^C}{2}\not{q}_2 u(p^-)\ , \quad (16.38)$$

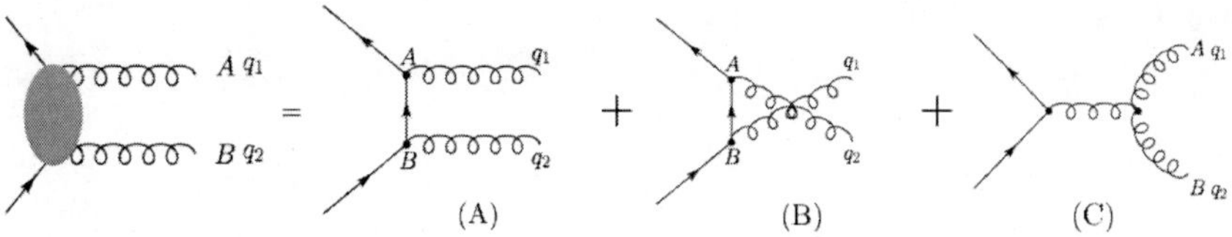

Figure 16.3 Contribution of two gluon intermediate states to the unitarity relation for the amplitude of fermion–antifermion annihilation.

and, similarly,

$$\mathcal{N}_{DE}^{\rho\sigma}(q_1, q_2)(q_1)_\rho = +(-ig)^2 f_{DEL} \frac{1}{q^2} (q_2)_\sigma \bar{u}(p^-) \frac{\lambda^C}{2} \slashed{q}_1 v(p+) , \quad (16.39)$$

$$\mathcal{N}_{DE}^{\rho\sigma}(q_1, q_2)(q_2)_\sigma = -(-ig)^2 f_{DEL} \frac{1}{q^2} (q_1)_\rho \bar{u}(p^-) \frac{\lambda^C}{2} \slashed{q}_1 v(p^+) . \quad (16.40)$$

We can now discuss the unitarity relation, illustrated in Figure 16.4. We calculate the discontinuity in the elastic amplitude due to the two-gluon intermediate state, Figure 16.4(a). Applying the Cutkosky rule to the gluon propagators, we have

$$-Disc \, \mathcal{M} = -\frac{1}{2} \int d\Omega \, \mathcal{M}_{AB}^{\mu\nu}(-g_{\mu\rho})(-g_{\nu\sigma})\mathcal{N}^{\rho\sigma AB} , \quad (16.41)$$

where we have denoted with $d\Omega$ the integration measure of the phase space of the two gluons

$$\frac{1}{2}d\Omega = \frac{1}{2} \frac{d^3 q_1}{2q_1^0 (2\pi)^3} \frac{d^3 q_2}{2q_2^0 (2\pi)^3} (2\pi)^4 \delta(q - q_1 - q_2) , \quad \int d\Omega = \frac{1}{8\pi} , \quad (16.42)$$

and included a factor $\frac{1}{2}$ for the identical particles.

We furthermore decompose each factor $-g_{\mu\nu}$ into transverse, longitudinal and time components according to (16.29)

$$-g_{\mu\nu} = P_{\mu\nu} + P_{\mu\nu}^q + P_{\mu\nu}^n ,$$

with

$$P_{\mu\nu}^q = \frac{q_\mu q_\nu}{|\mathbf{q}|^2} , \qquad P_{\mu\nu}^n = -\frac{q_\mu n_\nu + n_\mu q_\nu}{|\mathbf{q}|} , \quad (16.43)$$

where P is the projector on the transverse states and n_μ is the unit vector in the time direction.

The term with $P \otimes P$ reproduces the right-hand side of the optical theorem, as shown in Figure 16.4. Therefore we must check if the sum of the other terms vanishes or not. On the basis of (16.37)–(16.40) we note that:

- terms with $P(1) \otimes [P^q(2) + P^n(2)]$, $[P(1)^q + P^n(1)] \otimes P(2)$ give zero contributions, because the amplitudes vanish if they are multiplied by $\epsilon(1)q_2$, or $q_1\epsilon(2)$;

- the terms in $P^q(1) \otimes P^q(2)$ or $[P^q(1) \otimes P^n(2) + P^n(1) \otimes P^q(2)]$ similarly vanish because the amplitudes give zero if multiplied by $q_1 q_2$;

- potentially damaging terms arise from the product $P^n(1) \otimes P^n(2)$ and give exactly $[n(1)q_2 \otimes q_1 n(2) + q_1 n(2) \otimes n(1)q_2]$.

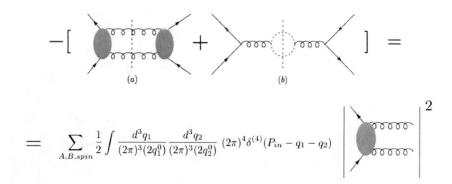

$$= \sum_{A.B.spin} \frac{1}{2} \int \frac{d^3q_1}{(2\pi)^3(2q_1^0)} \frac{d^3q_2}{(2\pi)^3(2q_2^0)} (2\pi)^4 \delta^{(4)}(P_{in} - q_1 - q_2) \left| \begin{matrix} \\ \end{matrix} \right|^2$$

Figure 16.4 The contribution of the intermediate ghost state must be added to the intermediate state of two gluons in order for the fermion–antifermion annihilation amplitude to satisfy the unitarity condition.

The potential anomaly is therefore

$$\Delta = -\frac{1}{2} \int d\Omega \left\{ [\mathcal{M}_{AB}^{\mu\nu} n_\mu(q_2)_\nu] \cdot [(q_1)_\rho n_\sigma \mathcal{N}_{AB}^{\rho\sigma}] \right.$$
$$\left. + [\mathcal{M}_{AB}^{\mu\nu}(q_1)_\mu n_\nu] \cdot [n_\rho(q_2)_\sigma \mathcal{N}_{AB}^{\rho\sigma}] \right\} . \quad (16.44)$$

The two terms differ in the exchange of $1 \leftrightarrow 2$ and are therefore equal; we can take one of them and cancel the factor $\frac{1}{2}$ for identical particles.

Using the explicit expressions (16.38) and (16.39), we find

$$\Delta = - \int d\Omega \, \frac{g^4}{q^4} j_i^C j_j^L f_{ABC} f_{ABL} (q_2)_i (q_1)_j = \frac{g^4}{32\pi} \frac{T(A)}{3q^2} j^A j^{A\star} \neq 0 . \quad (16.45)$$

The discontinuity due to the ghosts, however, is also associated with the cut of the two gluons, Figure 16.4(b). For the contribution of this discontinuity we find

$$\Delta_{ghost} = \frac{g^2}{q^4} j_\mu^A j_\nu^{B\star} [-i\mathrm{Disc}(\Pi^{(b)})_{AB}^{\mu\nu}]$$
$$= +\frac{g^4}{32\pi} \frac{T(A)}{3q^2} (j_\mu^A g^{\mu\nu} j_\nu^{A\star}) \quad (16.46)$$
$$= -\frac{g^4}{32\pi} \frac{T(A)}{3q^2} \mathbf{j}^A \cdot \mathbf{j}^{A\star} ,$$

and so we arrive at the expected result

$$\Delta + \Delta_{ghost} = 0 . \quad (16.47)$$

CHAPTER 17

EFFECTIVE CONSTANTS AT HIGH ENERGY AND GRAND UNIFICATION

CONTENTS

In this chapter, we would like to deepen the concept of the *running coupling constant* and discuss how this has led to the first ideas for a complete unification of the gauge forces, excluding gravity, or what is known as *grand unification.*

17.1   THE DETERMINATION OF α_s

The constant α_s is introduced in analogy with QED, starting from (15.48), to characterise the strength of the QCD interaction

$$\alpha_s = \frac{g_{eff}^2}{4\pi} \ . \tag{17.1}$$

Equation (15.51) determines the variation of α_s as a function of the momentum scale. We define

$$t = \log q^2 \ , \quad \beta(g) = \frac{b}{4\pi} \, g^3 \ , \quad b = -\frac{1}{4\pi} \left[\frac{11}{3} T(A) - \frac{4}{3} n_f T(R) \right] \ , \tag{17.2}$$

and rewrite the equation as

$$\frac{2dg_{eff}}{g_{eff}^3} = \frac{b}{4\pi} \ dt \ , \tag{17.3}$$

or

$$d(-\frac{4\pi}{g_{eff}^2}) = d(-\frac{1}{\alpha_s}) = b \, dt \ , \tag{17.4}$$

from which, integrating from a reference momentum, μ^2, up to q^2,

$$\frac{1}{\alpha_s(q^2)} = \frac{1}{\alpha_s(\mu^2)} - b \ln \frac{q^2}{\mu^2} \ , \tag{17.5}$$

or

$$\alpha_s(q^2) = \frac{\alpha_s(\mu^2)}{1 - b\alpha_s(\mu^2) \ln \frac{q^2}{\mu^2}} \ . \tag{17.6}$$

Equation (17.6) defines the *running coupling constant*, the coupling constant appropriate to describe phenomena at the scale q^2, in terms of the constant defined at a fixed scale μ^2. As q^2 grows, the running constant *decreases or increases* according to whether $b < 0$ (QCD, asymptotic freedom) or $b > 0$ (QED).

In QCD, because the coupling can be small but not vanishing, we can use perturbation theory to determine the deviations from the scaling rules for $q^2 > 1$ GeV^2 where the scaling laws begin to be approximately valid. Measurements of these deviations have been carried out at high-energy accelerators, in particular at SLAC, CERN, the DESY laboratory in Hamburg and at Fermilab, near Chicago. In this way, it has been possible to determine the behaviour of α_s as a function of q^2, to be compared with the predictions of QCD, represented by the lowest-order result (17.6) or more precise estimates at higher orders; see Figure 15.1 [23, 24]).

17.2   THE LANDAU POLE AND THE CONTINUUM LIMIT

The case $b > 0$ applies to QED,[1] as seen from (15.50) by setting $T(A) = 0$, $n_f T(R) = \sum Q_i^2$, where Q_i are the electric charges of the charged fermions, quarks and leptons, in units of the proton charge

$$b_{QED} = +\frac{1}{3\pi} \sum_i Q_i^2 \ . \tag{17.7}$$

As noted by Landau, equation (17.6) in the case $b = b_{QED} > 0$ presents a paradoxical aspect of the growth of q^2: the denominator vanishes for a finite value of q^2. For this value, the running constant encounters a singularity, the *Landau pole*, beyond which $\alpha(q^2)$ would become negative and, correspondingly, the theory would become unstable. The value of q^2 corresponding to

---

[1] We restrict ourselves for the present to energies below the threshold for production of charged intermediate bosons.

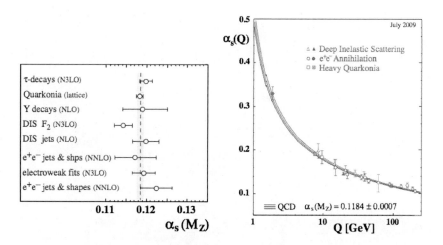

Figure 17.1 On the left, a compilation of values of α_s at the fixed scale of the mass of the Z^0, obtained from several different processes. On the right is shown the theoretical variation of $\alpha_s(q^2)$ compared with available experimental information (Dissertori and Salam in the Particle Data Group 2014 [23]; the figure is a redrawing of the original in [24]).

the Landau pole therefore represents a limit to the validity of the theory, commonly known as a *cutoff*, beyond which the integrations over momentum cannot be continued.

Another way of expressing the same problem is by writing the low-energy constant in terms of the scale limit of validity of QED at high energy, $q^2 = \Lambda^2$. This should be the case for a QED formulated on a space-time lattice with finite spacing, $a = 1/\Lambda$:

$$\alpha(\mu^2) = \frac{\alpha(\Lambda^2)}{1 + \alpha(\Lambda^2) b_{QED} \ln \frac{\Lambda^2}{\mu^2}} . \qquad (17.8)$$

The continuum theory is obtained in the limit of zero lattice spacing, $\Lambda = 1/a \to +\infty$.

In the neighbourhood of the Landau pole, $\alpha(\Lambda^2)$ is large and equation (17.8) predicts

$$\alpha(\mu^2) \approx \frac{1}{b_{QED} \ln \frac{\Lambda^2}{\mu^2}} . \qquad (17.9)$$

Thus, in the $\Lambda \to \infty$ limit, the coupling tends to zero, implying that *in the continuum limit, QED exists only as a free theory*.

The opposite case is that of an asymptotically free theory, like QCD. If we formulate QCD at the level of a finite lattice, the lattice coupling constant, $\alpha_s(\Lambda^2)$, is obtained from (17.6):

$$\alpha_s(\Lambda^2) = \frac{\alpha_s(\mu^2)}{1 + |b|\alpha_s(\mu^2)\ln\frac{\Lambda^2}{\mu^2}} \sim \frac{1}{|b|\ln(\frac{1}{(a\mu)^2})} , \qquad (17.10)$$

or

$$(a\mu)^2 = e^{\left(-\frac{1}{|b|\alpha_s(\Lambda^2)}\right)} . \qquad (17.11)$$

The continuum limit does not give rise to pathologies, at least in this approximation, and it corresponds to allowing the lattice constant to tend to zero, with μ constant.

17.3   EFFECTIVE CONSTANTS OF THE STANDARD THEORY

The Standard Theory is based on the gauge group $SU(3)_{colour} \otimes SU(2)_L \otimes U(1)_Y$, characterised by three coupling constants, g_s, g_W and g', respectively; see [2]. The electroweak constants are connected to the electric charge by the relations

$$e = g_W \sin\theta = g'\cos\theta , \qquad (17.12)$$

where θ is the electroweak mixing angle introduced by Glashow, Weinberg and Salam [2], see Chapter 18.

The fermion interactions are identified by the quantum numbers shown in the schemes of (14.21) and (14.23).

Considering one generation, we have

- colour: six quark triplets;

- $SU(2)_L$: a quark doublet (e.g. u, d) in three colours and one left-handed doublet of leptons.

The charge associated with the $U(1)_Y$ interaction is conventionally denoted as $\frac{1}{2}Y$ and is connected to the electric charge and weak isospin by the relation

$$Q = T_3 + \frac{1}{2}Y , \qquad (17.13)$$

valid separately for left- and right-handed fields.

To calculate the β function of the three interactions, we must take into account:

- some results valid for an $SU(N)$ group with fundamental representation $\boldsymbol{N}$ (see problems 1 and 2);

$$T(A) = N , \quad (SU(N), N \geq 2) ,$$

$$T(\boldsymbol{N}) = \frac{1}{2} , \quad \text{(Dirac fermions) ;} \qquad (17.14)$$

- that the contribution to the β function of a chiral, left- or right-handed, fermion is $\frac{1}{2}$ the contribution of a Dirac fermion

$$T(\mathbf{N})_{chiral} = \frac{1}{4} , \quad \text{(chiral fermions)} . \tag{17.15}$$

Putting everything together, starting from the formula (17.2), we find:

$$\beta(g_s) = \frac{g_s^3}{4\pi} b_s , \qquad b_s = -\frac{1}{4\pi}(11 - \frac{4}{3}n_{gen}) , \tag{17.16}$$

$$\beta(g_W) = \frac{g_W^3}{4\pi} b_W , \qquad b_W = -\frac{1}{4\pi}(\frac{22}{3} - \frac{4}{3}n_{gen}) , \tag{17.17}$$

$$\beta(g') = \frac{(g')^3}{4\pi} b_Y , \qquad b_Y = +\frac{1}{4\pi}\frac{2}{3}(\sum |\frac{Y_L}{2}|^2 + \sum |\frac{Y_R}{2}|^2)n_{gen}$$

$$= +\frac{1}{4\pi}(\frac{20}{9})n_{gen}, \tag{17.18}$$

where, for each generation

$$\sum |\frac{Y_L}{2}|^2 = \frac{2}{3} , \qquad \sum |\frac{Y_R}{2}|^2 = \frac{8}{3} . \tag{17.19}$$

Repeating, with obvious redefinition of the constants, the procedure of (17.6) and using M_Z^2 as the subtraction point, we find, for $q^2 \gg M_Z^2$

$$\frac{1}{\alpha_s(q^2)} = \frac{1}{\alpha_s(M_Z^2)} + \frac{1}{4\pi}(11 - \frac{4}{3}n_{gen}) \ln(\frac{q^2}{M_Z^2}) , \tag{17.20}$$

$$\frac{1}{\alpha_W(q^2)} = \frac{1}{\alpha_W(M_Z^2)} + \frac{1}{4\pi}(\frac{22}{3} - \frac{4}{3}n_{gen}) \ln(\frac{q^2}{M_Z^2}) , \tag{17.21}$$

$$\frac{1}{\alpha'(q^2)} = \frac{1}{\alpha'(M_Z^2)} - \frac{1}{4\pi}(\frac{20}{9})n_{gen} \ln(\frac{q^2}{M_Z^2}) . \tag{17.22}$$

From relations (17.12) it follows that

$$\frac{1}{\alpha_W} + \frac{1}{\alpha'} = \frac{\sin^2\theta}{\alpha} + \frac{\cos^2\theta}{\alpha} = \frac{1}{\alpha} , \tag{17.23}$$

and hence the running electromagnetic constant is obtained simply by summing equations (17.21) and (17.22). Thus we obtain

$$\frac{1}{\alpha(q^2)} = \frac{1}{\alpha(M_Z^2)} - \frac{1}{3\pi}(-\frac{11}{2} + \frac{8}{3}n_{gen}) \ln(\frac{q^2}{M_Z^2}) . \tag{17.24}$$

The term proportional to n_{gen} reproduces the traditional formula for the QED β function, (17.7), with, in addition, the negative contribution of the charged vector boson, $W^{\pm}$.

Equations (17.20)–(17.24) allow extrapolation of the running constants to energies much higher than those explored experimentally, at least up to regions in which new particles, in addition to those required by the Standard Theory, could be produced. Strictly speaking, taking M_Z as the renormalisation point we are still in a non-asymptotic energy region compared to the masses of the Z, W and the top quark. Corrections can be made, by calculating the evolution of the constants with equations in which contributions of particles below their production threshold are suppressed and repositioning the subtraction point to an energy, e.g. $2m_{top}$, at which we enter the asymptotic region for all particles of the Standard Theory. These corrections are almost irrelevant, however, because of the logarithmic dependence which, to be significant, requires energy variations of orders of magnitude, rather than simple factors of $2-4$. A further correction is the contribution of the scalar Higgs electroweak doublet, which would be operative above the Higgs boson threshold. We leave as an exercise to the reader the computation of this numerically negligible contribution.

Figure 17.2 shows the high-energy behaviour of the effective constants, $\alpha, \alpha_W, \alpha_s$, starting from M_Z, extrapolating to $2m_{top}$ with the procedure described, and applying equations (17.20), (17.21) and (17.24) above $2m_{top}$, up to 10 TeV, which is the region that will be explored with the next generation of particle accelerators.

Fundamental conjugate representations of $SU(N)$. The generators of the fundamental representation of $SU(N)$ are constructed with Hermitian $N \times N$ matrices which generalise the Pauli and Gell–Mann matrices (see [2]) and which have matrix elements different from zero only for a given pair of row and column indices:

$$(\lambda^{(\bar{l}\bar{m})})_{ij} = \lambda\delta_{\bar{l}i}\delta_{\bar{m}j} + \lambda^*\delta_{\bar{l}j}\delta_{\bar{m}i} , \quad l \neq m, \ \lambda = 1, i , \qquad (17.25)$$

or are diagonal matrices with zero trace.

The number of independent diagonal matrices with zero trace is $N - 1$ while the number of non-diagonal elements with $i < j$ is $N(N - 1)/2$. The number of generators of $SU(N)$ is therefore equal to $N^2 - 1$. The λ matrices are usually normalised as

$$\text{Tr}(\lambda^A\lambda^B) = 2\delta^{AB} , \quad A, B = 1, \cdots, N^2 - 1 , \qquad (17.26)$$

and the generators of $SU(N)$ are

$$(t^A(\mathbf{N}))_{ij} = \frac{1}{2}\lambda^A , \quad A, B = 1, \cdots, N^2 - 1 , \quad \text{representation } \mathbf{N} . \ (17.27)$$

The complex conjugate representation, which for $N > 2$ is non-equivalent to the representation $\mathbf{N}$, is usually denoted as $\bar{\mathbf{N}}$. Its generators are given by

$$t^A(\bar{\mathbf{N}}) = -(t^A(\mathbf{N}))^T = -\frac{1}{2}(\lambda^A)^T , \quad A, B = 1, \cdots, N^2-1 , \quad \text{representation } \bar{\mathbf{N}} .$$
$$(17.28)$$

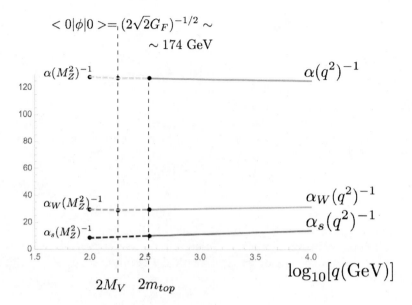

Figure 17.2 High-energy behaviour of the effective constants, α, α_W and α_s starting from M_Z, extrapolating to $2m_{top}$ with the procedure described in the text and applying the renormalisation group equations above $2m_{top}$. The points represent the values of the constants at M_Z, at the vector boson threshold, $2M_Z$, and for t–$\bar{t}$ pair production, $2m_t$. The vacuum expectation value of the Higgs field, which defines the energy scale of the Standard Theory, is also shown.

One way to derive this result is to start from the transformation rule for an element, q_i, of $\boldsymbol{N}$:

$$q_i \rightarrow (q')_i = (1 - i\epsilon^A t^A)_i^j q_j \,, \tag{17.29}$$

with ϵ^A infinitesimal parameters. We take the complex conjugate and obtain

$$q_i^* \rightarrow (q')_i^* = [(1 - i\epsilon^A t^A)_i^j]^* q_j^* = (1 + i\epsilon^A t^A)_j^i q_j^* \,, \tag{17.30}$$

where we have used the hermiticity of t^A. Therefore, if we define by $\bar{q}^i$ the element of $\boldsymbol{N}$ which transforms as q_i^*, we find

$$(\bar{q}')^j = (1 + i\epsilon^A t^A)_j^i \bar{q}^j = \left\{ \left[1 - i\epsilon^A (-t^A)^T \right] \bar{q} \right\}^j \,, \tag{17.31}$$

which leads to (17.28).

We now consider the tensor product $\boldsymbol{N} \otimes \bar{\boldsymbol{N}}$, the tensors with two indices, T_i^j, which transform like the product $\bar{q}^j q_i$:

$$(T')_i^j = (1 - i\epsilon^A t^A)_i^k (1 + i\epsilon^A t^A)_h^j T_k^h = T_k^h (1 + i\epsilon^A t^A)_h^j \, (1 - i\epsilon^A t^A)_i^k \,. \tag{17.32}$$

If we set $i = j$ and sum, the equation becomes

$$(T')^i_i = T^h_k[(1 + i\epsilon^A t^A)(1 - i\epsilon^A t^A)]^k_h = T^h_h , \qquad (17.33)$$

which proves that the trace of the tensor is invariant.

The $N \otimes N$ representation, with dimensions equal to N^2, decomposes into a singlet, dimension 1, and an irreducible representation of dimension $N^2 - 1$, which is easily recognised to be the regular representation of the group. Using the previous result, we can therefore obtain an explicit representation of the generators of the regular representation:

$$T^A = (\frac{1}{2}\lambda^A) \otimes 1 + 1 \otimes (-\frac{1}{2}\lambda^A)^T . \qquad (17.34)$$

Problem 1. In the $SU(2)$ group, the infinitesimal generators in the fundamental representation (spinor representation, spin $\frac{1}{2}$) and in the regular representation (spin 1) are constructed with the Pauli matrices and with the completely antisymmetric tensor:

$$(t^i)_{ab} = \frac{1}{2}(\tau^i)_{ab} , \quad a, b = 1, 2, \ i = 1, 2, 3 ,$$

$$(T^i)_{jk} = -i\epsilon^{jik} , \quad i, j, k = 1, 2, 3 . \qquad (17.35)$$

Prove that

$$\text{Tr}(t^i t^j) = T(\mathbf{2})\delta_{ij} = \frac{1}{2}\delta_{ij}; \quad \text{Tr}(T^i T^j) = T(A)\delta_{ij} = 2\delta_{ij} . \qquad (17.36)$$

Problem 2. Prove that for the $SU(N)$ group, the relations shown in (17.14) hold:

$$T(A) = N , \quad T(\mathbf{N}) = \frac{1}{2} . \qquad (17.37)$$

17.4   GRAND UNIFICATION AND OTHER HYPOTHESES

The success of the Standard Theory has produced a first, very important, unification of the forces acting at nuclear and subnuclear level; the three interactions, strong, electromagnetic and weak, are all derived from gauge symmetries. The differences are due to a phenomenon of spontaneous symmetry breaking, which introduces a mass scale below which the weak interactions freeze into Fermi interactions, and gives a real difference in strength between the electroweak constants, g, g' and g_s, Figure 17.2.

After the pioneering works of Pati and Salam [25], the hypothesis that the gauge group of the Standard Theory arises from a simple group, and that the differences between the three interactions are due to a spontaneous breaking of symmetry at energies higher than the electroweak level, was put forward in 1974 by Georgi and Glashow, who proposed [26] the first theory of grand unification based on the $SU(5)$ group (for a summary of grand unification theories see, for example, [27]).

The group $SU(5)$ is the minimal symmetry group which contains the Standard Theory, according to the scheme

$$SU(5) \supset SU(3)_{color} \otimes SU(2)_L \otimes U(1)_Y , \qquad (17.38)$$

and which allows complex representations capable of describing the observed chiral fermions.

It is useful to describe all the fermions with chiral left-handed fields. In this case, instead of fields associated, for example, with the right-handed electron, we consider[2] the field associated with its antiparticle, the left-handed positron $(e_c)_L$ and similarly the fields associated with u_R and d_R are substituted by $(u_c)_L$ and $(d_c)_L$. The extraordinary feature of $SU(5)$ is that we can include all the observed fermions from one generation in two irreducible representations:

$$\mathbf{5} : \begin{pmatrix} \nu \\ e \\ (d_c)^{1,2,3} \end{pmatrix} , \quad 1,2,3 = \text{colour indices} , \qquad (17.39)$$

$$\mathbf{\bar{10}} = [\mathbf{5} \otimes \mathbf{5}] \quad \text{antisymmetric tensor} : \begin{pmatrix} 0 & e_c & d_1 & d_2 & d_3 \\ & 0 & u_1 & u_2 & u_3 \\ & & 0 & (u_c)^3 & -(u_c)^2 \\ & & & 0 & (u_c)^1 \\ & & & & 0 \end{pmatrix} . \qquad (17.40)$$

The multiplet $\mathbf{5} \oplus \mathbf{\bar{10}}$ contains only the left-handed neutrino. Among the numerous extensions of $SU(5)$, the simplest one is based on the group $O(10) \supset SU(5) \otimes U(1)$, which has the complex representation

$$\mathbf{16}_{O(10)} = [\mathbf{5} \oplus \mathbf{\bar{10}} \oplus \mathbf{1}]_{SU(5)} , \qquad (17.41)$$

allowing us to add a right-handed neutrino.

The $SU(5)$ group predicts 24 vector bosons, which include:

- eight $SU(3)_{colour}$ gluons;

- four $SU(2)_L \otimes U(1)_Y$ vector bosons: $W^{\pm}, Z^0$ and the photon;

- six additional vector bosons, $SU(2)_L$ doublets and colour triplets: $X_i^{(-1/3)}, X_i^{(-4/3)}$ and their antiparticles, $\bar{X}_i^{(+4/3,+1/3)}$.

---

[2]The suffix c denotes the charge conjugate field.

The X bosons must be very heavy, because they mediate transitions, inside **5** and **10**, which *violate baryon number conservation*. The transitions shown in Figure 17.3 lead to the reactions

$$u + d \to e^+ + \bar{u} , \quad \text{or} \quad (uud) \to e^+ + \bar{u}u , \tag{17.42}$$

which would manifest themselves in proton decay

$$p \to e^+ + \pi^0 . \tag{17.43}$$

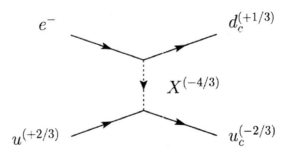

Figure 17.3 In the grand unification theory based on the $SU(5)$ group, amplitudes involving the exchange of intermediate bosons X (see text) do not obey baryon number conservation.

The decay rate of (17.43) is inversely proportional to M_X^4 and hence can be made sufficiently small to satisfy the very strict limits which already existed in 1974, requiring mass values $M_X \sim 10^{15}$ GeV, which are extremely high compared to the scale of electroweak energies. Given the logarithmic dependence of (17.20)–(17.24), such an elevated mass is consistent with the vast energy interval which must exist between the electroweak and grand unification scales, to allow the constants g, g' and g_s to converge to the common value required by grand unification.

To obtain a quantitative estimate, we must examine the form of the electroweak interactions at the grand unification scale, $q^2 \sim M_X^2$.

We consider the fermions of **5**. The GUT interaction takes the standard Yang–Mills form

$$\begin{aligned}
\mathcal{L}_I &= \bar{F} \left[g_{GUT} \frac{\lambda^A}{2} \gamma^\mu V_\mu^A \right] F \\
&= \bar{F} \left[g_s \frac{\lambda^i}{2} \gamma^\mu g_\mu^i + g \frac{\tau^3}{2} \gamma^\mu W_\mu^3 + g' \frac{Y}{2} \gamma^\mu B_\mu + \cdots \right] F , \\
& \qquad A = 1, \cdots, 24 , \quad i = 1, \cdots, 8 ,
\end{aligned} \tag{17.44}$$

where we have separated the interactions with the vector bosons of the Standard Theory, gluons, W^3 and B, and the dots denote interactions with all the other vector bosons. We see from (17.44) that the gluons and W^3 are coupled with matrices which have the same normalisation as the $SU(5)$ interactions. Therefore we expect that

$$\alpha_s(q^2),\ \alpha_W(q^2) \to \alpha_{GUT}, \quad \text{for} \quad q^2 \to M_{GUT}^2 . \tag{17.45}$$

Conversely, on **5**,

$$\text{Tr}\left(\frac{Y}{2}\right)^2 = \frac{5}{6} = \frac{5}{3} \times \frac{1}{2} = \frac{5}{3}\,\text{Tr}\left(\frac{\tau^3}{2}\right)^2 . \tag{17.46}$$

Hence, we expect

$$\alpha'(q^2) \to \frac{3}{5}\alpha_{GUT}\ , \quad \text{for} \quad q^2 \to M_{GUT}^2\ , \tag{17.47}$$

so as to reproduce an interaction normalised to the GUT interactions.[3] More simply, we can calculate the evolution in q^2 of α_s, α_W and $\frac{5}{3}\alpha'$ and see if they converge to the same value for any q^2, which would indicate the point of grand unification.

The continuation of Figure 17.2 to very high energy is shown in Figure 17.4 and shows a surprising convergence of the constants, for energies which are of the order required by the stability of the proton.

The tendency of the constants to converge towards similar values at very high energy is a striking piece of evidence in favour of grand unification.

The absence of convergence at a single point, shown by Figure 17.4, has given rise to numerous studies—for example, on the effect of higher orders on the evolution of the coupling constants, and on possible extensions to the grand unification group—as well as to conjectures about the existence of phenomena beyond the Standard Theory occurring at accessible energies, which would change the extrapolation of the constants to high energy.

Several authors, from about the year 2000 onward, noticed that convergence to a grand unification point would be considerably improved by assuming that, at energies of order[4] of 1 TeV, particles predicted by a new symmetry, *supersymmetry*, which connects particles of different spins, were to be produced. Up to now there has been no sign of the presence of such particles up to masses of 500–600 GeV. Supersymmetric particles will be searched for more intensively in the coming years in collisions at the CERN Large Hadron Collider up to the maximum energy and luminosity accessible, with a discovery potential of order 2–3 TeV.

---

[3]It is easily seen that the same scaling factor is obtained by considering $\overline{\mathbf{10}}$.

[4]1 TeV = 1000 GeV = 10^{12} eV.

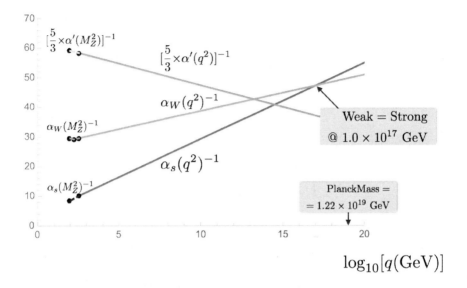

Figure 17.4 The extremely high-energy behaviour of the coupling constants of the Standard Theory with only the observed three generations of quarks and leptons shows an approximate convergence to the values required by $SU(5)$ grand unification at a mass scale $M_{GUT} \sim 10^{15}$ GeV.

The exploration of the whole region in which it is suspected that supersymmetric channels could open requires a further generation of particle accelerators, currently in the study phase in Europe, the United States, Japan and China.

The search for proton decay.   The search for proton instability was carried out, in the 1970s and 1980s, with huge detectors in underground laboratories, shielded by thick layers of rock from the background produced by cosmic rays, like Kamiokande and SuperKamiokande in Japan and IMB in the United States. These searches did not produce positive results and have extended the limits on the proton lifetime, in particular of the decay mode (17.43), to extreme values compared to what is possible with this type of experiment. The present limit from SuperKamiokande is [23]

$$\frac{\tau}{B(p \to e^+ \pi^0)} > 8.2 \cdot 10^{33} \text{ years} . \tag{17.48}$$

The problem of compatibility of such large values with the predictions of grand unified theories is still open (see again [27]).

LIMITS ON THE MASS OF THE HIGGS BOSON

CONTENTS

Scalar fields are a central element of the unified electroweak theory, in that they make possible the spontaneous breaking of the $SU(2)_L \otimes U(1)_Y$ gauge symmetry required to give rise to the masses of quarks, leptons and vector fields, cf. [2].

In this chapter we study the limits on the masses of these scalars, which are obtained from the condition that the theory remains stable up to a certain energy scale, for example, the grand unification energy or the energy characteristic of gravitation.

Max Planck was the first to observe that Newton's constant, which defines the strength of gravitational forces, combined with the universal constants of Relativity, the velocity of light c, and Quantum theory, the quantum of action, $\hbar$, leads to an energy scale which is extraordinarily large with respect to the typical energies encountered in atomic and nuclear physics. This is the so-called Planck mass, which, in natural units is $M_{\text{Planck}} = G_N^{-1/2} = 1.2 \cdot 10^{19}$ GeV.

Lev Landau speculated that the onset of quantum gravity at the Planck mass could naturally provide a finite cut-off to cure the so-called triviality problem of QED he had discovered; see the discussion of the Landau pole, Section 17.2.

18.1   SCALAR FIELDS IN THE STANDARD THEORY

To trigger the spontaneous symmetry breaking, a doublet of complex scalar fields with $Y = +1$ is inserted into the Standard Theory

$$\phi = \begin{pmatrix} \phi^+ \\ \phi^0 \end{pmatrix} = \begin{pmatrix} \frac{\phi_1 - i\phi_2}{\sqrt{2}} \\ \frac{\phi_0 - i\phi_3}{\sqrt{2}} \end{pmatrix} . \tag{18.1}$$

The self-interaction of the scalar fields is described by a renormalisable potential similar to that introduced in (8.1)

$$\mathcal{L}_s(\phi, \partial_\mu \phi) = \partial \phi_\mu^\dagger \partial^\mu \phi - V(\phi) , \tag{18.2}$$

with

$$V(\phi) = m^2 \, \phi^\dagger \phi + \frac{\lambda}{6} \, (\phi^\dagger \phi)^2 = \frac{1}{2} m^2 (\phi_i \phi_i) + \frac{\lambda}{4!} \, (\phi_i \phi_i)^2 , \tag{18.3}$$

where the sum over repeated indices, $i = 0, \cdots 3$, is understood. The Lagrangian (18.2) allows a global symmetry under $O(4) \supset SU(2)_L \otimes U(1)_Y$. The symmetry under $SU(2)_L \otimes U(1)_Y$ is promoted to a local symmetry with the minimal substitution

$$i\partial_\mu \to iD_\mu = i\partial_\mu + g\frac{\boldsymbol{\tau}}{2} \, \boldsymbol{W}_\mu + g'\frac{1}{2}B_\mu = i\partial_\mu + g^A t^A W_\mu^A , \tag{18.4}$$

and the addition of the relevant Yang–Mills tensors

$$\mathcal{L}_{YM-EW} = -\frac{1}{4}(\boldsymbol{W}_{\mu\nu}\boldsymbol{W}^{\mu\nu} + B_{\mu\nu}B^{\mu\nu}) . \tag{18.5}$$

Among the interactions of ϕ with the vector fields, we note the quartic (*seagull*) interaction:

$$\mathcal{L}^{(4)} = \frac{1}{2} \, \phi^\dagger \left\{ g^A t^A, g^B t^B \right\} \phi \, W^{A,\mu} W_\mu^B , \tag{18.6}$$

which, added to $\mathcal{L}_{YM}$, corresponds to giving the vector bosons a mass dependent on ϕ

$$M_{AB}(\phi) = \phi^\dagger \left\{ g^A t^A, g^B t^B \right\} \phi . \tag{18.7}$$

The interaction of scalar fields with quarks and leptons is described by couplings without derivatives (Yukawa couplings), made possible by the fact that the scalar doublet allows a symmetric coupling of the left-handed fermion doublets, quarks and leptons, to the right-handed singlets, cf. Section 14.2. In this way, in the presence of spontaneous breaking, the masses and mixing angles of fermions of the Standard Theory [2] are generated.

Yukawa couplings are proportional to fermion masses, so the most important one, by far, is the coupling of the heaviest quark, the t quark, which is the only one we will consider.

$$\mathcal{L}_t = g_t \, \bar{t}_R \left(\phi^0 t_L - \phi^+ b_L \right) + \text{h.c.} \tag{18.8}$$

In what follows, we will focus on the gauge theory based on $SU(3)_c \otimes SU(2)_L \otimes U(1)_Y$, restricted to scalar fields (colour singlets) and the set of $(t, b)_L$, t_R and b_R (colour triplets), interacting with each other and with the gauge fields.[1] The corresponding Lagrangian, cf. [2], is obtained from (14.17) and (14.20), specialised to the gauge group of the Standard Model, to the scalar matter fields and to third-generation quarks, (14.21), with the addition of (18.3) and (18.8)

$$\mathcal{L} = \mathcal{L}_{gauge} - V(\phi) + \mathcal{L}_t \ . \tag{18.9}$$

Spontaneous symmetry breaking. For $m^2 < 0$, the potential (18.3) has a minimum for $\phi \neq 0$ and the Higgs field acquires a non-zero vacuum expectation value. With suitable definitions, we can choose the minimum of the potential for real ϕ^0 with the other components equal to zero

$$\langle 0|\phi^0|0\rangle = \langle 0|\frac{\phi_0}{\sqrt{2}}|0\rangle = \sqrt{\frac{-3m^2}{\lambda}} = \eta \neq 0 \ . \tag{18.10}$$

Under these conditions, we write

$$\phi = \begin{pmatrix} \frac{\phi_1 - i\phi_2}{\sqrt{2}} \\ \eta + \frac{\sigma}{\sqrt{2}} - i\frac{\phi_3}{\sqrt{2}} \end{pmatrix} \ , \tag{18.11}$$

and expand the total Lagrangian (scalar, vector and quark fields) in powers of the quantum fields, σ and ϕ_i $(i = 1, 2, 3)$.

The expansion of the potential (18.3) around the minimum, (18.10), gives

$$V = -\frac{\lambda}{6}\eta^4 + \frac{1}{2}(\frac{2}{3}\lambda\eta^2)\sigma^2 + \frac{\lambda\eta}{3\sqrt{2}}\ \sigma(\sigma^2 + \sum_i \phi_i^2) + \frac{\lambda}{4!}\ (\sigma^2 + \sum_i \phi_i^2)^2 \ . \tag{18.12}$$

We note:

- the absence of terms linear in the fields, since we have expanded around a minimum;

- the absence of quadratic terms of the form $\sum_i \phi_i^2$, which is a consequence of the Goldstone theorem; $\phi_{1,2,3}$, which correspond to the three spontaneously broken generators, are massless;

- the quartic terms respect the $O(4)$ symmetry as in (18.3).

The gauge symmetry changes the situation.

For local transformations, the fields ϕ_i transform in a non-homogeneous way and we can choose a gauge, the *unitary gauge*, in which the Goldstone

---

[1]For consistency it is necessary to add the third lepton generation τ, ν_τ, cf. (14.23), required to cancel the quantum anomalies which prevent the conservation of currents associated with the gauge fields; see for example [22]; however, the effect of the lepton doublet on the considerations which follow is completely negligible.

fields are identically zero in all space. At the same time, the corresponding vector bosons acquire the mass given by (18.7), with $\phi = \eta$. In the unitary gauge, the only physical scalar field is the *Brout–Englert–Higgs boson*, corresponding to the σ field in (18.11)

$$\phi(x) = \begin{pmatrix} 0 \\ \eta + \frac{\sigma(x)}{\sqrt{2}} \end{pmatrix} \quad \text{(unitary gauge)}, \tag{18.13}$$

with mass

$$M_\sigma^2 = \frac{2}{3}\lambda\eta^2 . \tag{18.14}$$

Correspondingly, cf. [2], the interaction of the scalar field with the vector bosons reduces to the seagull term

$$\mathcal{L}^{(4)} = \frac{1}{2} \phi^T \{g^A t^A, g^B t^B\} \phi \, (W^A)^\mu W_\mu^B \quad \text{(unitary gauge)} \tag{18.15}$$

$$= M_W^2 \left[1 + \frac{\sigma(x)}{\sqrt{2}\eta}\right]^2 (W^\mu)^\dagger W_\mu + \frac{1}{2} M_Z^2 \left[1 + \frac{\sigma(x)}{\sqrt{2}\eta}\right]^2 Z^\mu Z_\mu ,$$

where

$$M_W^2 = \frac{1}{2}g^2\eta^2, \qquad M_Z^2 = \frac{1}{2}\frac{g^2}{\cos^2\theta}\eta^2 , \tag{18.16}$$

and θ is the Glashow–Weinberg–Salam angle, which connects the fields W^3 and B to the mass eigenstates:

$$W_\mu^3 = \cos\theta Z_\mu + \sin\theta A_\mu ,$$
$$B_\mu = -\sin\theta Z_\mu + \cos\theta A_\mu . \tag{18.17}$$

The photon, coupled to the conserved electromagnetic current, naturally remains massless and does not interact directly with the σ field.

From (18.16) and from the relation which links the Fermi constant to the mass of the W we determine the value of η

$$\frac{G_F}{\sqrt{2}} = \frac{g^2}{8M_W^2} = \frac{1}{4}\eta^{-2} \quad , \quad \text{or} \quad \eta = (2\sqrt{2}G_F)^{-1/2} \approx 174 \text{ GeV} . \tag{18.18}$$

Still in the unitary gauge, the Yukawa interaction takes the form

$$\mathcal{L}_t = g_t \left[\eta + \frac{\sigma(x)}{\sqrt{2}}\right] (\bar{t}_L t_R) + \text{h.c.} = m_t \, \bar{t}t + g_t \frac{\sigma}{\sqrt{2}}\bar{t}t \quad \text{(unitary gauge)}, \tag{18.19}$$

from which it follows that

$$g_t = \frac{m_t}{\eta} . \tag{18.20}$$

The massless theory. To study the high energy limit of the effective constants λ and g_t, we must consider the massless theory, $m^2 \to 0$, as we did for QED. In this limit the spontaneous breaking disappears and the theory reduces to Yang–Mills and Yukawa interactions of massless scalars and fermions.

The vertices of the massless theory are obtained from (18.1) and (18.3), with $m^2 = 0$, and from (18.8) for the Yukawa interaction. The interaction of the scalar fields with the gauge fields is obtained from the minimal substitution (18.4).

Starting from (18.15) and setting $\eta = 0$, the quartic (seagull) term takes the form:

$$\mathcal{L}^{(4)} = \frac{1}{4}g^2 \left(\frac{\phi_0^2}{2}\right)(W_\mu^1 W^{1,\mu} + W_\mu^2 W^{2,\mu}) + \frac{1}{4}(g^2 + g'^2)\left(\frac{\phi_0^2}{2}\right) Z_\mu^\dagger Z^\mu + \cdots , \quad (18.21)$$

where we have extracted the relevant terms when the external lines are of the ϕ_0 type.

In the spontaneously broken theory, the scalars ϕ_i provide the longitudinal components of the vector fields with mass. Instead, in the limit of zero mass, they are coupled, for example, to the top quark with a constant g_t, independent of the constants g, g' with which the transverse components of the vector fields are coupled.

To clarify the situation, we recommence from the unitary gauge and consider W exchange between the fermion lines in the $b + t \to t + b$ transition in the limit of high momentum transfer, $q^2 \to -\infty$, Figure 18.1.

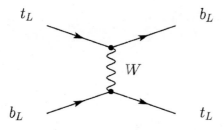

Figure 18.1 In the massless theory, the longitudinal component of W becomes an independent field.

In the unitary gauge, cf. [2], W exchange is described by the propagator

$$D_{\mu\nu} = \frac{i}{q^2 - M_W^2}\left(-g_{\mu\nu} + \frac{q_\mu q_\nu}{M_W^2}\right) . \quad (18.22)$$

In the limit $q^2 \to -\infty$, the first term gives a genuine order g^2 contribution, due to the transverse components[2]

$$A_{transv} = \left(\bar{b}_L \gamma^\mu t_L\right) \times \left(\bar{t}_L \gamma^\nu b_L\right) \frac{1}{2} g^2 \frac{-i g_{\mu\nu}}{q^2} . \tag{18.23}$$

In the longitudinal term, we use the equation of motion on the fermion lines (neglecting m_b) to find, in the same limit

$$A_{long} = \left(\bar{b}_L t_R\right) \times \left(\bar{t}_R b_L\right) \frac{i}{q^2} \frac{1}{2} \frac{m_t^2 g^2}{M_W^2} = \left(\bar{b}_L t_R\right) \times \left(\bar{t}_R b_L\right) \frac{i}{q^2} g_t^2 , \tag{18.24}$$

which corresponds to the amplitude due to the exchange of the scalar ϕ^+ coupled according to (18.8). A similar argument connects the exchange of the longitudinal Z to the exchange of the scalar ϕ^0 in the diagonal transition, $t \to t$.

18.2 LIMITS ON THE MASS OF THE HIGGS BOSON

The β functions of the electroweak constants up to the two-loop (NL) approximation are calculated in [29]. Referring to Appendix F for the details of the calculation, we give the one loop $\beta(\lambda)$ function which is obtained from the diagrams of Figure F.1

$$16\pi^2 \beta(\lambda) = 4\lambda^2 + 12\, g_t^2 \lambda - 36\, g_t^4 - 9\, g^2 \lambda - 3\, g'^2 \lambda + \frac{27}{4}\, g^4 + \frac{9}{2}\, g^2 g'^2 + \frac{9}{4}\, g'^4 . \tag{18.25}$$

In a similar way the various terms which constitute the $\beta(g_t)$ function are calculated, starting from the Feynman diagrams in Figure F.3. This leads to

$$16\pi^2 \beta(g_t) = \left(\frac{9}{2}\, g_t^2 - 8\, g_s^2 - \frac{9}{4}\, g^2 - \frac{17}{12}\, g'^2\right) g_t . \tag{18.26}$$

The renormalisation group equations which determine the behaviour of the effective constants $\lambda(q)$ and $g_t(q)$ as functions of the momentum scale q are a straightforward generalisation of (15.51). To leading logarithmic order (LL, Section 13.4)

$$q \frac{\partial \lambda(q)}{\partial q} = \beta(\lambda) = \frac{1}{16\pi^2} \left[4\lambda^2 + 12\, g_t^2\, \lambda - 36\, g_t^4 \right.$$
$$\left. - \left(9\, g^2 \lambda + 3\, g'^2 \lambda - \frac{27}{4}\, g^4 - \frac{9}{2}\, g^2 g'^2 - \frac{9}{4}\, g'^4\right) \right] , \tag{18.27}$$

---

[2]The factor $\frac{1}{2}$ in the formula arises from the fact that the fields $W^{1,2}$ are coupled through the Pauli matrices $\frac{1}{2}\tau^{1,2}$, while the vertices in Figure 18.1 correspond to raising and lowering operators, $\tau^\pm$. The factor $\frac{1}{2}$ follows from the relation $\frac{1}{2}\tau^1 \otimes \frac{1}{2}\tau^1 + \frac{1}{2}\tau^2 \otimes \frac{1}{2}\tau^2 = \frac{1}{2}\tau^+ \otimes \tau^- + \cdots$.

$$q\frac{\partial g_t(q)}{\partial q} = \beta(g_t) = \frac{1}{16\pi^2}\left[\frac{9}{2}g_t^2 - 8g_s^2 - \left(\frac{9}{4}g^2 + \frac{17}{12}g'^2\right)\right]g_t\,, \qquad (18.28)$$

where we have put the subdominant terms due to the electroweak interactions in parentheses.

In the absence of other interactions, (18.27) shows that $\beta(\lambda) > 0$. The $\lambda\phi^4$ interaction, *per se*, has the same pathology encountered in QED, a Landau pole at a finite energy scale.

According to the discussion in Section 17.2, in the continuum limit the $\lambda\phi^4$ theory should tend to the free theory (or be *trivial*). Non-perturbative arguments in this direction have been given by Kogut and Wilson for the lattice $\lambda\phi^4$ theory [30].

In the case of the Standard Theory, the coupling constant λ is determined by the mass of the Higgs boson according to (18.14) and could be large enough at presently achievable energies to give rise to the Landau pole at energies not astronomically large, as happens in QED. The requirement that this does not occur below a given energy, Λ, leads to an upper limit on the Higgs boson mass, as a function of Λ. The same happens for the Yukawa interaction constant, giving an upper limit to the mass of the *top* quark.

Furthermore, equation (18.27) leads to a lower limit on the mass of the Higgs boson. Actually, for too small values of $\lambda(\mu)$, the negative term of order g_t^4 in the right-hand side of (18.27) can push $\lambda(q)$ to negative values, for which the theory is unstable.

An upper limit on the Higgs boson mass of the order of 150 GeV, dependent on the number of generations of quarks and leptons, was obtained in [31] from the requirement that λ does not develop the Landau pole for energies below the Planck mass, $M_{\text{Planck}} = 1.2 \cdot 10^{19}$ GeV.

A first estimate of the stability of the Higgs boson mass as a function of the *top* quark mass, for $\Lambda = M_{\text{GUT}} = 10^{15}$ GeV, was obtained by Cabibbo *et al.* [32]; see also [33]. At the time of this work, the mass of the *top* quark was not known. Inserting the actual value, $m_t = 174$ GeV, leads to

$$150\text{ GeV} < M_H < 180\text{ GeV}\,,$$

$$(\Lambda = 10^{15}\text{ GeV}\,,\ m_t = 174\text{ GeV}\,,\ \text{LL approximation})\,. \qquad (18.29)$$

The behaviour of λ for the extreme values of this interval, obtained by numerically integrating equations (18.27) and (18.28), is shown in Figure 18.2. For comparison, the same figure shows the LL behaviour of the curve corresponding to the observed value of $M_H = 125$ GeV. The figure shows clearly that the energy value at which λ becomes negative can be appreciably influenced by higher-order corrections.

The calculation of the two-loop β function [29] and a preliminary determination of the mass of the *top* quark, allowed an approximate estimate to next

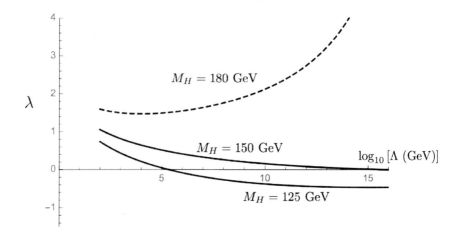

Figure 18.2 Behaviour of the effective constant $\lambda(q)$ in the Standard Theory to leading logarithm (LL) approximation, [32]. The values indicated ($M_H = 150$ and 180 GeV) are lower and upper limits on the Higgs boson mass from the conditions that the theory does not become unstable and does not develop a Landau pole below 10^{15} GeV. The curve with $M_H = 125$ GeV corresponds to the observed mass of the Higgs boson. The rather flat behaviour around $\lambda \leq 0$ suggests that the lower limit is sensitive to higher-order corrections, as confirmed in [34] and [36]; see Figure 18.3.

to leading log (NLL) order of the lower limit [34]

$$M_H \geq 135 \text{ GeV} \ ,$$

$$(\Lambda = 10^{15} \text{ GeV} \ , \ m_t = 174 \text{ GeV} \ , \text{ NLL approximation}) \ . \quad (18.30)$$

The argument received considerable attention after the discovery of the Higgs boson in 2012. Figure 18.3 shows the behaviour of λ, estimated in [36] in the next successive approximation (NNLL) with the more precise values of α_s and mass of the *top* quark available today.

Within errors, the observed value of M_H is such that the Standard Theory could be extended to the Planck mass without encountering instabilities, a noteworthy result which has given rise to some speculation (see for example [35]), but that in substance leaves completely open the possibility of new physics regimes at energies intermediate between those presently achievable and Planck's energy.

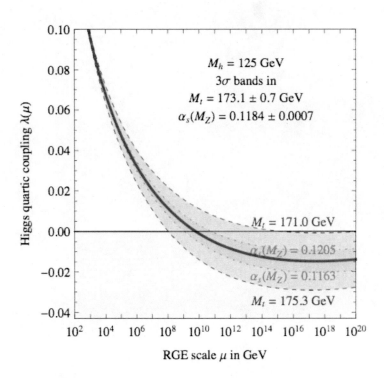

Figure 18.3 Behaviour of the effective constant $\lambda(q)$ in the Standard Theory in the next to next to leading log (NNLL) approximation. The dependence on uncertainties with which the colour constant, α_s, and *top* quark mass are known is shown. The figure is derived from [36].

THE WEAK MUON ANOMALY

CONTENTS

The advent of the renormalisable, electroweak Weinberg–Salam theory made it possible to compute the one-loop weak corrections to the muon anomaly, completing Schwinger's calculation of the pure electromagnetic correction illustrated in Chapter 12.

Previous calculations of the weak anomaly were made in the intermediate vector boson theory [37] and were plagued by uncertainties related to the divergent structure of the theory. The calculation of the weak anomaly in the Weinberg–Salam theory, reported in Ref. [38], provided a first example of how a renormalisable, spontaneously broken gauge theory works (see the paper of Fujikawa, Lee and Sanda for a particularly clear exposition).

The expected weak corrections are of order g^2, similar to the order α corrections considered in Section 12.5. However, since the intermediate vector boson masses are much larger than the muon mass, the actual order of the weak corrections is determined by the Fermi constant to be

$$(a_\mu)_W \sim \frac{1}{\pi^2} \frac{g^2 m^2}{8 M_W^2} \times \text{numerical factor} \sim \frac{Gm^2}{\sqrt{2}\pi^2} \sim 10^{-8}, \qquad (19.1)$$

m being the muon mass. At this level, weak corrections are hidden by the hadronic corrections and, as we shall discuss later, it is not yet possible to meaningfully compare the theory with the experimental data. The corresponding corrections to the electron anomaly are completely invisible, for the time being.

In this chapter we introduce the main features of renormalisable R_ξ gauges in a spontaneously broken theory and then present the calculation of the weak muon anomaly in the renormalisable Feynman–'t-Hooft gauge, following the strategy illustrated in Section 12.5.

The starting point is the Lagrangian describing scalar and vector fields introduced in the previous section, equations (18.1)–(18.5), which we summarise here:

$$\mathcal{L}(\phi, V) = (D_\mu \phi)^\dagger D^\mu \phi - V(\phi) - \frac{1}{4}(\boldsymbol{W}_{\mu\nu}\boldsymbol{W}^{\mu\nu} + B_{\mu\nu}B^{\mu\nu}) \quad (19.2)$$

$$D_\mu = \partial_\mu - ig\frac{\boldsymbol{\tau}}{2}\,\boldsymbol{W}_\mu - ig'\frac{1}{2}B_\mu \quad (19.3)$$

and define:

$$S = \frac{\phi_1 - i\phi_2}{\sqrt{2}}; \quad S_3 = \phi_3 \quad (19.4)$$

$$W = \frac{W_\mu^1 - iW_\mu^2}{\sqrt{2}}\,. \quad (19.5)$$

We expand the scalar field Lagrangian to obtain the vector boson masses and their couplings to the Goldstone bosons. We organize the expansion in order of the powers of η, the vacuum value of the scalar field, equation (18.10), and the number of derivatives of the scalar fields, restricting to terms necessary for our calculation.

Two η s, no derivatives. This is the same as in the unitary gauge and it gives us masses and mixing angles of the vector bosons, equations (18.16, 18.17),

One η, one derivative. One finds:

$$M_Z S_3 \partial_\mu Z^\mu + iM_W(S^\dagger \partial_\mu W^\mu - S\partial_\mu W^{\mu\dagger})\,. \quad (19.6)$$

One η, no derivative. These terms provide the trilinear couplings of the Goldstone to the vector bosons. We shall need only the S–W–A coupling, which is

$$\mathcal{L}_{SWA} = +\frac{eg\eta}{2}A^\mu(W_\mu S^\dagger + W_\mu^\dagger S)\,. \quad (19.7)$$

19.1   THE R_ξ GAUGE

't-Hooft's has extended to the spontaneously broken theory the choice of the gauge-fixing terms made in (14.45). With reference to the vector fields that take definite masses, one introduces the Lagrange multipliers (for a more complete discussion of gauge fixing and ghosts in a spontaneously broken gauge theory, see e.g. [39])

$$f_A = \partial_\mu A, \qquad (19.8)$$
$$f_W = \partial_\mu W^\mu + i\xi M_W S, \qquad (19.9)$$
$$f_Z = \partial_\mu Z^\mu + \xi M_Z S_3, \qquad (19.10)$$

and adds to the Lagrangian (19.2) the gauge-fixing terms:

$$\mathcal{L}_{g.f.} = -\frac{1}{2\xi} \left(f_A^2 + f_Z^2 \right) - \frac{1}{\xi} f_W f_W^\dagger . \qquad (19.11)$$

Explicitly, taking e.g. the Z term:

$$-\frac{1}{2\xi} f_Z^2 = -\frac{1}{2\xi} (\partial_\mu Z^\mu)^2 - M_Z S_3 (\partial_\mu Z^\mu) - \frac{1}{2} (\xi M_Z^2) S_3^2 , \qquad (19.12)$$

one obtains (i) a gauge-fixing term to the Z quadratic Lagrangian; (ii) a mixed term $M_Z S_3(\partial_\mu Z^\mu)$, which cancels the similar term arising from the covariant derivatives in the Lagrangian of the scalar fields, equation (19.6); (iii) a mass term for S_3, providing to the would-be Goldstone boson a mass $M_3^2 = \xi M_Z^2$.

After adding the gauge-fixing terms to the Yang–Mills Lagrangian, one easily finds the vector boson propagators (see e.g. [2]) in the generic gauge ($V = W, Z$):

$$D^{\mu\nu}(p) = \frac{i}{p^2 - M_V^2} \left[-g^{\mu\nu} + (1-\xi) \frac{p^\mu p^\nu}{p^2 - \xi M_V^2} \right] . \qquad (19.13)$$

The gauge parameter ξ is arbitrary. Special values are:

$\xi = 0$:

$$\text{(Landau gauge)} : D^{\mu\nu}(p) = \frac{i}{p^2 - M_V^2} \left[-g^{\mu\nu} + \frac{p^\mu p^\nu}{p^2} \right] \qquad (19.14)$$

$\xi = 1$

$$(\text{'t} - \text{Hooft} - \text{Feynman gauge}) : D^{\mu\nu}(p) = \frac{-ig^{\mu\nu}}{p^2 - M_V^2} \qquad (19.15)$$

$\xi = +\infty$

$$\text{(Unitary gauge)} : D^{\mu\nu}(p) = \frac{i}{p^2 - M_V^2} \left[-g^{\mu\nu} + \frac{p^\mu p^\nu}{M_V^2} \right] . \qquad (19.16)$$

The propagator of the corresponding would-be Goldstone boson is:

$$D(p) = \frac{i}{p^2 - \xi M_V^2} . \qquad (19.17)$$

For $0 < \xi$ both propagators go asymptotically like p^{-2}, making the theory renormalisable by power counting, like QED. However, to fix the gauge, we have produced unphysical poles in the scalar and in the longitudinal parts of vector propagators, which have to disappear from physical amplitudes. The cancellation has to be guaranteed by the Ward identities derived from gauge invariance.

The general argument goes, schematically, as follows. Consider a vector boson exchange between two vertices, V_A and V_B. Multiplication of the vertex by the W momentum makes the divergence of the current, which is proportional to the coupling of the corresponding scalar, would-be Goldstone boson:

$$ig(\partial_\mu J_V^\mu)_A = igm_A J_{SA} = ig\eta \, g_{SA} J_{SA} = M_V \left(g_{SA} J_{SA} \right) \ . \tag{19.18}$$

Writing $m_A = g_{SA}\eta$, we took into account that symmetry breaking originates from the vacuum value of the Higgs field via the coupling of A to the scalar fields.

The exchange of the longitudinal gauge boson then gives:

$$\begin{aligned}
A_a &= \frac{i}{p^2 - M_V^2} \, ig(\partial_\mu J_V^\mu)_A \, \frac{1 - \xi}{p^2 - \xi M_V^2} ig(\partial_\mu J_V^\mu)_B = \\
&= \frac{(1 - \xi)M_V^2}{p^2 - M_V^2} \, ig_{SA} J_{SA} \, \frac{i}{p^2 - \xi M_V^2} ig_{SB} J_{SB} \ .
\end{aligned} \tag{19.19}$$

On the other hand, the exchange of the scalar field between the same vertices is

$$A_b = ig_{SA} J_{SA} \, \frac{i}{p^2 - \xi M_V^2} ig_{SB} J_{SB} \ , \tag{19.20}$$

and the sum of the two amplitudes is

$$\begin{aligned}
A_a + A_b &= ig_{SA} J_{SA} \, \frac{i}{p^2 - \xi M_V^2} ig_{SB} J_{SB} \left[\frac{(1 - \xi)M_V^2}{p^2 - M_V^2} + 1 \right] = \\
&= ig_{SA} J_{SA} \, \frac{i}{p^2 - M_V^2} ig_{SB} J_{SB} \ ,
\end{aligned} \tag{19.21}$$

which is indeed independent of ξ.

The argument would be spoiled by the presence of Adler–Bell–Jackiw anomalies [40] in the partial conservation of the axial currents, which are implied by the V–A structure of the weak interactions. Bouchiat, Iliopoulos and Meyer [41], have shown however that the anomalies do cancel if the electric charges of the fermions sum to zero, as happens in the Standard Theory with the composition of quarks and leptons introduced in Section 14.2.

A useful check.   To see how gauge invariance works, we consider the tree-level amplitudes shown in Figure 19.1, which receive contributions from the vector and scalar propagator.

We consider the W exchange first, Figure 19.1(a). To be specific, we fix the external W to be W^2, then the exchanged W is W^1 and the amplitude, see equations (14.51), (14.52), is:

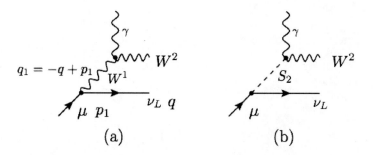

Figure 19.1 For external W^2, the exchanged vector boson in (a) is W^1. In the R_ξ gauge, the longitudinal, gauge-dependent, part of the propagator in (a) combines with the Goldstone boson propagator in (b), to yield a ξ independent result; see text.

$$A_a = \bar{u}_{\nu_L}(q)\frac{ig}{2}\tau_1\gamma^\rho\frac{1-\gamma_5}{2}u_\mu(p_1)D^{\rho\mu}(q_1)(-e)\epsilon_{132}V^{\mu\nu\lambda}(q_1,q_3,q_2) \quad (19.22)$$

with $q_1 = p_1 - q$ and $q_1 + q_2 + q_3 = 0$. We insert the longitudinal, gauge-dependent, part of the vector propagator, (19.13). Multiplication by q_1 in the fermion and Yang–Mills vertices gives:

$$q_1^\rho\, \bar{u}_{\nu_L}(q)\frac{ig}{2}\tau_1\gamma_\rho\frac{1-\gamma_5}{2}u_\mu(p_1) =$$

$$= m\, \bar{u}_{\nu_L}(q)\frac{ig}{2}\frac{1+\gamma_5}{2}u_\mu(p_1); \quad (19.23)$$

$$(q_1)_\mu\, V^{\mu\nu\lambda}(q_1,q_3,q_2) = g^{\nu\lambda}(q_2^2 - q_3^2) + \cdots = g^{\nu\lambda}M_W^2 , \quad (19.24)$$

where the ellipsis denotes terms that vanish when multiplied by the external polarisations of the on-shell photon and W^2.

Putting everything together, we find

$$A_a = -g_Y\frac{ge\eta}{2}\bar{u}_{\nu_L}(q)\frac{1+\gamma_5}{2}u_\mu(p_1)\,\frac{(1-\xi)M_W^2}{(q_1^2 - M_W^2)(q_1^2 - \xi M_W^2)}\,g^{\nu\lambda} , \quad (19.25)$$

where we have introduced the Yukawa, scalar-lepton coupling related to the muon mass $m = g_Y\eta$.

We turn now to diagram (b) in Figure 19.1, in which the exchanged scalar is ϕ_2. Using the result in (19.7), we find[1]

$$A_b = i\frac{-ig_Y}{\sqrt{2}}\,\bar{u}_{\nu_L}(q)\frac{1+\gamma_5}{2}u_\mu(p_1)\,\frac{i}{q_1^2 - \xi M_W^2}\,\frac{ige\eta}{\sqrt{2}}\,g^{\nu\lambda} =$$

$$= -g_Y\frac{ge\eta}{2}\,\bar{u}_{\nu_L}(q)\frac{1+\gamma_5}{2}u_\mu(p_1)\,\frac{1}{q_1^2 - \xi M_W^2}\,g^{\nu\lambda} . \quad (19.26)$$

---

[1] The first $-i/\sqrt{2}$ comes from the normalisation of ϕ_2 in equation (19.4), when inserted in the Yukawa coupling $\bar{\nu}_L S\mu_R$.

The total amplitude is then:

$$A_a + A_b =$$

$$-g_Y \frac{ge\eta}{2} \bar{u}_{\nu_L}(q) \frac{1+\gamma_5}{2} u_\mu(p_1) \frac{1}{q_1^2 - \xi M_W^2} g^{\nu\lambda} \left(\frac{(1-\xi)M_W^2}{q_1^2 - M_W^2} + 1 \right) =$$

$$= -g_Y \frac{ge\eta}{2} \bar{u}_{\nu_L}(q) \frac{1+\gamma_5}{2} u_\mu(p_1) \frac{1}{q_1^2 - M_W^2} g^{\nu\lambda} =$$

$$= \text{gauge independent.} \tag{19.27}$$

19.2 MUON ANOMALY: W EXCHANGE

We are now ready to compute the weak muon anomaly, starting from the diagrams which describe the effect of W exchange and of the accompanying S exchange, Figure 19.2. We set ourselves in the 't-Hooft–Feynman gauge, $\xi = 1$ and follow the strategy described in Section 12.5.

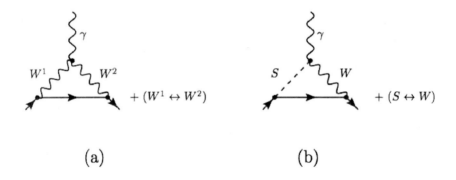

Figure 19.2 Diagrams contributing to the weak muon anomaly to order g^2: charged boson exchange.

We use the following notations for the momenta:

$$p_1, p_2 = \text{incoming and outgoing muon};$$

$$q = \text{virtual neutrino};$$

$$q_3 = p_2 - p_1 = \text{incoming photon}; \quad Q = \frac{p_1 + p_2}{2}$$

$$q_1 = p_1 - q = W^1/S; \quad q_2 = -p_2 + q = W^2/S .$$

$$\tag{19.28}$$

We write the amplitude corresponding to diagram (a) as:

$$A_a = \int \frac{d^4q}{(16\pi)^4} \frac{N_a^\nu(p_1, p_2, q)}{(q^2 + i\epsilon)z \left[(q-p_1)^2 - M_W^2 + i\epsilon\right] \left[(q-p_2)^2 - M_W^2 + i\epsilon\right]} .$$

$$\tag{19.29}$$

Discarding terms of $\mathcal{O}(q_3^2)$ and $\mathcal{O}(m^2/M_W^2)$, one finds

$$\left[(q - p_1)^2 - M_W^2 + i\epsilon\right]\left[(q - p_2)^2 - M_W^2 + i\epsilon\right] = \left[(q - Q)^2 - zM_W^2\right]^2 \tag{19.30}$$

so that, using the Feynman formula, equation (12.73), we obtain

$$\begin{aligned}
A_a &= \int_0^1 dz\, 2z \int \frac{d^4q}{(16\pi)^4} \frac{N_a^\nu(p_1, p_2, q)}{[(q - Qz)^2 - zM_W^2]^3} = \\
&= \int_0^1 dz\, 2z \int \frac{d^4k}{(16\pi)^4} \frac{N_a^\nu(p_1, p_2, k + Qz)}{(k^2 - zM_W^2)^3}
\end{aligned} \tag{19.31}$$

after the change of variables $q = k + zQ$.

We consider now the function in the numerator of (19.29).

Omitting initial and final spinors for simplicity of notation, we find

$$\begin{aligned}
N_a^\nu &= \frac{ig\tau_2}{2}\gamma_\lambda \frac{1 - \gamma_5}{2} i\slashed{q} \frac{ig\tau_1}{2}\gamma_\mu \frac{1 - \gamma_5}{2} \times \\
&\quad \times (-i)^2(-e)\epsilon_{213}V^{\lambda\mu\nu}(q_2, q_1, q_3) + \\
&\quad \cdots \frac{ig\tau_1}{2} \cdots \frac{ig\tau_2}{2} \cdots \epsilon_{123} \cdots
\end{aligned} \tag{19.32}$$

The two terms produce the commutator $[\frac{\tau_2}{2}, \frac{\tau_1}{2}]$, which, taken between initial and final muons, gives an additional factor of $+i/2$.

Simplifying and using the previous expression for V, equation (14.52), we find

$$\begin{aligned}
N_a^\nu &= e\frac{g^2}{2}\frac{1 - \gamma_5}{2} \gamma_\lambda \slashed{q} \gamma_\mu \times \\
&\quad \times V^{\lambda\mu\nu}(q_2, q_1, q_3) = \\
&= e\frac{g^2}{2}\frac{1 - \gamma_5}{2}\left[\gamma_\alpha \slashed{q}\gamma^\alpha(q_2 - q_1)_\nu + (\slashed{q_1} - \slashed{q_3})\slashed{q}\gamma_\nu + \gamma_\nu \slashed{q}(\slashed{q_3} - \slashed{q_2})\right].
\end{aligned} \tag{19.33}$$

The numerator N is at most quadratic in k, but the quadratic terms contribute a logarithmic divergence to the form factor proportional to γ^ν and $\gamma^\nu\gamma_5$. Linear terms integrate to zero and we are reduced to computing $N_a^\nu(p_1, p_2, zQ)$.

Note that there is no parity violating term of first order in q_3, so we can drop the γ_5 together with the terms proportional to γ^ν and $\gamma^\nu\gamma_5$.

Using equations (19.28) and the equations of motion on the external spinors, we can express the numerator in terms of q_3 and Q. After some algebra, we arrive at:

$$N_a^\nu = \frac{eg^2}{2}m\left\{Q^\nu z(1 - z) - \frac{3}{4}z[\slashed{q_3}, \gamma^\nu]\right\}. \tag{19.34}$$

The required result is obtained by the substitutions

$$[q\!\!\!/ , \gamma^\nu] = 4m(\frac{i}{2m}\sigma^{\nu\rho}(q_3)_\rho);$$

$$Q^\nu = m(\frac{i}{2m}\sigma^{\nu\rho}(q_3)_\rho) + \text{terms} \propto \gamma^\nu \qquad (19.35)$$

to get

$$N_a^\nu = \frac{eg^2}{4}m^2\,(2z + 4z^2)\cdot\sigma^{\nu\rho}(q_3)_\rho + \text{ terms} \propto \gamma^\nu. \qquad (19.36)$$

The integration indicated in (19.31) is done with the help of Appendix E, which gives

$$\int \frac{d^4k}{[k^2 - zM_W^2 + i\epsilon]^3} = \frac{-i\pi^2}{2}\frac{1}{zM_W^2}\,. \qquad (19.37)$$

Collecting all results and factorizing the factor $-ie$, which goes with the tree-level Lagrangian, we find:

$$a(\mu)_a = \frac{g^2m^2}{8M_W^2}\frac{1}{8\pi^2}\frac{7}{3} = \frac{Gm^2}{\sqrt{2}\pi^2}\frac{1}{8}(\frac{7}{3})\,. \qquad (19.38)$$

The calculation of diagram (b) is considerably simpler. The starting point is the expression

$$A_a = \int \frac{d^4q}{(16\pi)^4}\frac{N_b^\nu(p_1, p_2, q)}{(q^2 + i\epsilon)z\,[(q - p_1)^2 - M_W^2 + i\epsilon]\,[(q - p_2)^2 - M_W^2 + i\epsilon]}\,, \qquad (19.39)$$

which has the same denominator as (19.29) but a different numerator. For the exchange of S_1 and omitting the external spinors, one finds

$$(N_b^\nu)_1 = \frac{ig\tau_1}{2}\gamma^\nu\frac{1 - \gamma_5}{2}i q\!\!\!/ \frac{1 + \gamma_5}{2}\frac{ig_Y}{\sqrt{2}} \times \frac{ie g\eta}{\sqrt{2}} =$$

$$= \frac{eg^2\eta}{4}\frac{m}{\eta}\frac{1 + \gamma_5}{2}\gamma^\nu q\!\!\!/ \,. \qquad (19.40)$$

Adding the exchange of S_2 gives a factor of 2. The diagram with $W \leftrightarrow S$ simply sends $\frac{1+\gamma_5}{2}\gamma^\nu q\!\!\!/ \to q\!\!\!/ \gamma^\nu\frac{1-\gamma_5}{2}$.
Dropping the terms with γ_5, we find

$$N_b^\nu = e\frac{g^2}{4}m\cdot(2q^\nu)\,. \qquad (19.41)$$

The shift in the integration variable $q = k + zQ$ simply amounts to replacing q by zQ, since the term linear in k integrates to zero, and

$$N_b^\nu = e\frac{g^2}{2}m\cdot zQ^\nu = e\frac{g^2}{2}m^2z\,(\frac{i}{2m}\sigma^{\nu\rho}(q_3)_\rho) + \text{terms} \propto \gamma^\nu\,. \qquad (19.42)$$

Carrying out the integration in k and z, we find finally

$$a(\mu)_b = \frac{Gm^2}{\sqrt{2}\pi^2} \frac{1}{8} (\frac{3}{3}),$$

(19.43)

and the total from Figure 19.2 is

$$a(\mu)_W = \frac{Gm^2}{\sqrt{2}\pi^2} \frac{1}{4} (\frac{5}{3})$$

(19.44)

as reported in [38].

19.3 Z AND HIGGS BOSON EXCHANGE

The contributions from the exchange of neutral bosons arise from the diagrams of Figure 19.3, including the routing of momenta, and

$$q_3 = p_2 - p_1; \quad Q = \frac{p_2 - p_1}{2}.$$

(19.45)

We first consider diagram (c). The denominators, to first order in q_3 and neglecting m with respect to M_Z, can be written as

$$\frac{1}{(q^2 - M_Z^2 + i\epsilon)\left[(q - q_1)^2 - m^2 + i\epsilon\right]\left[(q - q_2)^2 - m^2 + i\epsilon\right]}$$
$$= \int dz \, \frac{2z}{\left[(q - zQ)^2 - z^2 m^2 - (1 - z)M_Z^2 + i\epsilon\right]^3}.$$

(19.46)

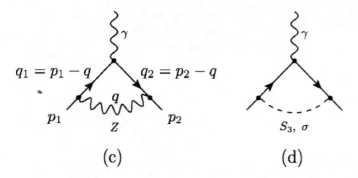

(c)

(d)

Figure 19.3 Diagrams contributing to the weak muon anomaly to order g^2: neutral boson exchange.

The numerator, after obvious simplifications, is

$$N_Z^\nu(p_1, p_2, q) =$$
$$= e\left[(g_L\gamma_\rho a^- + g_R\gamma_\rho a^+)(q\!\!\!/_1 + m)\gamma^\nu(q\!\!\!/_2 + m)(g_L\gamma^\rho a^- + g_R\gamma^\rho a^+)\right],$$

(19.47)

with $a^{\pm} = \frac{1 \pm \gamma 5}{2}$. Making the shift $q = k + zQ$ and setting $k = 0$ as before, after obvious steps, we get

$$N_Z^{\nu}(p_1, p_2, q) = -eP^{\nu}(p_1, p_2, q);$$
$$P^{\nu} = 8m(1 - z)Q^{\nu} g_L g_R +$$
$$+ (g_R^2 + g_L^2) \left[-p\!\!\!/\gamma^{\nu}p\!\!\!/ + z(p\!\!\!/\gamma^{\nu}Q\!\!\!\!/ + Q\!\!\!\!/\gamma^{\nu}p\!\!\!/) - z^2 Q\!\!\!\!/\gamma^{\nu}Q\!\!\!\!/ \right] \ . \ (19.48)$$

Eliminating $p_{1,2}$ in favour of Q and q_3, using the equation of motion of the external spinor and equations (19.35), we find[2]:

$$N_Z^{\nu} = -e \cdot \left(\frac{i}{2m} \sigma^{\nu \rho}(q_3)_{\rho} \right) \cdot 2m^2 \cdot (1 - z) \left[(g_L^2 + g_R^2)(2 - z) - 4g_L \cdot g_R \right] \ . \ (19.49)$$

Finally, performing the $dz\, d^4k$ integrations and dividing by the tree-level value, $-ie$, we find

$$a(\mu)_Z = \frac{Gm^2}{\sqrt{2}\pi^2} \frac{\cos^2 \theta}{g^2} \cdot \left[-\frac{2}{3}(g_L^2 + g_R^2) + 2g_L\, g_R \right] \ . \qquad (19.50)$$

We leave to the reader the task of proving that the exchange of S_3 gives a result smaller by a factor of m^2/M_Z^2 and the exchange of the Higgs boson gives

$$a(\mu)_{\sigma} = -\frac{Gm^2}{\sqrt{2}\pi^2} \frac{m^2}{4M_{\sigma}^2} \log \frac{M_{\sigma}^2}{m^2} \ , \qquad (19.51)$$

largely negligible given the mass of the Higgs boson.

19.4   COMPARISON WITH DATA

The chiral couplings are found directly from the weak assignments in (14.23)

$$g_L = \frac{g}{\cos \theta} \left(-\frac{1}{2} + \sin^2 \theta \right); \quad g_R = \frac{g}{\cos \theta} \sin^2 \theta \qquad (19.52)$$

with $\cos \theta$ related to the masses in (18.16). We find

$$a(\mu)_Z = \frac{Gm^2}{\sqrt{2}\pi^2} \frac{1}{6} \left(1 - 6\frac{M_W^2}{M_Z^2} + 4\frac{M_W^4}{M_Z^4} \right) =$$
$$= \frac{Gm^2}{\sqrt{2}\pi^2} \frac{1}{6} \left(-1 - 2\sin^2 \theta + 4\sin^4 \theta \right) \qquad (19.53)$$

---

[2]The denominator in equation (19.46) and the numerator given here allow us to recover the QED limit by setting $M_Z \to 0$ and $g_L = g_R = e$. It is easy to see that one reobtains Schwinger's result (12.82).

and

$$a(\mu)_{\text{weak}} = a(\mu)_W + a(\mu)_Z =$$
$$= \frac{Gm^2}{\sqrt{2}\pi^2} \frac{1}{8} \left[\frac{5}{3} + \frac{1}{3}(1 - 4\sin^2\theta)^2 \right] = 195 \cdot 10^{-11} . \quad (19.54)$$

The authors of [42] compare the most recent data on the anomaly with the full Standard Theory prediction, which includes QED corrections up to order α^4, the hadronic correction estimated via the cross section of the inelastic process $e^+e^- \to$ hadrons, and the weak corrections (19.54) supplemented by the two-loop calculation.

The comparison displays a discrepancy at the level of 2–3 standard deviations:

$$\Delta a(\mu) = a^{\text{exp}} - a^{\text{SM}} = 288(63)(49) \cdot 10^{-11} , \quad (19.55)$$

which is at present unexplained.

CHAPTER 20

EFFECTIVE POTENTIAL AND NATURALNESS

CONTENTS

The considerations about spontaneous symmetry breaking up to now are based on the classical potential introduced in (18.3). The quantum effects of the potential can be calculated by an expansion in powers of Planck's constant, $\hbar$, starting from the functional methods introduced originally by Schwinger [43], applied to spontaneous symmetry breaking by Jona–Lasinio and other authors [44, 45], and more recently by Coleman and Weinberg [28] in the context of the Standard Theory. For an extended review, see [46].[1] After analysing the ultraviolet divergences which characterise the effective potential, we illustrate the conceptual difficulties which arise from the presence of elementary scalar fields in a renormalisable field theory.

20.1 EFFECTIVE POTENTIAL

The starting point for the construction of the effective potential is the definition of the generating functional $Z(J)$ introduced in (3.20), which we recall here for convenience

$$Z[J] = \int d[\xi]d[\phi] \ \exp\left[\frac{i}{\hbar}\left(S(\xi,\phi) - \int d^4z \sum_k \phi_k(z)J_k(z)\right)\right]. \quad (20.1)$$

The action (20.1) may include spinor and gauge fields, which we denote collectively by $\xi(x)$. The functional integration denoted by $d[\xi]$ refers to them.

---

[1]LM is grateful to Massimo Testa for some illuminating discussions on the subject.

However, our attention is focused on the scalar fields, and the classical currents, J, are coupled only to them.

To carry out explicit calculations, it is useful to perform the functional integrations in four-dimensional Euclidean space.[2] In the integral which defines the action, we substitute $x_0 \to -ix_4$ (cf. Section 2.4)

$$i \int d^4x \, \mathcal{L}(\xi, \phi) \to + \int d^4x_E \mathcal{L}(\partial_0 \to -i\partial_4) = -S_E(\xi, \phi) \; , \qquad (20.2)$$

and the exponent is such as to guarantee the convergence of the functional integration.[3] In the case of scalar fields

$$S_E(\phi) = \int d^4x_E \left[\frac{1}{2}\phi(-\Box_E)\phi + V(\phi) \right] \; , \qquad (20.3)$$

where $\Box_E$ is the sum of the squares of the derivatives with respect to all coordinates.

With these substitutions, we have

$$Z_E[J] = \int d[\xi]d[\phi] \, \exp\left[-\frac{1}{\hbar}\left(S_E(\xi, \phi) + \int d^4z_E \sum_k \phi_k(z)J_k(z) \right) \right]$$
$$= e^{-\frac{1}{\hbar}W_E(J)} \; , \qquad (20.4)$$

and the exponential now converges when the fields and their derivatives go to infinity. $W_E(J)$ is the generator of connected diagrams (cf. Section 8.3).

We define a classical scalar field corresponding to each current:

$$\phi_k(x) = \frac{\delta W_E(J)}{\delta J_k(x)} \; , \qquad (20.5)$$

and we expect to be able to obtain J in terms of ϕ from (20.5). Thus we define the effective potential, a functional of $\phi_k(x)$, as the Legendre transform of $W(J)$:

$$\Gamma(\phi) = W_E(J) - \int d^4z_E \sum_k J_k(z)\phi_k(z) \; , \qquad (20.6)$$

and from now on, for notational simplicity, we omit the index k from the scalar fields, and the suffix E from the spatial coordinates.

The central property of Γ is

$$\frac{\delta\Gamma}{\delta\phi(x)} = \int d^4z \, \frac{\delta W_E}{\delta J(z)} \frac{\delta J(z)}{\delta\phi(x)} - \int \frac{\delta J(z)}{\delta\phi(x)}\phi(z) - J(x) = -J(x) \; . \qquad (20.7)$$

---

[2] This step is equivalent to carrying out the integrations in the Feynman diagrams by performing a Wick rotation.

[3] For a finite number of degrees of freedom, $L = T - V$, while $L_E = T_E + V$ with $T_E > 0$ and hence also $L_E > 0$.

For $J \to 0$:

$$\frac{\delta \Gamma}{\delta \phi(x)}\bigg|_{J=0} = 0 \, , \tag{20.8}$$

$$\phi(x)|_{J=0} = \frac{\delta W_E(J)}{\delta J(x)}\bigg|_{J \to 0}$$

$$= \frac{1}{Z(0)} \int d[\xi]d[\phi] \ \phi(x) \exp\left[-\frac{1}{\hbar}S_E(\xi,\phi)\right]$$

$$= < 0|\phi^{(q)}(x)|0 > \, . \tag{20.9}$$

In words:

- *the extreme of the potential, for zero external currents, corresponds to the vacuum expectation value of the quantum field $\phi^{(q)}$.*

Furthermore it can be seen that, to have limited fluctuations around the vacuum (a stable vacuum)

- *the extreme must correspond to the absolute minimum of the potential.*

The study of Γ, a functional of the classical variables $\phi_k(x)$, allows an examination of the possible quantum vacuums of the theory.

If we require that the stable vacuum should be invariant under the proper Poincarè group (translations and proper orthochronous Lorentz transformations), the values of the fields $\xi(x)$ in the vacuum must vanish and we can limit our considerations to scalar fields independent of x.

The potential $\Gamma(\phi)$ has a simple diagrammatic interpretation. Using the expansion in powers of ϕ:

$$\Gamma(\phi) = \sum_n \frac{1}{n!}\Gamma^{(n)}(x_1, x_2, \cdots, x_n)\phi(x_1)\phi(x_2)\cdots\phi(x_n) \, , \tag{20.10}$$

we can show that

- $\Gamma^{(n)}(x_1, x_2, \cdot, x_n)$ *is the sum of connected 1-particle irreducible (1PI) diagrams with n external lines, deprived of the corresponding propagator: in short the n-point vertices, connected and 1PI.*

In the case of constant fields, the external lines have vanishing momenta.

20.2 EXPANSION AROUND THE CLASSICAL LIMIT

In the limit $\hbar \to 0$, the exponent in (20.1) must be stationary and we recover the solution of the classical equation of motion:

$$\frac{\delta S_E}{\delta \phi_k(x)} + J_k(x) = 0 \, , \quad \frac{\delta S_E}{\delta \xi(x)} = 0 \, . \tag{20.11}$$

In the absence of external currents, the classical solution for the spinor and vector fields is simply $\xi(x) = 0$. For the scalar fields, the equations of motion with constant fields reduce to

$$\frac{\partial V(\phi)}{\partial \phi} + J = 0 \ . \tag{20.12}$$

We denote the solution of (20.12) by ϕ_J and we set $\phi = \phi_J + \sigma$ in the functional integral (20.1).

According to the saddle point method, we must expand the exponent of (20.1) up to terms quadratic in deviations from the classical solution, obtaining

$$Z(J) = \exp -\frac{1}{\hbar} \left[S_E(0, \phi_J) + \int d^4z \, \phi_J(z) J(z) \right]$$
$$\times \int d[\xi] d[\sigma] \exp \left[-\frac{1}{\hbar} \int d^4x \mathcal{L}_E^{(2)}(\xi, \sigma, \phi_J, x) \right]. \tag{20.13}$$

We have omitted terms linear in the σ fields which are of the form

$$\int d^4x \ \sigma(x) \left[\frac{\delta S_E(0, \phi)}{\delta \phi(x)} \bigg|_{\phi_J} + J(x) \right] \ , \tag{20.14}$$

and vanish from the equation of motion (20.11).

$\mathcal{L}_E^{(2)}$ is the (Euclidean) Lagrangian truncated to second order in ξ and σ, which depends on ϕ_J, the starting point of the expansion of the scalar fields. The dependence on the ξ fields is at least quadratic, hence we can write

$$\mathcal{L}_E^{(2)} = \mathcal{L}_{E\xi}^{(2)}(\xi, \phi_J) + \mathcal{L}_{E\sigma}^{(2)}(\sigma, \phi_J) \ . \tag{20.15}$$

The first term on the right-hand side is the Lagrangian of the spinor and gauge fields, restricted to quadratic terms, which can depend on the scalar fields only through ϕ_J. The second term is simply the mass term of the σ fields, calculated in the field $\phi_J = constant$

$$\mathcal{L}_{E\sigma}^{(2)} = \frac{1}{2} \frac{\partial V(\phi)}{\partial \phi_k \partial \phi_h} |_{\phi_J} \ \sigma_k(x) \sigma_h(x) \ . \tag{20.16}$$

We can write symbolically

$$\mathcal{L}_E^{(2)} = \bar{\zeta} K_\zeta(\phi_J) \zeta + V_\mu K_\xi^{\mu\nu}(\phi_J) V_\nu + \sigma K_\sigma(\phi_J) \sigma \ , \tag{20.17}$$

where ζ and V_μ denote the various fermion and gauge fields present in the theory. The integral in (20.13) reduces to the product of Gaussian integrals, which each give a contribution equal to

$$[\det \frac{K}{\hbar}]^{\mp 1/2} \ , \tag{20.18}$$

for every degree of freedom, with exponents equal to ∓ 1 according to whether boson or fermion degrees of freedom are being treated (cf. Section 6.1.3). From here we find

$$W_E(J) = S_E(0, \phi_J) + \int d^4z\, \phi_J(z) J(z) \pm \frac{N_i}{2} \hbar \sum_i [\log \det K_i(\phi_J)] + \text{const} .$$

$$(20.19)$$

The factor $1/\hbar$ in (20.18) is absorbed into the constant and disappears with the normalisation of Z, hence the term additional to the classical action is effectively of order $\hbar$, compared to the classical term.

The exponent of the determinant of K, N_i, is the number of degrees of freedom of the field. On the basis of the formulae (6.40) and (6.43), $N_i = 2, 1, 2$ for, respectively, complex fermion, real scalar and complex scalar fields. The sign $+$ $(-)$ holds for boson (fermion) fields.

We can now derive ϕ_J in terms of ϕ from (20.19)

$$\phi(x) = i\hbar \frac{W(J)}{\delta J(x)}$$

$$= \int d^4z \left[-\frac{\delta S(0, \phi)}{\delta \phi(z)} \Big|_{\phi_J} \frac{\delta \phi_j(z)}{\delta J(x)} \right] + \int d^4z\, \frac{\delta \phi_J(z)}{\delta J(x)} J(z) + \phi_J(x) + O(\hbar)$$

$$(20.20)$$

$$= \phi_J + O(\hbar) .$$

We denote with $\hbar\psi$ the difference defined by

$$\phi_J = \phi + \hbar\psi .$$

$$(20.21)$$

Finally, we can calculate $\Gamma(\phi)$

$$\Gamma(\phi) = S_E(0, \phi) + \int d^4z\, \phi(z) J(z) - \int d^4z\, \psi(z) \left[\frac{\delta S_E(\phi)}{\delta \phi(z)} \Big|_{\phi_J} + J(z) \right]$$

$$+ \hbar \sum_i [\frac{\pm N_i}{2} \log \det K_i(\phi)] - \int d^4z J(z) \phi(z) ,$$

$$(20.22)$$

or

$$\Gamma(\phi) = S_E(0, \phi) + \hbar \sum_i [\frac{\pm N_i}{2} \log \det K_i(\phi)] .$$

$$(20.23)$$

We have again used the fact that the first derivative of the action in the presence of J vanishes at the saddle point, equation (20.11).

The method can be extended to the calculation of the corrections of order $\hbar^2$ and beyond, as we show in the following section; cf. [46].

Calculation of log det K.   In cases of interest, the operators K_i are traced back to the operator

$$K = -\Box_E + M^2(\phi) \ . \tag{20.24}$$

If ϕ is independent of the coordinates, the eigenvalues in four-dimensional Euclidean space can immediately be calculated. The eigenfunctions are plane waves with Euclidean momentum p and eigenvalues

$$\lambda(p) = p^2 + M^2 \ , \quad p^2 \geq 0 \ . \tag{20.25}$$

In a finite 4-volume $\mathcal{V}_E$ with periodic boundary conditions, the eigenvalue density is obtained by extending the well-known three-dimensional formulae, cf. [1]:

$$dn(p) = \mathcal{V}_E \frac{d^4 p}{(2\pi)^4} \ . \tag{20.26}$$

In terms of the eigenvalues λ_i of K, $\det K = \prod_i \lambda_i$, or

$$\log \det K = \log \prod_i \lambda_i = \text{Tr} \log K \ , \tag{20.27}$$

from which

$$
\begin{aligned}
\log \det K &= \mathcal{V}_E \int \frac{d^4 p}{(2\pi)^4} \text{Tr} \log [p^2 + M^2(\phi)] \\
&= \frac{\mathcal{V}_E}{16\pi^2} \text{Tr} \int dp^2 \, p^2 \log [p^2 + M^2(\phi)] \ ,
\end{aligned}
\tag{20.28}
$$

where the Tr operation is carried out over the possible additional indices of K, due, for example, to $M^2(\phi)$.

We perform the integration over momentum with an ultraviolet cutoff, Λ^2. We find

$$
\log \det K(\phi) = \frac{\mathcal{V}_E}{16\pi^2} \left\{ \left(\Lambda^2 \text{Tr} \, M^2(\phi) - \frac{1}{2} \log \Lambda^2 \text{Tr} \, M^4(\phi) \right) \right.
$$
$$
\left. + \frac{1}{2} \text{Tr} \left[M^4(\phi) \log M(\phi)^2 \right] \right\} \ . \tag{20.29}
$$

Constants independent of the fields disappear in the renormalisation and are systematically omitted.

For constant fields, the action in (20.23) is proportional to $\mathcal{V}_E$ and we can therefore, for corrections due to a scalar boson field, simplify (20.23) into

$$V_{corr}(\phi) = V(\phi) + \delta V(\phi) \ ,$$

$$
\delta V = \hbar \frac{\pm N_i}{2} \frac{1}{16\pi^2} \left\{ \left(\Lambda^2 \text{Tr} \, M^2(\phi) - \frac{1}{2} \log \Lambda^2 \, \text{Tr} \, M^4(\phi) \right) \right.
$$
$$
\left. + \frac{1}{2} \text{Tr} \left[M^4(\phi) \, \log M(\phi)^2 \right] \right\} \ . \tag{20.30}
$$

20.3   LOOP EXPANSION OF THE POTENTIAL

As we have seen, the action divided by $\hbar$ appears in the exponent of the generating functional and we can ask how to organise the amplitudes in a series of increasing powers of Planck's constant. Expansion in powers of $\hbar$ was proposed by several authors as an alternative to the perturbative expansion (expansion in powers of the interaction Lagrangian) and has been shown to be of great utility in the analysis of gauge theories [28, 46].

If we deal with fundamental particles, the expansion is not necessarily useful for calculations: generally we end up with setting $\hbar = 1$. The conceptual importance of the expansion lies in the fact that the sum of terms up to a certain order gives an approximation at least compatible with the symmetries of the system, and such that there cannot be systematic cancellations or conspiracies between different orders.

Furthermore, the expansion to a certain order in $\hbar$ corresponds to the sum of an infinite number of diagrams of the perturbative series and provides some insight into the properties of the exact result, better than do the first terms of the perturbative series.

We can show that the powers of $\hbar$ are linked to the number of independent integrations over internal momenta of the Feynman diagrams. This number, L, can be visualised as the number of closed loops of the diagram, in which flows a momentum not determined by the momenta of the external particles. The expansion in powers of $\hbar$ reorders the terms of the perturbative series as an expansion in the number of loops.

To obtain the result, we can restrict the diagrams to those connected and *1-particle irreducible* (1PI). A diagram with V vertices of the interaction Lagrangian and I internal lines corresponds to carrying out a number of independent (four-dimensional) integrations equal to

$$L = I - (V - 1) \, . \tag{20.31}$$

The reason for V–1 is that at each vertex there is a Dirac δ-function of momentum, but invariance under translations requires that a δ-function expresses the conservation of energy and overall momentum, and hence only V–1 δ-functions are used to reduce the number of integrations.

The power of $\hbar$ of the overall result is linked to L. Every vertex carries a factor $\hbar^{-1}$, every propagator (the inverse of the free Lagrangian) carries a factor $\hbar$. Therefore, the diagram that we considered has a power $\hbar^P$, with P equal to

$$P = -V + I = L - 1 \, . \tag{20.32}$$

The expansion of the potential in loops has a simple diagrammatic interpretation of the terms of order $\hbar^0$ and $\hbar$ in (20.23) (it is necessary to take into account that in the definition (20.6) we have rescaled by one power of $\hbar$)

- $\hbar^0$: diagrams with zero loops, no integration, are the diagrams called *tree diagrams*; it is the part of the classical Lagrangian, $V(\phi)$, without derivatives;

- $\hbar$: are the corrections calculated with the saddle point method, equation (20.23); they correspond to diagrams with one loop and an arbitrary number of outgoing zero-momentum lines.

The argument shows also that the subsequent corrections in $\hbar$ are associated with diagrams with two loops, three loops, etc.

20.4 ONE LOOP POTENTIAL IN THE STANDARD THEORY

We apply the results of the previous section to the calculation of the effective potential of the Higgs field in the Standard Theory. As shown by (20.23), we must identify the Lagrangians which are quadratic in the quantum fields and calculate the determinant of the differential operator which characterises them.

Scalar fields. The quadratic Lagrangian arising from (18.2) and (18.3) is written (in the Standard Theory $a, b = 0, \cdots, N-1$, $N = 4$)

$$
\begin{aligned}
\mathcal{L}_{E\sigma}^{(2)}(\sigma, \phi) &= \frac{1}{2}\partial_\mu \sigma^a \partial^\mu \sigma^a + \frac{1}{2}\,\sigma^a M(\phi)_{ab}^2 \sigma^b \\
&= \frac{1}{2}\left[\sigma^a \delta_{ab}\left(-\Box + M_{ab}^2(\phi)\right)\sigma^b\right] = \frac{1}{2}\,\sigma^a\,K_\sigma(\phi)\,\sigma^b\,,
\end{aligned}
\tag{20.33}
$$

where ϕ is the classical field and

$$
\begin{aligned}
M^2(\phi)_{ab} &= \frac{\partial V(\phi)}{\partial \phi^a \partial \phi^b} = m^2\,\delta_{ab} + \frac{\lambda}{6}\,\left(\delta_{ab}\phi^2 + 2\phi_a\phi_b\right)\,, \\
\phi^2 &= \sum_{i=1,N} \phi_i \phi_i\,.
\end{aligned}
\tag{20.34}
$$

From (20.30) we find

$$
\frac{1}{2}\left[\log \det K_\sigma(\phi)\right] = \frac{1}{16\pi^2}\,\frac{1}{2}\left\{\left(\Lambda^2 \mathrm{Tr}\,M^2(\phi) - \frac{1}{2}\log\Lambda^2 \mathrm{Tr}\,M^4(\phi)\right)\right.
$$

$$
\left. + \frac{1}{2}\mathrm{Tr}\left[M^4(\phi)\,\log M(\phi)^2\right]\right\}\,.
\tag{20.35}
$$

The terms which depend on the cutoff are a second-order polynomial in ϕ^2 and are balanced by counterterms generated by the potential expressed in terms of the physical mass and coupling constant:

$$
V(\phi) = \frac{1}{2}m^2\phi^2 + \frac{\lambda}{4!}\,(\phi^2)^2 - \frac{1}{2}\delta m^2\phi^2 - \frac{\delta\lambda}{4!}\,(\phi^2)^2\,.
\tag{20.36}
$$

If we require that the final correction vanishes for a value of the field such that $M^2(\phi) = \mu^2$, we obtain

$$\Gamma(\phi) = \frac{1}{2}m^2\phi^2 + \frac{\lambda}{4!}(\phi^2)^2 + \frac{1}{64\pi^2}\text{Tr}\left[M^4(\phi)\log\frac{M(\phi)^2}{\mu^2}\right] . \qquad (20.37)$$

To diagonalise the mass matrix we use the projector

$$P_{ab} = \phi_a\phi_b\,(\phi^2)^{-1} \quad , \quad P^2 = P , \qquad (20.38)$$

and write

$$M^2 = (1 - P)(m^2 + \frac{1}{6}\phi^2) + P(m^2 + \frac{1}{2}\phi^2) = (1 - P)G(\phi) + PH(\phi) , \quad (20.39)$$

from which we find the contribution of the scalar fields to $\Gamma(\phi)$ [29]

$$\delta\Gamma_\sigma = +\frac{1}{16\pi^2}\left[\frac{1}{4}H^2\log\frac{H}{\mu^2} + \frac{3}{4}G^2\log\frac{G}{\mu^2}\right] . \qquad (20.40)$$

In the $m = 0$ limit

$$\delta\Gamma_\sigma|_{m=0} = \frac{1}{16\pi^2}\left(2\lambda^2\,\log\frac{\phi^2}{\mu^2}\right)\frac{\phi_0^4}{24} . \qquad (20.41)$$

Fermions. We choose the classical field with $\phi_0 \neq 0$ and all the other components vanishing. From (18.19) we obtain

$$\mathcal{L}_t^{(2)} = \bar{t}\,(i\slashed{\partial} - M_t(\phi))\,t \quad , \quad M_t(\phi) = \frac{g_t}{\sqrt{2}}\phi_0 , \qquad (20.42)$$

or[4]

$$K_t = \slashed{p} - M_t(\phi) . \qquad (20.43)$$

We write

$$\begin{aligned}
-\text{Tr}\log K_t &= -\int\frac{d^4p}{(2\pi)^4}\text{Tr}\log\left[\slashed{p} - M_t(\phi)\right] \\
&= -\frac{1}{2}\int\frac{d^4p}{(2\pi)^4}\text{Tr}\left[\log\left(\slashed{p} - M_t(\phi)\right) + \log\left(-\slashed{p} - M_t(\phi)\right)\right] \quad (20.44) \\
&= -\frac{6}{16\pi^2}\int dp^2\,p^2\log\left[p^2 + M_t(\phi)^2\right] + \text{const} ,
\end{aligned}$$

where a factor $4 \cdot 3$ arises from the trace over the Dirac and colour indices. In conclusion

$$\begin{aligned}
\delta\Gamma_t &= -\frac{1}{16\pi^2}\left[3\,M_t(\phi)^4\log\frac{M_t(\phi)^2}{\mu^2}\right] \\
&= \frac{1}{16\pi^2}\left[-18\,g_t^4\log\frac{M_t(\phi)^2}{\mu^2}\right]\frac{\phi_0^4}{24} . \qquad (20.45)
\end{aligned}$$

---

[4]The physical mass of the quarks is obtained from $\phi_0 = \sqrt{2}\eta$; in Euclidean space, $\slashed{p} = -p_4\gamma_4 - \boldsymbol{p}\cdot\boldsymbol{\gamma}$ with $\gamma_4 = i\gamma_0, \gamma_4^2 = -1$ and $\slashed{p}^2 = -p^2$.

Gauge fields. The relevant Lagrangian is

$$\mathcal{L}_V^{(2)} = -\frac{1}{4}\left(W_{\mu\nu}^1 W^{1\mu\nu} + W_{\mu\nu}^2 W^{2\mu\nu}\right) - \frac{1}{4}Z_{\mu\nu}Z^{\mu\nu}$$

$$+ \frac{1}{2}M_W(\phi)^2\left(W_\mu^1 W^{1\mu} + W_\mu^2 W^{2\mu}\right) + \frac{1}{2}M_Z(\phi)^2\, Z_\mu Z^\mu \qquad (20.46)$$

$$- \frac{1}{2\alpha^2}\left[(\partial^\mu W_\mu^1)^2 + (\partial^\mu W_\mu^2)^2 + (\partial^\mu Z_\mu)^2\right]\,,$$

where we have introduced a gauge-fixing term. As before, we choose the classical field with $\phi_0 \neq 0$, all the other components vanish and[5]

$$M_W(\phi)^2 = \frac{1}{4}g^2\phi_0^2, \qquad M_Z(\phi)^2 = \frac{1}{4}(g^2 + g'^2)\phi_0^2\,. \qquad (20.47)$$

Proceeding as in Section 5.2 we calculate the inverse of K (in Euclidean space) for each of the vector fields

$$K_{\mu\nu}^{-1} = \frac{1}{p^2 + M(\phi)^2}\left[\delta_{\mu\nu} - (1-\alpha)\frac{p_\mu p_\nu}{p^2 - \alpha M(\phi)^2}\right]\,. \qquad (20.48)$$

In the Landau gauge, which corresponds to $\alpha = 0$, the numerator is the projector of the states which satisfy the Lorenz condition, with

$$P_{\mu\nu} = \left[\delta_{\mu\nu} - \frac{p_\mu p_\nu}{p^2}\right]\,, \quad \text{Tr}\, P = 3\,, \qquad (20.49)$$

and we can write the contribution of a neutral vector boson to the potential as

$$\delta\Gamma_V = \frac{1}{2}\text{Tr}\log K = -\text{Tr}\log K^{-1} = \frac{3}{2}\text{Tr}\int\frac{d^4p}{(2\pi)^4}\log[p^2 + M(\phi)^2]\,, \quad (20.50)$$

from which, specialising to the case of the Standard Theory

$$\delta\Gamma_V = \frac{1}{16\pi^2}\left\{\frac{3}{2}\left[2\frac{M_W(\phi)^4}{2}\log\frac{M_W(\phi)^2}{\mu^2} + \frac{M_Z(\phi)^4}{2}\log\frac{M_Z(\phi)^2}{\mu^2}\right]\right\}$$

$$= \frac{1}{16\pi^2}\left[\frac{9}{4}g^4\log\frac{M_W(\phi)^2}{\mu^2} + \frac{9}{8}(g^2 + g'^2)^2\log\frac{M_Z(\phi)^2}{\mu^2}\right]\frac{\phi_0^4}{24}\,. \quad (20.51)$$

Renormalisation group for the potential. The methods used to study the variation of the scattering amplitudes at high energy can be used to characterise the behaviour of the effective potential at large values of the field. For simplicity, in what follows we will limit ourselves to consider in detail the purely massless scalar theory with a single coupling constant λ, in which the potential becomes finite without needing to consider the renormalisation of

---

[5]The physical mass of W and Z is obtained from $\phi_0 = \sqrt{2}\eta$.

the field, because $Z_2 = 1$ to one loop. Later we will comment on the extension to the Standard Theory.

To keep the corrections (20.35) finite, avoiding the singularity for $\phi \to 0$, we must define the renormalised constant at a value $\phi^2 = \mu^2 \neq 0$ and write, schematically

$$
\begin{aligned}
V &= [\lambda_0 - C \, \lambda_0^2 \, \log\frac{\Lambda^2}{\phi^2}] \, \frac{\phi^4}{24} \\
&= [\lambda_0 - C \, \lambda_0^2 \, \log\frac{\Lambda^2}{\mu^2}] \, \frac{\phi^4}{24} + C \, \lambda_0^2 \, \log\frac{\phi^2}{\mu^2} \, \frac{\phi^4}{24} \\
&= [\lambda(\mu) + C \, \lambda^2 \, \log\frac{\phi^2}{\mu^2}] \, \frac{\phi^4}{24} \ ,
\end{aligned}
\tag{20.52}
$$

from which it follows that

$$
\beta(\lambda) = \mu\frac{\partial\lambda(\mu)}{\partial\mu} = 2C\lambda^2 \ , \qquad \delta\Gamma = C \, \lambda^2 \, \log\frac{\phi^2}{\mu^2} \, \frac{\phi^4}{24} \ ,
\tag{20.53}
$$

with $C = 2/(16\pi^2)$, as shown by a comparison with (18.27) and (20.41).

As there is no renormalisation of the field, the proportionality of the potential to the fourth power of the field is dictated by purely dimensional reasons. In what follows, we define

$$
V = \lambda_{eff}(t, \lambda) \, \frac{\phi^4}{24} \ , \qquad t = \log\frac{\phi}{\mu} \ ,
\tag{20.54}
$$

and focus our attention on the dependence of t on λ_{eff}.

The construction that leads to (20.52) shows that λ_{eff} is independent of the value of μ, on condition of compensating a variation of μ with an appropriate redefinition of λ. In formulae, we find the Callan–Symanzik equation [47]

$$
\frac{d}{d\mu}[\lambda_{eff}(t, \lambda)] = \left(\mu\frac{\partial}{\partial\mu} + \beta(\lambda)\frac{\partial}{\partial\lambda}\right)\lambda_{eff} = 0 \ .
\tag{20.55}
$$

The solution of (20.55) is obtained starting from the function $\bar\lambda(t)$ with

$$
\frac{d\bar\lambda(t)}{dt} = \beta(\bar\lambda) \ , \qquad \bar\lambda(0) = \lambda(\mu) \ .
\tag{20.56}
$$

Actually, up to higher orders

$$
\beta(\lambda)\frac{\partial\lambda_{eff}}{\partial\lambda} = \beta(\lambda) \ ,
$$

and therefore

$$
\frac{\partial\lambda_{eff}}{\partial t} = -\mu\frac{\partial\lambda_{eff}}{\partial\mu} = \beta(\lambda)\frac{\partial\lambda_{eff}}{\partial\lambda} = \beta(\lambda_{eff}) \ ,
\tag{20.57}
$$

from which it follows that

$$\lambda_{eff} = \bar{\lambda}(t) \ . \tag{20.58}$$

Summarising, with $\beta(\lambda) = 2C\lambda^2$, we have

$$V = \lambda_{eff} \ \frac{\phi^4}{24} \ , \tag{20.59}$$

with

$$\lambda_{eff} = \frac{\lambda(\mu)}{1 - C\lambda(\mu) \log \frac{\phi^2}{\mu^2}} \quad , \quad C = \frac{2}{16\pi^2} \ . \tag{20.60}$$

The result has been derived for values of ϕ not too different from μ, when $\lambda \log \frac{\phi}{\mu} << 1$, so that perturbation theory can be used. However, by integrating the equation for increasing values of μ, we can reconstruct the variation in ϕ beyond what is predicted by perturbation theory. The solution (20.59) sums the leading powers in the parameter $\lambda \log \frac{\phi}{\mu}$, as in the case of QED and QCD (cf. Section 13.4 and Section 15.3).

As expected from the positive sign of the β function, the potential exhibits a Landau pole for large values of ϕ.

Standard Theory. The presence of non-irreducible divergent diagrams, shown in Figure F.1(d) and (g), ensures that the potential in the Standard Theory should also be finite for non-constant fields only after renormalisation of the wave function. In the potential, this implies an anomalous dimension for the ϕ fields and it is necessary to add the corresponding term to the Callan–Symanzik equation, leading to its general form for a renormalisable theory

$$\left(\mu \frac{\partial}{\partial \mu} + \sum_i \beta_i \frac{\partial}{\partial g_i} - \gamma \phi \frac{\partial}{\partial \phi} \right) V = 0 \ , \tag{20.61}$$

where we have denoted the coupling constants of the various interactions, λ, g_t, g, g', generically as g_i and the relevant β functions with β_i. However, we can omit the terms in which the β function does not refer to the constant λ, because they should give higher-order contributions, resulting in

$$\left(\mu \frac{\partial}{\partial \mu} + \beta(\lambda, g_t, \cdots) \frac{\partial}{\partial \lambda} - \gamma \phi \frac{\partial}{\partial \phi} \right) V = 0 \ . \tag{20.62}$$

To lowest order, the steps shown in (20.52) can be repeated by adding the contributions originating from the diagrams of Figure F.1 which correspond to corrections to the potential via the coupling constant and the normalisation of the field. It is useful to distinguish the contributions from irreducible diagrams,

Figures F.1(c), (e) and (h), from those of reducible diagrams, Figures F.1(d) and (g):

$$V = \left[\lambda_0 - (\sum C_{irr} + 4\lambda \sum C_{red}) \log\frac{\Lambda^2}{\phi^2}\right] \frac{1}{24}\phi^4 \left[1 + 4\sum C_{red} \log\frac{\Lambda^2}{\phi^2}\right]$$

$$= \left[\lambda_0 - (\sum C_{irr} + 4\lambda \sum C_{red}) \log\frac{\Lambda^2}{\mu^2} + (\sum C_{irr} + 4\lambda \sum C_{red}) \log\frac{\phi^2}{\mu^2}\right]$$

$$\times \frac{1}{24}\phi^4 \left[1 - 4\sum C_{red} \log\frac{\Lambda^2}{\mu^2}\right] \times \left[1 + 4\sum C_{red} \log\frac{\phi^2}{\mu^2}\right] . \quad (20.63)$$

The divergent corrections are absorbed into $\lambda(\mu)$ and $\phi(\mu)$, with

$$\beta(\lambda, g_t, \cdots) = +2(\sum C_{irr} + 4\lambda \sum C_{rid}) , \quad (20.64)$$

$$\gamma(\lambda, g_t, \cdots) = +2\sum C_{rid} , \quad (20.65)$$

and we can write

$$V = \left[\lambda(\mu) + \beta(\lambda, g_t, \cdots) \log\frac{\phi}{\mu}\right] \frac{1}{24} \left[\phi(\mu)(1 + \gamma \log\frac{\phi}{\mu})\right]^4 . \quad (20.66)$$

By construction, (20.66) fulfils the condition

$$\frac{dV}{d\mu} = 0 . \quad (20.67)$$

It can immediately be seen that the form of V which satisfies (20.62) is

$$V = V(t, \lambda, g_t, \cdots, \phi) = V(t, \bar{\lambda}(t), \bar{g}_t(t), \cdots, \phi(t)) , \quad (20.68)$$

where t is defined in (20.54), and $\bar{\lambda}(t), \bar{g}_t(t), \cdots$ are the solutions of the equations

$$\frac{d\bar{\lambda}}{dt} = \beta(\bar{\lambda}, \bar{g}_t, \cdots) \text{ etc.} \quad (20.69)$$

and

$$\phi(t) = \phi(\mu) \exp\left\{\int_0^t dt' \ \gamma[\bar{\lambda}(t'), \bar{g}_t(t'), \cdots]\right\} . \quad (20.70)$$

To one-loop order, explicitly

$$V = \frac{\bar{\lambda}(t)}{24} \left\{[\phi(\mu)]^{\int_0^t \gamma(t') \ dt'}\right\}^4 , \quad (20.71)$$

implying

$$\mu\frac{\partial V}{\partial \mu} = -\frac{d\bar{\lambda}}{dt} + 4\gamma\bar{\lambda} ,$$

$$\left[\beta(\lambda, g_t, \cdots) \frac{\partial}{\partial \lambda} - 4\gamma\right] V = \beta(\bar{\lambda}, \cdots) - 4\gamma\bar{\lambda} ,$$

$$(20.72)$$

and the Callan–Symanzik equation is satisfied by virtue of (20.69).

The form of the potential in (20.71) is connected to the considerations discussed in a previous chapter, Section 18.2, where we studied the equations (20.69) in an explicit form. If (18.25) leads to a negative value of $\bar{\lambda}$ at a certain mass scale Λ, we must expect that the potential becomes negative for $\phi \sim \Lambda$, as shown in Figure 20.1, with the destabilisation of the minimum from the minimum at low energy, where the Standard Theory resides, in favour of a new minimum, corresponding to a Higgs boson with a mass of the order of Λ.

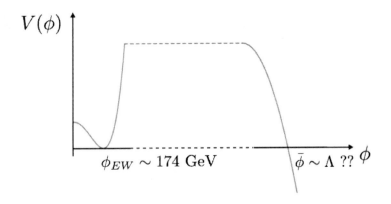

Figure 20.1 Qualitative behaviour of the potential of equation (20.71). The dotted lines mark the regions where we do not have a detailed calculation of the potential. If the running coupling $\bar{\lambda}$ becomes negative at a large energy scale, Λ, the potential becomes negative for values of the field $\bar{\phi} \sim \Lambda$ much larger than the electroweak vacuum value ϕ_{EW} determined by the Fermi constant. The theory is unstable, possibly towards a new minimum at Λ.

20.5 NON-NATURALNESS OF THE STANDARD THEORY

We have many times mentioned the necessity of a completion of the Standard Theory at high energy. Leaving aside motives of elegance which would require a unified theory, at least of the interactions associated with the $SU(2)_L \otimes U(1)_Y$ gauge group, it does not seem possible to contemplate a world in which gravity should be separated from the other forces of nature and not quantised. As we noted earlier, the constant which characterises gravity, the Planck mass, has an extraordinarily large value compared to the mass scale of the Standard Theory, fixed by the Fermi constant

$$< 0|\phi|0 >= \eta = (2\sqrt{2}G_F)^{-1/2}$$

$$\approx 274 \text{ GeV} << M_{Planck} = G_N^{-1/2} \approx 10^{19} \text{ GeV} . \quad (20.73)$$

It is therefore reasonable to think that the Standard Theory should be the "low" energy limit of a theory whose effects would show up at energies $\Lambda > \eta$ and to believe that in any case $\Lambda \leq M_{Planck}$, the equality applying in the case in which there are no other interactions below quantum gravity.

But how large can Λ be? Or, can we find clues in the Standard Theory that there should be new physical phenomena at high energy? And what estimates can we propose for the value of Λ?

A question of this type was posed at the beginning of the Fermi theory of weak interactions. In this case, to state the problem was simpler. The theory proposed by Fermi is not renormalisable and we can imagine that the new phenomena would come into play at the scale fixed by the Fermi constant, namely energies of the order of $G_F^{-1/2} \sim 300$ GeV. There, the increasing amplitudes calculated with perturbation theory exceed the unitarity condition on the S-matrix. In practice, somewhat before this point, the vector bosons make the amplitudes less singular and the Fermi theory is merged into the Standard Theory, of which it represents the low energy limit.

A similar situation appears in the theory of grand unification in which, at our energies, the effects of the exchange of additional bosons would be manifested by Fermi interactions, for example those supposedly responsible for proton decay.

In the Standard Theory, since it is renormalisable, the problem is posed in a similar way to that of QED. The Landau pole sets a limit to the extension of integrations over momentum and makes it necessary to complete QED with new interactions, to absorb it into a consistent theory in continuous space-time. In this situation, QED, and with it the Standard Theory, would be integrated directly with gravity. This would also be compatible with the considerations made on the Higgs boson mass in Section 18.2; see Figure 18.3.

New signs emerge if we follow the idea that the scale of possible new physics provides the cutoff on the momentum integrations of the amplitudes of QED or the Standard Theory.

Generally in QED there are amplitudes quadratically or linearly divergent, in the vacuum polarisation and the electron propagator. However, the quadratic divergence vanishes by virtue of gauge invariance, see Section 11.1, and the linear divergence vanishes because of chiral symmetry (see the discussion on the electron mass in Section 12.3). Therefore, in QED, the ultraviolet cutoff is frozen inside the logarithmic divergences and does not produce appreciable effects on quantum corrections to the mass of the particles which appear at low energy, even if we attribute to it a value equal to the Planck mass or grand unification mass.

An elegant criterion for theory to be *natural* even in the presence of two very different mass scales as in equation (20.73), is that the limit $m \to 0$ of the "light" particles corresponds to an increase of symmetry [48].

The cases of the spin $\frac{1}{2}$ fermion and of the spin 1 boson clearly conform to this picture: either chiral symmetry or gauge symmetry are recovered as $m \to 0$. For spin 0 particles, the light mass is natural if the particle is the Goldstone boson of a spontaneously broken global symmetry. This is the case, for example, of the pion which, in the limit of massless quarks, becomes the Goldstone boson of chiral symmetry.

It is easy to verify that none of these circumstances applies to the case of the Higgs field of the Standard Theory. The most direct way is to consider the structure of the quadratic divergences of the corrections to the effective potential, starting from equations (20.30) and (20.35). If we confine ourselves to terms quadratic in the fields, we can write:

$$V^{(2)} = \frac{1}{2} m_0^2 \phi^2 + \frac{\Lambda^2}{32\pi^2} \sum_J (-1)^{2J} (2J+1) \, M_J(\phi)^2 \, , \qquad (20.74)$$

where the summation runs over the particles of the Standard Theory (counting particles and antiparticles separately), and $M_J(\phi)$ is the mass that the particle of spin J takes in the Higgs field ϕ.

In the Standard Theory, the constants which enter into the expression for $M_J(\phi)$ are independent of each other and there is no natural relation which makes the coefficient of Λ^2 zero. If we set

$$M_H^2 \sim m_0^2 - \text{constant} \cdot \Lambda^2 \sim (100 \text{ GeV})^2 \, , \qquad (20.75)$$

for $\Lambda = M_{GUT}$ or M_{Planck} it is clearly necessary to have an extraordinary conspiracy between the starting mass and the correction, to make the resultant mass of the order of the Fermi scale, as observed.

However, the fact that in the sum of (20.74) terms enter with opposite signs suggests that some appropriate extra symmetry could force the summation to be zero. Clearly a symmetry of this type should connect fermions and bosons, i.e. must be a *supersymmetry*, the symmetry discovered by Wess and Zumino and by Volkov and Akulov in the 1970s [49]. In practice, as noted by several authors [50], in a supersymmetric theory there is exact equality between the number of fermion and boson degrees of freedom and the rule

$$\sum_J (-1)^{2J} (2J+1) M_J^2 = 0 \, , \qquad (20.76)$$

holds exactly. In the presence of explicit or spontaneous supersymmetry breaking, the masses of the particles satisfy the 't-Hooft criterion [48]; the symmetry becomes exact in the limit of mass differences equal to zero, and the Standard Theory completed in a supersymmetric theory would be natural, with respect to both grand unification and quantum gravity, if the breaking of supersymmetry would not occur orders of magnitude above the Fermi scale. From this it follows that the particles implied by supersymmetry must appear at a mass scale of order 1 to a few TeV.

Realistic models of supersymmetric electroweak interactions have been constructed starting from the 1970s and 1980s [51] [52]; for more recent reviews see for example [53].

Supersymmetry today appears as the only reasonable possibility to have a spin 0 particle as elementary as quarks, leptons and gauge fields. The alternative is that the Higgs boson is not elementary, but rather a bound fermion-antifermion state, just as the pion is composed of a quark and antiquark, and that it should be, approximately, the Goldstone boson of a symmetry that will appear at a higher energy [54]. In this picture, the only elementary particles should be spin $\frac{1}{2}$ and spin 1 particles, from which it is possible to construct natural theories, based on gauge and chiral symmetries. In these schemes, the cutoff Λ in (20.75) would be the energy at which we can resolve the Higgs boson into its elementary constituents (i.e. new types of quark bound by new forces not yet observed). The naturalness of the theory requires, also in this case, new phenomena to be observed in the region from 1 to several TeV.

The observations made at the CERN with the Large Hadron Collider during 2011–2013 did not find evidence for supersymmetric particles within mass limits that extend from 600 GeV to a little above 1 TeV, nor of new particles associated with hypothetical constituents of the Higgs boson.

A new series of measurements has just begun at higher energy and luminosity, which will explore the mass range from about 1–2.5 TeV.

New accelerators will be needed to explore the multiTeV region and give an answer to the issues raised by the mere existence of the Higgs scalar boson.

Transition Amplitude Calculation

CONTENTS

In this appendix, we describe the calculation of the matrix element $\langle q_2 | e^{-i\frac{p^2}{2m}T} | q_1 \rangle$ corresponding to the absence of a potential, discussed in Chapter 2.

A.1   TRANSITION AMPLITUDE FOR ZERO POTENTIAL

We assume that the eigenstates q and p are normalised so that

$$\langle q'|q \rangle = \delta(q' - q) \quad , \quad \int dq |q\rangle\langle q| = \mathbf{1} \ .$$

If we normalise the states $|p\rangle$ so that

$$\langle q|p \rangle = \frac{1}{\sqrt{2\pi}} e^{ipq} \ ,$$

we find that

$$\langle p'|p \rangle = \delta(p' - p) \quad , \quad \int dp |p\rangle\langle p| = \mathbf{1} \ .$$

We therefore have

$$\begin{aligned}
\langle q_2 | e^{-i\frac{p^2}{2m}T} | q_1 \rangle &= \int dk \langle q_2 | e^{-i\frac{p^2}{2m}T} | k \rangle \langle k | q_1 \rangle \\
&= \int dk e^{-i\frac{k^2}{2m}T} \langle q_2 | k \rangle \langle k | q_1 \rangle \\
&= \frac{1}{2\pi} \int dk e^{-i\frac{k^2}{2m}T} e^{i(q_2 - q_1)k} \ .
\end{aligned}$$

The integration is simplified by making the exponent a perfect square. In this way we obtain the expression

$$\langle q_2 | e^{-i\frac{p^2}{2m}T} | q_1 \rangle = \frac{1}{2\pi} e^{i\frac{m(q_2-q_1)^2}{2T}} \int dk \; e^{-i\frac{(k-k_{cl})^2}{2m}T} \quad , \quad k_{cl} = \frac{m(q_2-q_1)}{T} \quad ,$$

which is convergent for $\operatorname{Im} T < 0$. For real values of T we can define, with a change of variable, $Q = k - k_{cl}$:

$$\int dQ e^{-i\frac{Q^2}{2m}T} = \lim_{\eta \to 0^+} \int dQ e^{-i\frac{Q^2}{2m}(T-i\eta)} = \sqrt{\frac{2\pi m}{iT}} \quad ,$$

where the notation $\eta \to 0^+$ denotes that the limit is taken starting from positive values of η. The necessity to approach the limit of real values of time starting from complex values in the lower half-plane is reflected, as we saw in Chapter 3, in the famous "rule of $i\epsilon$" in the calculation of propagators and Feynman diagrams. Substituting into the previous expression, the result of equation (2.3) is obtained.

Connected Diagrams

CONTENTS

B.1   GENERATING FUNCTIONAL OF CONNECTED DIAGRAMS

We want to prove that the generating functional $Z[J]$ can be written in the form

$$Z[J] = \exp(W[J]) = \sum_{k=0}^{\infty} \frac{1}{k!} W[J]^k \ , \tag{B.1}$$

where $W[J]$ is the sum of all the connected diagrams. The proof, which we will discuss for the case of the $\lambda\phi^4$ theory, applies similarly to the perturbative expansion of any field theory.

We can write $Z[J]$ in terms of the "vertex" operator, V, defined by

$$Z[J] = e^V Z^0[J] = \sum \frac{V^k}{k!} Z^0[J] \ , \tag{B.2}$$

where the operator V is obtained directly from the interaction Lagrangian

$$V = i \int d^4x \mathcal{L}^1 \left(i \frac{\delta}{\delta J(x)} \right) \ , \tag{B.3}$$

and depends on the theory. In the case we are considering, this operator is [see equation (8.6)]

$$V = \frac{-i\lambda}{4!} \int d^4x \left(i \frac{\delta}{\delta J(x)} \right)^4 \ ,$$

while in different theories it can take a more complex form, with more J functions corresponding to several fields. The generating functional of the free theory, $Z^0[J]$, can be written in the form

$$Z^0[J] = \exp W^0[J] \ , \tag{B.4}$$

where $W^0[J]$ is the sum of connected diagrams not containing vertices. In the

$\lambda\phi^4$ theory the only diagram of this type is diagram (d) of Figure 8.2, and we find (see equation 8.5)

$$W^0[J] = \frac{i}{2} \iint d^4x\, d^4y\; J(x)\, \Delta_F(x-y)\, J(y) \; . \tag{B.5}$$

Each functional derivative of Z [see for example equations (8.9) and (8.10)] contains a factor $Z^0[J]$, hence we can write

$$Z[J] = \widetilde{Z}[J]\, Z^0[J] = \widetilde{Z}[J]\, \exp(W^0[J]) \; , \tag{B.6}$$

and note that $\widetilde{Z}[J]$ can be expressed by means of the sum of all diagrams G—connected and unconnected—*in which each connected component has at least one vertex*

$$\widetilde{Z} = 1 + \sum G \; . \tag{B.7}$$

To prove (B.1) it is therefore necessary to prove that

$$\widetilde{Z} = \exp(\widetilde{W}[J]) = \sum_{n=0}^{\infty} \frac{1}{n!} \widetilde{W}[J]^n \; , \tag{B.8}$$

where $\widetilde{W}[J]$ is the sum of all *connected diagrams, with one or more vertices,* which we imagine to be ordered in a series $\{D_1, D_2, \dots\}$

$$\widetilde{W}[J] = \sum_{i=1}^{\infty} D_i[J] \; . \tag{B.9}$$

Denoting the diagrams with one vertex by $v_i = 1$, those with two vertices by $v_i = 2$ and so on, we can therefore write

$$\exp(\widetilde{W}[J]) = e^{D_1} e^{D_2} \cdots e^{D_k} \cdots = \sum_{n_1, n_2 \dots n_k \dots} \frac{D_1^{n_1}}{n_1!} \cdots \frac{D_k^{n_k}}{n_k!} \cdots \; . \tag{B.10}$$

The term $V^k/k!$ in (B.2) will produce connected diagrams with k vertices (a subset of the list $\{D_1, D_2, \dots\}$) as well as unconnected diagrams which we will now denote with G

$$\frac{V^k}{k!} Z^0[J] = \left[\sum_i D_i\, \delta_{k\, v_i} + (\text{unconnected diagrams } G) \right] Z^0[J] \; . \tag{B.11}$$

We consider an unconnected diagram, i.e. of type G, which contains n_1 copies of connected diagrams D_1, n_2 copies of D_2 and so on. We can therefore write its contribution in the form

$$G = K_G\, (D_1)^{n_1} (D_2)^{n_2} \cdots \; , \tag{B.12}$$

where K_G is a combinatorial coefficient. To prove (B.8) we must prove that K_G is the same coefficient with which this term appears in (B.10), i.e. that

$$K_G = \frac{1}{n_1!} \cdots \frac{1}{n_k!} \cdots . \tag{B.13}$$

To calculate K_G we must start from (B.2). If $v_i \geq 1$ is the number of vertices in diagram D_i, the total number of components and of vertices in G, n and v, will be respectively[1]

$$n = \sum_{i=1}^{\infty} n_i \quad , \quad v = \sum_{i=1}^{\infty} n_i v_i .$$

Therefore G will be produced by the term $V^v/v!$ in (B.2). In this term we must choose the v_1 factors V which produce each of the n_1 copies of D_1, the v_2 factors which produce the copies of D_2 and so on [see equation (B.11)]. This choice can be made in

$$\frac{1}{\prod n_i!} \frac{v!}{\prod (v_i!)^{n_i}}$$

different ways, since there are $v!$ permutations of the factors V, but this number is divided by the number of permutations of V which contribute to each connected component of G. Therefore we divide by $\prod (v_i!)^{n_i}$, and by the number of permutations between the n_1 groups which give the n_1 copies of D_1, and so on. Finally, we divide by $\prod n_i!$. The factor $v!$ is simplified with the factor $1/v!$ which accompanies the term V^v in the expansion of $Z[J]$, equation (B.2). Similarly, each of the factors $v_i!$ in the denominator are combined [see equation (B.11)] with a V^{v_i} to generate the components D_i.

In conclusion, the value of the diagram G, composed of n_1 copies of D_1, n_2 copies of D_2, and so on, is given by

$$G[J] = \prod_{i=1}^{\infty} \frac{(D_i[J])^{n_i}}{n_i!} . \tag{B.14}$$

The coefficient K_G is hence that of (B.13), and this completes the proof.

---

[1] We note that even if the sums extend up to infinity, we are considering diagrams with a finite number of components, for which only a few n_k will be different from zero.

Lorentz invariance and one-particle states

C.1   RENORMALISATION CONSTANTS

In Section 3.4 we saw that the matrix elements of a scalar field between the vacuum and the one-particle states are given, in a theory without interactions, by expressions of the type of equation (3.75) in which a characteristic factor $1/\sqrt{2\omega}$ appears. In this appendix we will prove that this factor is determined by the invariance of the field ϕ under Lorentz transformations, and by the fact that we have chosen for the one-particle states the normalisation

$$\langle \mathbf{p}'|\mathbf{p}\rangle = \delta^{(3)}(\mathbf{p}' - \mathbf{p}) \, . \tag{C.1}$$

In the presence of interactions, the form of the matrix element between the vacuum and the one-particle states is determined to within a multiplicative constant, known as the *renormalisation constant*. In the case of a real scalar field, we must have

$$\langle 0| \, \phi(\mathbf{x}, t) \, |\mathbf{p}\rangle = \frac{\sqrt{Z}}{(2\pi)^{3/2}\sqrt{2\omega_p}} e^{i(\mathbf{p}\cdot\mathbf{x} - \omega_p t_x)} \, . \tag{C.2}$$

This result, which is the basis of the KL representation of the two-point Green's function, equation (8.39), was used in Chapter 7 to establish the relation between Green's functions and S-matrix elements.

We note that the dependence on $\mathbf{x}, t$ is fixed by the value of the momentum and energy of the particle, so it is sufficient to verify equation (C.2) for $|\mathbf{x}| = t = 0$

$$\langle 0| \, \phi(0) \, |\mathbf{p}\rangle = \frac{\sqrt{Z}}{(2\pi)^{3/2}\sqrt{2\omega_p}} \, . \tag{C.3}$$

For $|\mathbf{p}| = 0$, (C.3) can be considered as a definition of the *renormalisation constant Z*

$$\langle 0| \, \phi(0) \, ||\mathbf{p}| = 0 \rangle = \frac{\sqrt{Z}}{(2\pi)^{3/2}\sqrt{2m}} \, . \tag{C.4}$$

Therefore it remains only to prove that a Lorentz transformation leads from (C.4) to (C.3).

We consider a Lorentz transformation with velocity v along the x-axis (a *boost*) applied to the 4-vector $q \equiv \{E, \mathbf{q}\}$

$$q'_x = \frac{Ev + q_x}{\sqrt{1 - v^2}} \, , \quad q'_y = q_y \, , \quad q'_z = q_z \, , \quad E' = \frac{E + q_x v}{\sqrt{1 - v^2}} \, , \tag{C.5}$$

which is represented by a unitary transformation $\mathcal{B}_v$ on the Hilbert space. The action of $\mathcal{B}_v$ on the one-particle states can be represented in the form

$$\mathcal{B}_v |\mathbf{q}\rangle = h(\mathbf{q}, \mathbf{q}')|\mathbf{q}'\rangle \, , \tag{C.6}$$

while the vacuum state $|0\rangle$ must remain invariant

$$\mathcal{B}_v |0\rangle = |0\rangle \, . \tag{C.7}$$

Starting from zero momentum, a momentum $p = \{\omega_p = m/\sqrt{1 - v^2}, \mathbf{p}\}$, is obtained, i.e.

$$\mathcal{B}_v ||\mathbf{p}| = 0 \rangle = k(p)|\mathbf{p}\rangle \, , \tag{C.8}$$

where $(k(p) \equiv h(0, \mathbf{p}))$. Rotational invariance guarantees that $k(p)$ depends only on the magnitude of the vector $\mathbf{p}$, and we can choose the phase of the state $|\mathbf{p}\rangle$ so that $k(p)$ is real and positive.

To determine the value of $k(p)$ we note that [see equation (C.1)]

$$\delta^{(3)}(\mathbf{q}) = \langle \mathbf{q}||\mathbf{p}| = 0 \rangle = \langle \mathbf{q}|\mathcal{B}_v^\dagger \mathcal{B}_v ||\mathbf{p}| = 0 \rangle$$
$$= h^*(\mathbf{q}, \mathbf{q}')k(p)\langle \mathbf{q}'|\mathbf{p}\rangle = k^2(p)\delta^{(3)}(\mathbf{q}' - \mathbf{p}) \, ,$$

where in the final step we have used the first $\delta^{(3)}(\mathbf{q})$, which guarantees that $|\mathbf{q}| = 0$, and therefore we have substituted $h(\mathbf{q}, \mathbf{q}')$ with $h(0, \mathbf{p}) = k(p)$ which we know to be a real quantity. The vector $\mathbf{q}'$ present in the argument of the last δ-function depends on $\mathbf{q}$ via the Lorentz transformation. We therefore have

$$\delta^{(3)}(\mathbf{q}) = k^2(p)\delta^{(3)}\left(\mathbf{q}'(\mathbf{q}) - \mathbf{p}\right) = k^2(p)\left|\frac{\partial q'_i}{\partial q_k}\right|^{-1}_{|\mathbf{q}|=0} \delta^{(3)}(\mathbf{q}) \, ,$$

and with a simple calculation of the Jacobian of the Lorentz transformation, we find that

$$k^2(p) = \left|\frac{\partial q'_i}{\partial q_k}\right|_{|\mathbf{q}|=0} = \frac{1}{\sqrt{1 - v^2}} = \frac{\omega_p}{m} \, ,$$

from which it follows that (C.8) can be rewritten

$$\mathcal{B}_v |\mathbf{p} = 0 \rangle = \sqrt{\frac{\omega_p}{m}}|\mathbf{p}\rangle \, . \tag{C.9}$$

Because $\phi(x)$ is a scalar field, it must be true that

$$\mathcal{B}_v^\dagger \phi(0) \mathcal{B}_v = \phi(0) \ , \tag{C.10}$$

which implies

$$\langle 0 | \, \phi(0) \, | |\mathbf{p}| = 0 \rangle = \langle 0 | \mathcal{B}_v^\dagger \phi(0) \mathcal{B}_v | |\mathbf{p}| = 0 \rangle = \sqrt{\frac{\omega_p}{m}} \, \langle 0 | \, \phi(0) \, |\mathbf{p}\rangle \ .$$

We have therefore proven that (C.3)—and hence, for arbitrary values of $\mathbf{x}$ and t, also (C.2)—is obtained from (C.4).

Reduction formulae

CONTENTS

D.1   REDUCTION FORMULAE FOR THE COMPTON SCATTERING AMPLITUDE

As we saw in Chapter 10, the generating functional that is obtained to first order from the QED perturbative expansion can be written in the form[1]

$$Z_1[J^\rho, \bar{J}, J] = ie \int d^4u \, \frac{-i\delta}{\delta J(u)} \gamma^\mu \frac{i\delta}{\delta \bar{J}(u)} \frac{i\delta}{\delta J^\mu(u)} Z_0[J^\rho, \bar{J}, J]$$

$$= ie \int d^4u \, C_{1\mu}[J^\rho] C_2^\mu[\bar{J}, J] Z_0[J^\rho, \bar{J}, J] \, , \qquad (D.1)$$

where $Z_0[J^\rho, \bar{J}, J] = Z_0[J^\rho] Z_0[\bar{J}, J]$ is the generating functional of the free theory, equation (10.9),

$$C_{1\mu}[J^\rho] = - \int d^4x J^\rho(x) \Delta_{\rho\mu}(x - u) \, , \qquad (D.2)$$

$$C_2^\mu[\bar{J}, J] = -i\gamma^\mu S_F(u-u) - \int d^4y \, d^4z \bar{J}(y) S_F(y-u) \gamma^\mu S_F(u-z) J(z) \, , \quad (D.3)$$

and we have introduced the notation $\Delta_{\lambda\mu}(x - u) = g_{\lambda\mu} \Delta_F(x - u; 0)$.

To obtain the generating functional to second order, we must calculate the expression

$$\frac{1}{2} ie \int d^4v \, \frac{-i\delta}{\delta J(v)} \gamma^\nu \frac{i\delta}{\delta \bar{J}(v)} \frac{i\delta}{\delta J^\nu(v)} Z_1[J^\rho, \bar{J}, J] \, . \qquad (D.4)$$

The result of this operation is the sum of a large number of terms, which correspond to the different physical processes discussed in Section 10.4. Here, we only consider the contribution corresponding to the generating functional from which the Compton scattering amplitude is obtained, denoted $Z_2^C[J^\rho, \bar{J}, J]$.

---

[1] The presence of mass counterterms in the interaction Lagrangian is not relevant for the result that we propose to derive, and will therefore be neglected.

In this context the relevant functional derivative with respect to $J^\nu(v)$ can operate only on $Z_0[J^\rho]$, with the result

$$\frac{i\delta}{\delta J^\nu(v)} Z_0[J^\rho] = -\int d^4x' \, \Delta_{\sigma\nu}(v-x')J^\sigma(x')Z_0[J^\rho] , \qquad (D.5)$$

while the functional derivatives with respect to $\bar{J}(v)$ and $J(v)$ can operate, respectively, on $C_2[\bar{J}, J]$ and $Z_0[\bar{J}, J]$, or vice versa. In the first case we find (for clarity, we make the Dirac indices explicit)

$$\gamma^\nu_{\alpha\beta} \frac{i\delta}{\delta \bar{J}_\beta(v)} C_2[\bar{J}, J] = -i\int d^4z \, \gamma^\nu_{\alpha\beta} S_{F\beta\delta}(v-u)\gamma^\mu_{\delta\rho} S_{F\rho\sigma}(u-z)J_\sigma(z) , \quad (D.6)$$

and

$$\frac{-i\delta}{\delta J_\alpha(v)} Z_0[\bar{J}, J] = -\int d^4y \, \bar{J}_\xi(y)S_{F\xi\alpha}(y-v)Z_0[\bar{J}, J] . \qquad (D.7)$$

Combining equations (D.6)–(D.7) with the analogous expressions obtained by differentiating $Z_0[\bar{J}, J]$ with respect to $\bar{J}(v)$ and $C_2[\bar{J}, J]$ with respect to $J(v)$, we obtain

$$
\begin{aligned}
Z_2^C &= \frac{1}{2}(ie)^2 i \int d^4u\, d^4v\, d^4x\, d^4x'\, d^4y\, d^4z \, J^\lambda(x)\Delta_{\lambda\mu}(x-u)\Delta_{\sigma\nu}(v-x')J^\sigma(x') \\
&\quad \times \left[\bar{J}(y)S_F(y-v)\gamma^\nu S_F(v-u)\gamma^\mu S_F(u-z)J(z) \right. \\
&\quad\quad \left. + (y \leftrightarrows z, \ u \leftrightarrows v, \ \mu \leftrightarrows \nu) \right] .
\end{aligned}
\qquad (D.8)
$$

It is immediately seen that the two terms of (D.8) give identical contributions, which together eliminate the factor $\frac{1}{2}$. The final result can be written in the form

$$Z_2^C = (ie)^2 \int d^4u\, d^4v D_{1\mu\nu}[J^\rho]D_2^{\mu\nu}[\bar{J}, J]Z_0[J^\rho, \bar{J}, J] , \qquad (D.9)$$

with

$$D_{1\mu\nu}[J^\rho] = \int d^4x\, d^4x' \, J^\lambda(x)\Delta_{\lambda\mu}(x-u)\Delta_{\sigma\nu}(v-x')J^\sigma(x') , \qquad (D.10)$$

and

$$D_2^{\mu\nu}[\bar{J}, J] = i\int d^4y\, d^4z \, \bar{J}(y)S_F(y-v)\gamma^\nu S_F(v-u)\gamma^\mu S_F(u-z)J(z) . \qquad (D.11)$$

The Compton scattering amplitude is obtained, by means of the LSZ reduction formulae, from the four-point Green's function

$$G_{\alpha\beta}(x_1, x_2, x_1', x_2') = \langle 0|T\{A_\alpha(x_1)A_\beta(x_1')\bar{\psi}(x_2)\psi(x_2')\}|0\rangle , \qquad (D.12)$$

which in the path integral formalism can be rewritten in the form

$$G_{\alpha\beta}(x_1, x_2, x'_1, x'_2) =$$

$$\frac{1}{Z[0,0,0]} \frac{-i\delta}{\delta J(x_2)} \frac{i\delta}{\delta \bar{J}(x'_2)} \frac{i\delta}{\delta J^\alpha(x_1)} \frac{i\delta}{\delta J^\beta(x'_1)} Z[J^\rho, \bar{J}, J]\Bigg|_{J^\rho=\bar{J}=J=0} \quad . \quad (D.13)$$

Using the expression for the generating functional to second order, given by equations (D.9)–(D.11), it is immediately seen that the two functional derivatives with respect to $J^\alpha(x_1)$ and $J^\beta(x'_1)$ must operate on $D_1[J^\rho]$, with the result

$$\frac{i\delta}{\delta J^\alpha(x_1)} \frac{i\delta}{\delta J^\beta(x'_1)} D_1[J^\rho] =$$

$$- [\Delta_{\alpha\mu}(x_1 - u)\Delta_{\beta\nu}(v - x'_1) + \Delta_{\beta\mu}(x'_1 - u)\Delta_{\alpha\nu}(v - x_1)] \quad . \quad (D.14)$$

We now take the functional derivatives with respect to $\bar{J}(x_2)$ and $J(x'_2)$ which can operate only on $D_2[\bar{J}, J]$. The result obtained is

$$\frac{-i\delta}{\delta J(x_2)} \frac{i\delta}{\delta \bar{J}(x'_2)} D_2[\bar{J}, J]$$

$$= \frac{i\delta}{\delta J(x_2)} \int d^4z \, S_F(x'_2 - v)\gamma^\nu S_F(v - u)\gamma^\mu S_F(u - z)J(z)$$

$$= iS_F(x'_2 - v)\gamma^\nu S_F(v - u)\gamma^\mu S_F(u - x_2) \quad . \quad (D.15)$$

From equations (D.13)–(D.15) it follows that

$$G_{\alpha\beta}(x_1, x_2, x'_1, x'_2)$$

$$= -i(ie)^2 \int d^4u \, d^4v \, [\Delta_{\alpha\mu}(x_1 - u)\Delta_{\beta\nu}(v - x'_1) + \Delta_{\beta\mu}(x'_1 - u)\Delta_{\alpha\nu}(v - x_1)]$$

$$\times S_F(x'_2 - v)\gamma^\nu S_F(v - u)\gamma^\mu S_F(u - x_2) \quad . \quad (D.16)$$

Now we want to use the expression for the Green's function, (D.16), to obtain the amplitude of the process

$$\gamma(k, r) + e(p, s) \to \gamma(k', r') + e(p', s') \quad , \quad (D.17)$$

where (k, r), (k', r'), (p, s) and (p', s') are the 4-momenta and the electron and photon polarisation in the initial and final states. Using the reduction formulae discussed in Chapter [10], we can write the S-matrix element in the form

$$S_{if} = N \int dx_1 dx'_1 dx_2 dx'_2 e^{-i(kx_1 + px_2)} e^{i(k'x'_1 + p'x'_2)}$$

$$\times \epsilon^\beta(k', r')\overrightarrow{\Box}_{x'_1} \bar{u}(p', s')\overrightarrow{(i\partial\!\!\!/ - m)}_{x'_2} G_{\alpha\beta}(x_1, x_2, x'_1, x'_2)$$

$$\times \overleftarrow{(-i\partial\!\!\!/ - m)}_{x_2} u(p, s)\overleftarrow{\Box}_{x_1} \epsilon^\alpha(k, r) \quad , \quad (D.18)$$

where N is a normalisation factor that we will discuss later. The integrals over x_1, x_1', x_2 and x_2' are carried out by using the relations

$$\Box_x \Delta_F(x - y; 0) = -\delta^{(4)}(x - y) , \quad (i\slashed{\partial} - m)_x S_F(x - y) = \delta^{(4)}(x - y) , \quad \text{(D.19)}$$

with the result

$$S_{if} = -i(ie)^2 N \int d^4u \, d^4v \tag{D.20}$$

$$\times \left\{ e^{-i[(p+k)u-(p'+k')v]} \epsilon_\nu(k',r')\bar{u}(p',s')\gamma^\nu S_F(v-u)\gamma^\mu u(p,s)\epsilon_\mu(k,r) \right.$$

$$\left. + e^{-i[(p-k')u-(p'-k)v]} \epsilon_\mu(k',r')\bar{u}(p',s')\gamma^\nu S_F(v-u)\gamma^\mu u(p,s)\epsilon_\nu(k,r) \right\} .$$

For the last two integrations, the new variables $W = (v + u)/2$ and $w = v - u$ are used. Integrating over W, the δ-function expressing 4-momentum conservation is obtained and, using the concise notation

$$\epsilon_\mu(k,r) = \epsilon_\mu , \quad \epsilon_\nu(k',r') = \epsilon_\nu' , \quad u(p,s) = u , \quad \bar{u}(p',s') = \bar{u}' , \quad \text{(D.21)}$$

we can write the S-matrix element in the form

$$S_{if} = N \ (2\pi)^4 \delta(k + p - k' - p') \ ie^2 \int d^4w$$

$$\times \left[e^{i(p+k)w}\bar{u}'\slashed{\epsilon}'S_F(w)\slashed{\epsilon}u + e^{i(p-k')w}\bar{u}'\slashed{\epsilon}S_F(w)\slashed{\epsilon}'u \right]$$

$$= N \ (2\pi)^4 \delta(k + p - k' - p') \ ie^2\bar{u}' \left[\slashed{\epsilon}'S_F(p+k)\slashed{\epsilon} + \slashed{\epsilon}S_F(p-k')\slashed{\epsilon}' \right] u ,$$

$$\tag{D.22}$$

with

$$S_F(p) = \frac{\slashed{p} + m}{p^2 - m^2 + i\epsilon} . \tag{D.23}$$

The normalisation factor in equation (D.20) has the form

$$N = \frac{1}{\sqrt{(2\pi)^3 2\omega_k Z_3}} \frac{1}{\sqrt{(2\pi)^3 2\omega_p Z_2}} \frac{1}{\sqrt{(2\pi)^3 2\omega_{k'} Z_3}} \frac{1}{\sqrt{(2\pi)^3 2\omega_{p'} Z_2}} , \tag{D.24}$$

which contains the normalisations of the states describing the particles in the initial and final states, and we may set $Z_2 = Z_3 = 1$ to the present perturbative order.

Equation D.22 reproduces the Compton scattering amplitude to lowest order, well known in the literature (see e.g. [1], equations (14.82) and (14.83)).

Integrals

CONTENTS

E.1   INTEGRATION IN D DIMENSIONS

We derive the result of the following integral often encountered in perturbation theory

$$I(s, D, n) = \int \frac{d^D k}{[k^2 - s + i\epsilon]^n} = i\pi^{D/2}(-1)^n \frac{\Gamma(n - D/2)}{\Gamma(n)} \frac{1}{s^{n-D/2}} . \quad (E.1)$$

The integrals extend over a space of D dimensions, $k = \{k_0, k_1, ...k_{D-1}\}$, with a Minkowski metric $k^2 = k_0^2 - k_1^2 .... - k_{D-1}^2$. We will assume that s is real[1] and positive.

The Γ function is defined by

$$\Gamma(x) = \int_0^\infty dy \, y^{x-1} e^{-y} , \quad \Gamma(x) = (x - 1)\Gamma(x - 1) , \quad \Gamma(n) = (n - 1)! . \quad (E.2)$$

In reality, we are interested in the case $D = 4$, but we would also like to consider an analytic continuation to arbitrary values of D defined by the result in (E.1), which is an analytic function of D except for poles at $D/2 = n, n + 1, ...$.

Actually $\Gamma(x)$ is analytic for $\Re(x) > 0$, and can be continued to values $\Re(x) \leq 0$ by using the relation $\Gamma(x) = \Gamma(x + 1)/(x)$, with which it is easily proven that $\Gamma(x)$ is analytic in the whole of the complex plane with the exception of poles at $x = 0$, $x = -1$, $\cdots$. For example, starting from a Taylor series expansion in the vicinity of $x = 1$, where Γ is analytic (see for example [55])

$$\Gamma(1 + \eta) = 1 + \gamma\eta + \mathcal{O}(\eta^2) ,$$

---

[1] The value of $I(s, D, n)$ for complex values of s, if needed, can be obtained by analytic continuation.

where γ is Euler's constant, $\gamma = 0.5772\ldots$, we obtain, in the vicinity of $x = 0$

$$\Gamma(\eta) = \frac{1}{\eta} + \gamma + \mathcal{O}(\eta) \ .$$

To obtain the result in (E.1), we first rotate the integration path in the complex plane of k_0 from the real to the imaginary axis with a rotation in the anticlockwise direction. Due to the $+i\epsilon$ rule of the propagators, no singularities are encountered, as shown in Figure E.1. This operation is known as a Wick rotation. After the rotation we can take the limit $\epsilon \to 0$ since the two singularities at $k_0 = \pm\omega = \pm(s + k_1^2\ldots + k_{D-1}^2)^{1/2}$ are far away from the new integration path.

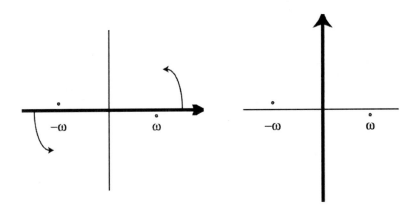

Figure E.1   Wick rotation.

We can therefore set $k_0 = ik_D$ and $dk_0 = i\,dk_D$, and rewrite the integral as

$$I(s, D, n) = i(-1)^n \int \frac{d^D p}{[p^2 + s]^n} \ , \tag{E.3}$$

where $p = \{k_1, \ldots k_{D-1}, k_D\}$ is a D-dimensional vector with Euclidean metric $p^2 = k_1^2 + k_2^2 \ldots + k_D^2$.

To calculate the integral we use polar coordinates in the space of D dimensions. Since the integrand does not depend on the angular variables, the latter can immediately be integrated and, with the change of variables $x = p^2/s$, implying $p\,dp = s\,dx/2$, we obtain

$$I(s, D, n) = i(-1)^n \int \frac{p^{D-1} dp\, d\Omega_D}{[p^2 + s]^n}$$

$$= i(-1)^n \frac{\Omega_D}{2s^{n-D/2}} \int_0^\infty \frac{x^{(D-2)/2} dx}{(1 + x)^n} \ . \tag{E.4}$$

The above integral is convergent as $x \to \infty$ if $n > D/2$. The solid angle in D dimensions is given by [2]

$$\Omega_D = \frac{2\pi^{D/2}}{\Gamma(D/2)} , \qquad (\text{E.5})$$

which reproduces the known results: $\Omega_2 = 2\pi$ and, since $\Gamma(3/2) = \frac{1}{2}\Gamma(1/2) = \pi^{1/2}/2$, $\Omega_3 = 4\pi$. In four dimensions $\Omega_4 = 2\pi^2$ is obtained.

The remaining integral is expressed by means of the Beta function (for the proof see Chapter 15 of [56], but here we use the notation of [55]),

$$B(z, w) = \int_0^\infty \frac{x^{z-1}dx}{(1+x)^{z+w}} = \frac{\Gamma(z)\,\Gamma(w)}{\Gamma(z+w)} , \qquad (\text{E.6})$$

and we finally recover the result of (E.1).

E.2   FEYNMAN PARAMETERS

A simple case.   We consider the integral

$$I(p) = \int d^D k \; \frac{1}{k^2 - m^2 + i\epsilon} \; \frac{1}{(k-p)^2 - m^2 + i\epsilon} . \qquad (\text{E.7})$$

$I(p)$ can be expressed in terms of a parametric integration over a real variable x ($0 \le x \le 1$) with a method introduced by Feynman. We start from the identity

$$\frac{1}{D_1 D_2} = \int_0^1 dx \frac{1}{[xD_1 + (1-x)D_2]^2} , \qquad (\text{E.8})$$

which is proven straightaway by carrying out the integration of the right-hand side

$$\int_0^1 dx \frac{1}{[xD_1 + (1-x)D_2]^2} = \int_0^1 dx \frac{1}{[x(D_1 - D_2) + D_2]^2} \qquad (\text{E.9})$$

$$= \frac{1}{D_1 - D_2}\left(\frac{1}{D_2} - \frac{1}{D_1}\right) = \frac{1}{D_1 D_2} . \qquad (\text{E.10})$$

Applying (E.8) we find

$$I(p) = \int_0^1 dx \int d^D k \; \frac{1}{[x(k-p)^2 + (1-x)k^2 - m^2]^2}$$

$$= \int_0^1 dx \int d^D k \; \frac{1}{[k^2 - 2xk \cdot p + xp^2 - m^2]^2} \qquad (\text{E.11})$$

$$= \int_0^1 dx \int d^D k \; \frac{1}{[(k - xp)^2 + x(1-x)p^2 - m^2]^2} .$$

---

[2] The result is easily obtained by evaluating the Gaussian integral in D dimensions $\int d^D x \; e^{-(x_1^2 + \cdots + x_D^2)}$ in Cartesian and polar coordinates.

After the change of variable $k \to k' = k - xp$, the integral (E.11) takes the form in (E.1) with

$$s = m^2 - x(1-x)p^2 \quad , \quad n = 2 , \tag{E.12}$$

from which it follows that

$$I(p) = i\pi^{D/2}\Gamma(2 - D/2) \int_0^1 dx \frac{1}{[m^2 - x(1-x)p^2]^{2-D/2}} . \tag{E.13}$$

We can now set $D = 4 - \eta$, with the idea of taking the limit $\eta \to 0^+$. We obtain

$$I(p) = i\pi^{2-\eta/2}\Gamma(\eta/2) \int_0^1 dx \; e^{-\frac{\eta}{2} \ln[m^2 - x(1-x)p^2]}$$

$$= i\pi^2 \left\{ \frac{2}{\eta} - \ln\pi - \int_0^1 dx \ln [m^2 - x(1-x)p^2] + \mathcal{O}(\eta) \right\} \tag{E.14}$$

$$= I(0) - i\pi^2 \int_0^1 dx \; \ln\left[1 - x(1-x)\frac{p^2}{m^2}\right] ,$$

with $I(0)$ a constant, diverging the limit $\eta \to 0^+$.

The general case. Equation (E.8) can be generalised to the case of n denominators. We start from the Schwinger representation

$$\frac{1}{D_1} = \int_0^{+\infty} d\alpha \; e^{-\alpha D_1} , \tag{E.15}$$

and write

$$I_n = \frac{1}{D_1 D_2 \cdots D_n} = \int_0^{+\infty} (\Pi_i d\alpha_i) \; e^{-\sum \alpha_i D_i} . \tag{E.16}$$

The right-hand side of the above equation can be rewritten introducing a δ-function and an integration over the additional variable λ

$$I_n = \int_0^{+\infty} d\lambda(\Pi_i d\alpha_i) \; \delta(\lambda - \sum \alpha_i) \; e^{-\sum \alpha_i D_i} . \tag{E.17}$$

Setting $\alpha_i = \lambda x_i$ we obtain

$$I_n \int (\Pi_i dx_i)\delta(1 - \sum x_i) \int_0^{+\infty} d\lambda \; \lambda^{n-1}e^{-\lambda(\sum x_i D_i)} , \tag{E.18}$$

where the integration over each x_i is now extended to the range $[0, 1]$. Finally, introducing the new variable $t = \lambda(\sum x_i D_i)$, we find

$$I_n = \int (\Pi_i dx_i)\delta(1 - \sum x_i)\frac{1}{(\sum x_i D_i)^n} \int_0^{+\infty} t^{(n-1)}e^{-t}$$

$$= \int [\Pi dx_i \; \delta(1 - \sum x_i)] \frac{(n-1)!}{(x_1 D_1 + x_2 D_2 + \cdots x_n D_n)^n} . \tag{E.19}$$

$\beta(\lambda)$ and $\beta(g_t)$ functions

CONTENTS

F.1   CALCULATION OF THE $\beta(\lambda)$ AND $\beta(g_t)$ FUNCTIONS

We describe here the details of the calculations of the $\beta(\lambda)$ and $\beta(g_t)$ functions, cf. [29].

F.2 $\beta(\lambda)$

The relevant Feynman diagrams are shown in Figure F.1. The integration over the internal lines is carried out with an upper limit on the momenta, an *ultraviolet cutoff* Λ, and with external momenta initially different from zero. For the simple cases in the figure, techniques of dimensional regularisation to identify the quadratically and logarithmically divergent terms are not needed.

We introduce the *superficial degree of divergence* which, for one-loop diagrams, is equal to

$$D = 4 + n - d , \tag{F.1}$$

where n and d are the powers of the momentum circulating in the loop, which occur in the numerator and the denominator of the function to be integrated. $D = 2$ and 0 correspond to quadratic and logarithmic divergences in the cutoff Λ.

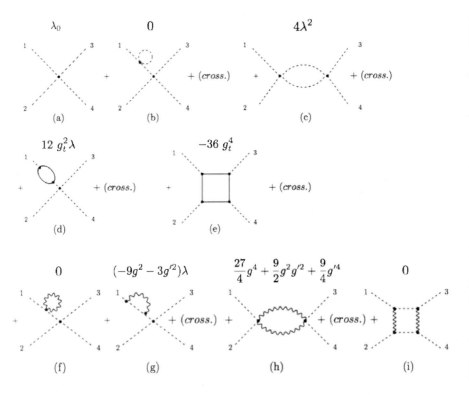

Figure F.1 Feynman diagrams for the $\beta(\lambda)$ function in the Standard Theory. The contributions shown for each diagram are multiplied by a common factor of $(16\pi^2)^{-1}$. The box diagrams with internal lines of alternate scalars and vectors, which are not divergent in the Landau gauge used in the text, are not shown.

The usefulness of the Landau gauge [28] is in the fact that this choice reduces the degree of divergence associated with the scalar-vector vertex in one-loop diagrams, see Fig. F.2. As a consequence, diagrams involving a loop in which scalars and vectors alternate become more convergent, or vanish if the external momentum vanishes, $q = 0$.

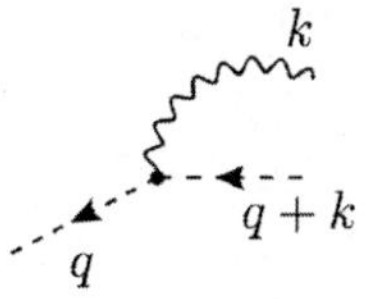

$$(2q + k)_\mu (g^{\mu\nu} - \tfrac{k^\mu k^\nu}{k^2}) = 2q^\nu$$

Figure F.2 Scalar-vector vertex in the Landau gauge. The momentum of the loop gives zero in the product of the scalar vertex with the propagator of the vector field, reducing the degree of divergence of the diagram by one unit.

Referring to the one-loop diagrams appearing in (a)–(i) of Figure F.1, we note that:

- (b) and (f) have $D = 2$ but contribute a constant independent of the external momentum, hence only to the renormalisation of the mass of the scalar (cf. Section 8.4), and we can ignore them;

- (d) is similar to vacuum polarisation in QED; the term with $D = 2$ is not zero but it contributes only to mass renormalisation; the derivative with respect to the external q^2 has $D = 0$ and it contributes to the β function, similar to what happens for renormalisation of the wave function, Z_2 in QED and QCD;

- (g) has in general $D = 2$, but has $D = 0$ in the Landau gauge, see Figure F.2; it contributes to the renormalisation of the wave function, as in the previous case; (i) has $D = 0$ in general, but is convergent in the Landau gauge and drops out ;

- (c), (e) and (h) have $D = 0$.

The logarithmically divergent contributions are calculated at an external momentum scale of order μ^2. The dependence on μ, for dimensional reasons, can only be of the type $\ln(\Lambda^2/\mu^2)$. Therefore, every logarithmically divergent contribution of the form $\delta\lambda^{(X)} = C^{(X)} \ln\Lambda^2$ contributes to the β function with a term

$$\mu\frac{\partial(\delta\lambda^{(X)})}{\partial\mu} = -2\Lambda^2\frac{\partial(\delta\lambda)^{(X)}}{\partial\Lambda^2} = -2C^{(X)} . \tag{F.2}$$

Diagrams (a), (c). We can restrict ourselves to the case in which the external states are all of type ϕ_0, making the potential (18.3) explicit as

$$V = \frac{\lambda_0}{4!}\left[\phi_0^4 + 2\phi_0^2(\sum_{1,N-1}\phi_i^2) + \cdots\right] = V_{00} + V_{0i} + \cdots . \qquad (F.3)$$

To first order, Figure F.1 (a), only V_{00} counts and the amplitude is equal to

$$A_0^{(4)} = FC \times (-i\frac{\lambda_0}{4!}) = -i\lambda_0 , \qquad (F.4)$$

where FC is a combinatorial factor which represents the number of ways in which we can connect the external states $1,\cdots,4$ to the vertex and is equal to $4\cdot3\cdot2\cdot1 = 4!$ as already noted with reference to equation (8.38). We shall write the higher-order corrections as $\delta A_0^{(4)} = -i\,\delta\lambda$.

In the first correction, Figure F.1 (c), the terms V_{00} (exchange of σ in the internal lines) and V_{0i} (exchange of ϕ_i in the internal lines) play a role. We find

$$-i\,\delta\lambda^{(c)} = FC \times (-i\frac{\lambda_0}{4!})^2 \int \frac{d^4k}{(2\pi)^4}\frac{i}{(k+q_1+q_2)^2}\frac{i}{(k)^2} , \qquad (F.5)$$

where FC is the appropriate combinatorial factor. The amplitude has $D = 0$ and we can calculate the logarithmically divergent term directly, neglecting the external momenta. After the Wick rotation, discussed in Appendix E, we obtain

$$-i\,\delta\lambda^{(c)} = FC \times (\frac{\lambda_0}{4!})^2 \times \frac{i}{16\pi^2}\ln\Lambda^2 + \cdots . \qquad (F.6)$$

To calculate FC, we consider first the diagram in which V_{00} appears and the states $1, 2$ are connected to the first vertex, Figure F.1(c). We note that:

- the factor $\frac{1}{2}$ from the Dyson formula is cancelled by the fact that there are two possible vertices which connect to $1, 2$;

- we can connect $1, 2$ to the first vertex in $4\cdot3$ different ways, and similarly $3, 4$ to the second vertex;

- we can connect internal lines from the first vertex to the second in $2\cdot1$ ways.

Therefore

$$(FC)_{00} = (4\cdot3)^2 \cdot 2 . \qquad (F.7)$$

We proceed in a similar way in the case of V_{0i} and obtain

$$(FC)_{0i} = 4\cdot(2)^2 \cdot 2 \cdot (N-1) , \qquad (F.8)$$

having taken account of the factor 2 in V_{0i} and that there are $N-1$ types of internal lines. Putting everything together and setting $N=3$ at the end, we find

$$-i\,\delta\lambda^{(c)} = \frac{9+(N-1)}{18}\frac{i\lambda_0^2}{16\pi^2}\ln\Lambda^2 + \cdots = \frac{2}{3}\frac{i\lambda_0^2}{16\pi^2}\ln\Lambda^2 + \cdots . \tag{F.9}$$

A further factor 3 is obtained from adding the crossed diagrams, with the exchanges $1,2 \to 1,3$ and $1,2 \to 1,4$.

Using (F.2), we finally find

$$(16\pi^2)\mu\frac{\partial(\delta\lambda^{(c)})}{\partial\mu} = (16\pi^2)\beta^{(c)} = +4\lambda^2 . \tag{F.10}$$

Diagram (e).   Considering external states of type ϕ_0, we have

$$-i\,\delta\lambda^{(e)} = (-1) \times FC \times (i\frac{g_t}{\sqrt{2}})^4 \int \frac{d^4k}{(2\pi)^4}\,\mathrm{Tr}\left(\frac{i}{\not k}\right)^4$$
$$= -4\cdot 3 \times FC \times (\frac{g_t}{\sqrt{2}})^4\frac{i}{16\pi^2}\ln\Lambda^2 . \tag{F.11}$$

The factor -1 corresponds to the fermion loop, and the factor $4\cdot 3$ follows from the trace over Dirac and colour indices. For the calculation of FC, we note that

- the Dyson factor $1/4!$ is cancelled by the number of ways in which the external states are associated with each of the four vertices;

- the vertices which follow the first vertex along the fermion line can be chosen in $3\cdot 2\cdot 1$ ways.

We therefore find

$$(16\pi^2)\mu\frac{\partial(\delta\lambda^{(e)})}{\partial\mu} = (16\pi^2)\beta^{(e)} = -36g_t^4 . \tag{F.12}$$

Diagram (h).   We first consider the W^1 exchange. In the Landau gauge and using (18.21), we find

$$-i\,\delta\lambda^{(h)} = (\frac{ig^2}{8})^2 FC \int \frac{d^4k}{(2\pi)^4}\frac{i}{k^2}\left(-g_{\mu\nu} + \frac{k_\mu k_\nu}{k^2}\right)\frac{i}{k^2}\left(-g^{\nu\mu} + \frac{k_\nu k_\mu}{k^2}\right)$$
$$= \frac{3}{64}FC\frac{ig^4}{16\pi^2}\ln\Lambda^2 . \tag{F.13}$$

The combinatorial factor is equal to $(2\cdot 1)^2\cdot 2 = 8$ corresponding to the ways of associating the external lines to $(\phi_0)^2$ at two vertices and to the ways of contracting the W^1 internal lines. Multiplying by $3\cdot 2$, to take account of the crossed diagrams and the exchange of W^2, and calculating in a similar way to the Z exchange, we obtain

$$-i\,\delta\lambda^{(h)} = \frac{i}{16\pi^2}\left[\frac{9}{4}g^4 + \frac{9}{8}(g^2 + g'^2)^2\right]\ln\Lambda^2 \,, \tag{F.14}$$

from which it follows that

$$(16\pi^2)\mu\frac{\partial(\delta\lambda^{(h)})}{\partial\mu} = (16\pi^2)\beta^{(h)} = \frac{27}{4}g^4 + \frac{9}{2}g^2g'^2 + \frac{9}{4}g'^4 \,. \tag{F.15}$$

Diagrams $(d), (g)$.   We denote the amplitude resulting from the integration over the closed fermion lines in (d) with

$$A^{(d)} = i(A + \delta Z\,q^2 + \cdots) \,, \tag{F.16}$$

where A corresponds to a mass renormalisation, which we can ignore.

If we imagine the diagrams of Figures F.1(a) and (d) inserted in a more complex diagram, hence with their external lines associated with propagators of the scalar fields, the sum of (a) and (d) is written

$$(a) + (d) = \frac{i}{q^2}\left[1 + i\delta Z q^2 \frac{i}{q^2}\right](-i\lambda_0) = \frac{i\sqrt{Z}}{q^2}(-i\sqrt{Z}\lambda_0) \,, \quad \sqrt{Z} = 1 - \frac{1}{2}\delta Z. \tag{F.17}$$

Therefore the charge λ_0 is renormalised as the effective charge $\sqrt{Z}\lambda_0 = \lambda_0 + \delta\lambda$. Taking into account that four lines merge at each vertex, the overall effective charge is obtained from

$$-i\,\delta\lambda^{(d)} = i\,2\,\delta Z\,\lambda_0 \,. \tag{F.18}$$

Putting $\delta Z = C \ln\Lambda^2$, we find from (F.18)

$$\beta^{(d)} = 4C \,, \tag{F.19}$$

and the same for diagram (g).

The amplitude resulting from the integration in diagram (d) is written[1]

$$A^{(d)} = (-1)\left(\frac{ig_t}{\sqrt{2}}\right)^2 \int \frac{d^4k}{(2\pi)^4}\,\mathrm{Tr}\left[\frac{i}{\not{k}}\frac{i}{\not{k}\not{+}\not{q}}\right]$$

$$= -\frac{g_t^2}{2}\,4\cdot 3 \int \frac{d^4k}{(2\pi)^4}\frac{k^2 + (k\cdot q)}{k^2\,(k+q)^2} \,.$$

Using the Feynman representation, Appendix E, easy steps lead to

$$A^{(d)} = -i\,6\frac{g_t^2}{16\pi^2}\left(-\Lambda^2 - \frac{3}{2}q^2\,\ln\Lambda^2 + \cdots\right)$$

$$= +i\,3\frac{g_t^2}{16\pi^2}q^2\,\ln\Lambda^2 + \cdots \,, \tag{F.20}$$

---

[1] As before, the factor $4\cdot 3$ arises from the trace over the Dirac indices and colour.

from which it follows that

$$16\pi^2\,\beta^{(d)} = +12\;g_t^2\lambda\;. \qquad (\text{F.21})$$

We leave to the reader the task of deriving the result:

$$16\pi^2\,\beta^{(g)} = -9\lambda g^2 - 3\lambda g'^2\;. \qquad (\text{F.22})$$

F.3 $\beta(g_t)$

The relevant Feynman diagrams are shown in Figure F.3, along with the results corresponding to each diagram.

Note the vanishing of the renormalisation of the wave function for exchange of electroweak vector bosons, analogous to what happens in QED in the Landau gauge.[2]

The calculations do not present particular difficulties, and we restrict our considerations to the contributions of diagrams (b) and (c).

The fermion inside the self-energy diagram in (b) and (c) can be t (two amplitudes $t \to t + \phi_0 \to t$ and $t \to t + \bar{\phi}_0 \to t$), or b (one amplitude $t \to b + \phi^+ \to t$). To isolate the logarithmic divergence, we proceed as for diagram (d) of Figure F.1. For the exchange of the neutral scalars and the charged scalar we find, respectively,

$$i\,\delta Z_n \slashed{q}(a^+ + a^-)\;,\quad (16\pi^2)\,\delta Z_n = \frac{1}{2}g_t^2\,\ln\Lambda^2\;, \qquad (\text{F.23})$$

$$i\,\delta Z_{ch}\slashed{q}\,a^+ \qquad ,\quad (16\pi^2)\,\delta Z_{ch} = \delta Z_n\;. \qquad (\text{F.24})$$

With these results we can construct the correct fermion propagator

$$
\begin{aligned}
S(q) &= \frac{i}{\slashed{q}}\left[1 + i\,\delta Z_L\,a^-\slashed{q} + i\,\delta Z_R\,a^+\slashed{q}\right]\frac{i}{\slashed{q}} + \cdots\\
&= \frac{i}{\slashed{q}}\left[1 - \delta Z_L\,a^- - \delta Z_L\,a^+\right] = \frac{i}{\slashed{q}}\,\mathcal{Z}\;,
\end{aligned} \qquad (\text{F.25})
$$

where

$$Z_L = 1 - \delta Z_n - \delta Z_{ch},\quad Z_R = 1 - \delta Z_n\;. \qquad (\text{F.26})$$

---

[2] Using the notation of Section 11.5, it is easy to confirm that $B = L = 0$ in the Landau gauge.

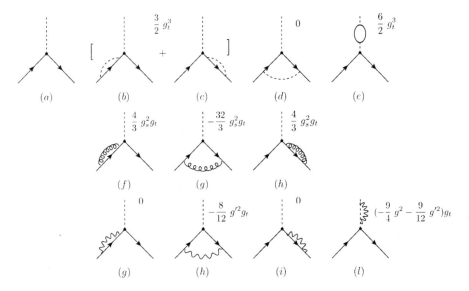

Figure F.3 Feynman diagrams for the $\beta(g_t)$ function in the Standard Theory. The contributions shown for each diagram are multiplied by a common factor of $(16\pi^2)^{-1}$. The wavy lines represent gluons in (f)–(h) and electroweak vector bosons in (g)–(l). Diagrams of the same order with the exchange of a vector boson between the fermion and the scalar are not divergent in the Landau gauge, and not shown.

We now sum diagrams (a), (b) and (c) and find

$$(a) + (b) + (c) = \frac{i}{\not{q}} \mathcal{Z}\, i\frac{g_t}{\sqrt{2}} \frac{i}{\not{q}} \mathcal{Z} = \frac{i}{\not{q}} \sqrt{\mathcal{Z}} \left[\sqrt{\mathcal{Z}}\, i\frac{g_t}{\sqrt{2}}\, \sqrt{\tilde{\mathcal{Z}}} \right] \frac{i}{\not{q}} \sqrt{\mathcal{Z}} , \qquad \text{(F.27)}$$

where

$$\mathcal{Z} = (1 - \delta Z_L)\, a^- + (1 - \delta Z_R)\, a^+ = Z_L a^- + Z_R a^+ \qquad \text{(F.28)}$$

$$\tilde{\mathcal{Z}} = Z_L a^+ + Z_R\, a^- . \qquad \text{(F.29)}$$

The renormalised charge is therefore

$$\left[\sqrt{\mathcal{Z}}\, i\frac{g_t}{\sqrt{2}}\, \sqrt{\tilde{\mathcal{Z}}} \right] = i\frac{g_t}{\sqrt{2}}\, \sqrt{Z_L Z_R} = i\frac{g_t}{\sqrt{2}} \left[1 - \frac{1}{2}\left(2\delta Z_n + \delta Z_{ch} \right) \right] , \qquad \text{(F.30)}$$

or

$$(16\pi^2)\, \delta g_t^{(b)+(c)} = -\frac{3}{4}\, g_t^2\, \ln\Lambda^2 = +\frac{3}{2}\, g_t^2\, \ln\mu , \qquad \text{(F.31)}$$

which gives the result shown in Figure F.3, (b) and (c).

Bibliography

[1] L. Maiani and O. Benhar, *Relativistic Quantum Mechanics*, CRC Press, 2015.

[2] L. Maiani, *Electroweak Interactions*, CRC Press, 2015.

[3] F. Mandl and G. Shaw, *Quantum Field Theory*, Wiley, 1984.

[4] R. P. Feynman and A. R. Hibbs, *Quantum Mechanics and Path Integrals*, McGraw-Hill, 1965.

[5] P. A. M. Dirac, Physikalische Zeitschrift der Sowjetunion, Band 3, Heft 1 (1933), Reprinted in *Quantum Electrodynamics*, ed. J. Schwinger, Dover, 1958.

[6] H. Kleinert, *Path Integrals in Quantum Mechanics, Statistics, Polymer Physics, and Financial Markets*, World Scientific, Singapore, 2006.

[7] J. J. Sakurai, *Modern Quantum Mechanics*, Addison-Wesley, 1994.

[8] J. D. Bjorken and S. Drell, *Relativistic Quantum Fields*, McGraw-Hill, 1965.

[9] C. Itzykson and J-B. Zuber, *Quantum Field Theory*, McGraw-Hill, 1980.

[10] L. Landau and E. Lifschitz, *Mecanique Quantique. Theorie Non Relativiste*, Mir, 1967.

[11] H. Lehman, K. Symanzik, and W. Zimmermann, *Nuovo Cimento* **1**, 205 (1955).

[12] F. Bloch and A. Nordsieck, *Phys. Rev.* **52**, 54 (1937).

[13] H. A. Bethe, *Phys. Rev.* **72**, 339 (1947).

[14] G. Abbiendi *et al.* [OPAL Collaboration], *Eur. Phys. J. C* **45** (2006) 1.

[15] M. Gell-Mann and F. E. Low, *Phys. Rev.* **95** (1954) 1300.

[16] C. N. Yang and R. L. Mills, *Phys. Rev.* **96** (1954) 191.

[17] L. Faddeev and V. Popov, *Phys. Lett.* **B**25 (1967) 29.

[18] V. N. Gribov, *Nucl. Phys.* **B 139** (1978) 1.

[19] R. P. Feynman, *Acta Phys. Polon.* **24** (1963) 697.

[20] B. S. DeWitt, *Phys. Rev.* **162** (1967) 1195.

[21] R. E. Cutkosky, *J. Math. Phys.* **1** (1960) 429.

[22] M. E. Peskin and D. V. Schroeder *An Introduction to Quantum Field Theory*, Perseus Books, 1995.

[23] K. A. Olive *et al.* (Particle Data Group), *Chin. Phys.* **C**, 38, 090001 (2014).

[24] S. Bethke, *Eur. Phys. J. C* **64** (2009) 689.

[25] J. C. Pati and A. Salam, *Phys. Rev. D* **10**, 275 (1974).

[26] H. Georgi and S. L. Glashow, *Phys. Rev. Lett.* **32**, 438 (1974).

[27] G. Ross, *Grand Unified Theories*, The Benjamin/Cummings Publishing Co. (Menlo Park, California), 1985; see also P. Ramond, *Journeys beyond the Standard Model*, Perseus Books (Cambridge, Massachusetts), 1999.

[28] S. R. Coleman and E. J. Weinberg, *Phys. Rev. D* **7** (1973) 1888.

[29] C. Ford, D. R. T. Jones, P. W. Stephenson and M. B. Einhorn, *Nucl. Phys. B* **395** (1993) 17.

[30] K. G. Wilson and J. B. Kogut, *Phys. Rep.* **12** (1974) 75.

[31] L. Maiani, G. Parisi and R. Petronzio, *Nucl. Phys. B* **136** (1978) 115.

[32] N. Cabibbo, L. Maiani, G. Parisi and R. Petronzio, *Nucl. Phys. B* **158** (1979) 295.

[33] M. Sher, *Phys. Rep.* **179** (1989) 273; M. Lindner, M. Sher and H. W. Zaglauer, *Phys. Lett. B* **228** (1989) 139.

[34] G. Altarelli and G. Isidori, *Phys. Lett. B* **337** (1994) 141; M. Sher, *Phys. Lett. B* **317** (1993) 159, *ibidem* **331** (1994) 448.

[35] A. Salvio and A. Strumia, *JHEP* **1406** (2014) 080.

[36] G. Degrassi, S. Di Vita, J. Elias-Miro, J. R. Espinosa, G. F. Giudice, G. Isidori and A. Strumia, *JHEP* **1208** (2012) 098.

[37] T. Burnett and J. J. Levine, *Phys. Lett.* **24** (1967) 467; S. J. Brodsky and J. Sullivan, *Phys. Rev.* **156** (1967)1644; T. Kinoshita *et al.*, *Phys. Rev.* **D 2** (1970) 910.

[38] R. Jackiw and S. Weinberg, *Phys. Rev.* **D5** (1972) 2396; G. Altarelli, N. Cabibbo and L. Maiani, *Phys. Lett.* **B 40** (1972) 415; I. Bars and M. Yoshimura, *Phys. Rev.* **D6** (1972) 374; K. Fujikawa, B. W. Lee and A. I. Sanda, *Phys. Rev.* **D6** (1972) 2923.

[39] M. Srednicki, *Quantum Field Theory*, Cambridge University Press, 2007.

[40] J. S. Bell and R. Jackiw, *Nuovo Cim.* A **60** (1969) 47; S. L. Adler, *Phys. Rev.* **177** (1969) 2426.

[41] C. Bouchiat, J. Iliopoulos and P. Meyer, *Phys. Lett.* B **38** (1972) 519.

[42] A. Hoecker, W. J. Marciano in: K. A. Olive *et al.* (Particle Data Group), *Chin. Phys.* **C 38** (2014) 090001.

[43] J. Schwinger, *Proc. Natl. Acad. Sci. U. S.* **37** (1951), 452; **37**, (1951), 455.

[44] G. Jona-Lasinio, *Nuovo Cimento* **34** (1964), 1790.

[45] J. Goldstone, A. Salam and S. Weinberg, *Nuovo Cimento* **127** (1962), 965.

[46] J. Iliopoulos, C. Itzykson and A. Martin, *Rev. Mod. Phys.* **47** (1975) 165.

[47] C. G. Callan, Jr., *Phys. Rev. D* **5** (1972) 3202; K. Symanzik, *Comm. Math. Phys.* **18** (1970) 227; *Commun. Math. Phys.* **23** (1971) 49

[48] G. 't Hooft, Naturalness, Chiral Symmetry and Spontaneous Chiral Symmetry Breaking. In *Recent Developments in Gauge Theories*. Plenum Press, 1978; previously, similar considerations were discussed in K. G. Wilson, *Phys. Rev. D* **3** (1971) 1818.

[49] J. Wess, B. Zumino, *Nucl. Phys.* **B70** (1974) 39; V. P. Akulov, D. V. Volkov, *Theor. Math. Phys.* **18** (1974) 28.

[50] L. Maiani, in *Proc. of the Summer School on Particle Physics*, Gif-sur-Yvette, 3–7 September 1979, Ed. M. Davier *et al.*, IN_2P_3, Paris, 1979; M. Veltman, *Acta Phys. Polon.* **B12** (1981) 437; E. Witten, *Nucl. Phys.* **B 188** (1981) 513 and *Phys. Lett.* **B 105** (1981) 267.

[51] P. Fayet, *Nucl. Phys. B* **90**, 104 (1975); *Phys. Lett. B* **64**, 159 (1976); B **69**, 489 (1977).

[52] S. Dimopoulos and H. Georgi, *Nucl. Phys. B* **193** (1981) 150.

[53] J. Gunion, H. Haber, G. Kane and S. Dawson, *The Higgs Hunter's Guide*, Reading, 1990; S. Heinemeyer, W. Hollik and G. Weiglein, *Phys. Rep.* **425** 265 (2006) ; A. Djouadi, *Phys. Rep.* **459** 1 (2008).

[54] See, for example, K. Agashe, R. Contino and A. Pomarol, *Nucl. Phys. B* **719** (2005) 165.

[55] M. Abramowitz and I. A. Stegun, *Handbook of Mathematical Functions*, Dover, 1972.

[56] H. Jeffreys and M. Jeffreys, *Methods of Mathematical Physics*, Cambridge University Press, 1972.

Index